Seventy-Five Years in Ecology:
The British Ecological Society

FRONTISPIECE 'The naturalists', painted by L. J. Watson. The figures are (left to right) E. B. Ford, A. G. Tansley, A. S. Watt and C. Diver, on an East Anglian heath in the summer of 1949.

Seventy-Five Years in Ecology:
The British Ecological Society

JOHN SHEAIL

NERC Institute of Terrestrial Ecology
Monks Wood Experimental Station
Huntingdon, Cambridgeshire

FOREWORD BY
CHARLES H. GIMINGHAM
President
British Ecological Society
1986–87

BLACKWELL SCIENTIFIC PUBLICATIONS

OXFORD LONDON EDINBURGH

BOSTON PALO ALTO MELBOURNE

© 1987 by
Blackwell Scientific Publications
Editorial offices:
Osney Mead, Oxford OX2 0EL
(*Orders*: Tel. 0865 240201)
8 John Street, London WC1N 2ES
23 Ainslie Place, Edinburgh EH3 6AJ
52 Beacon Street, Boston
  Massachusetts 02108, USA
667 Lytton Avenue, Palo Alto
  California 94301, USA
107 Barry Street, Carlton
  Victoria 3053, Australia

First published 1987

Set by Eta Services (Typesetters)
Ltd, Beccles, Suffolk, printed
by Redwood Burn Ltd,
Trowbridge, Wilts, and
bound by Butler & Tanner
Ltd, Frome, Somerset.

DISTRIBUTORS

USA and Canada
  Blackwell Scientific Publications Inc
  PO Box 50009, Palo Alto
  California 94303
  (*Orders*: Tel. (415) 965-4081)

Australia
  Blackwell Scientific Publications
  (Australia) Pty Ltd
  107 Barry Street,
  Carlton, Victoria 3053
  (*Orders*: Tel. (03) 347 0300)

British Library
Cataloguing in Publication Data
British Ecological Society
  Seventy five years in ecology: The
  British Ecological Society.—(British
  Ecological Society special publication,
  ISSN 0262-7027).
  1. British Ecological Society—History
  I. Title   II. Sheail, John   III. Series
  574.5'06'041   QH540
  ISBN 0-632-01911-5
  ISBN 0-632-01917-4 Pbk

Library of Congress
Cataloging-in-Publication Data
Sheail, John
  Seventy-five years in ecology: the British
  Ecological Society/John Sheail.
  p. cm.—(Special publication number . . .
  of the British Ecological Society)
  Bibliography: p.
  Includes index.
  ISBN 0-632-01911-5. ISBN 0-632-01917-4
(pbk.)
  1. British Ecological Society—
History.   I. Title. II. Title: 75 years in
ecology. III. Series: Special publication
. . . of the British Ecological Society: no.
QH137.S427 1987
574.5'06'041—dc19   87-16763   CP

Contents

List of Illustrations

ix

Foreword

In 1988 the British Ecological Society celebrates the 75th anniversary of its foundation. This is a significant moment in its history: a point at which it is instructive to survey what has been achieved in the past and in the light of this to develop an imaginative approach to the future.

All scientific progress is built on past achievement. This book is published in the belief that a knowledge of the history of the British Ecological Society will provide an important stimulus to the present generation of ecologists. For this reason all members of the Society are to receive copies, and I am confident not only that their interest will be aroused but also that the benefit and satisfaction they derive from the Society's activities will be enhanced. By providing such an admirably comprehensive account of the evolution of the Society during its first three-quarters of a century, this volume makes a notable contribution to the celebration of our anniversary. It will be complemented in due course by the published proceedings of two major symposium meetings, one seeking to identify and evaluate certain fundamental concepts which have emerged during this phase of the Society's life, and the other setting its sights on the future.

At the time of its foundation in 1913, the British Ecological Society was the only one of its kind in the world. Ecology was then just beginning to emerge as a distinctive approach to biological science. The history of the Society therefore amounts in large part to a history of the subject as it evolved in this country and responded to developments elsewhere. The Society owed its origin to a far-sighted group of pioneers who were convinced that ecology had a major role to play in the natural sciences and their application to practical problems, as well as in education. They could not have foreseen the extent to which these aspirations have been realized, nor that the membership would rise to more than 4300 within 75 years. John Sheail's study traces this progress, in terms of not only the growth of a 'learned society' but also the ways in which the varied personalities of many of its leading members affected the turn of events, often over the course of several generations. It also explores fascinating examples of the origins and interactions of scientific ideas, and of the continual changes of emphasis within an expanding scientific community. Present-day ecology has come a long way from its origins around the turn of the century, but its current and future thrusts will be strengthened and enlightened by a sound understanding of the road it has travelled to reach its present position.

CHARLES H. GIMINGHAM
President of the British
Ecological Society, 1986–87

xi

Preface

The British Ecological Society celebrates its 75th anniversary in 1988. The Society evolved out of the British Vegetation Committee, which was founded in 1904 to promote the survey and study of vegetation in the British Isles. This initiative was in turn the outcome of what many historians perceive to have been the emergence of modern ecology in the 1890s. In tracing the circumstances in which the British Ecological Society came to be established, and the course of events since 1913, this book is in effect celebrating a hundred years in ecology.

The first quarter-century of the British Ecological Society was celebrated by the election of its first president, A. G. Tansley, for a second term of office. In his presidential address of 1939, Tansley reviewed progress made in British ecology over that period. Celebrations for the golden jubilee took the form of a symposium, in the course of which 'the history of ecology in Britain' was outlined by W. H. Pearsall and E. J. Salisbury, who drew heavily on their own personal recollections of what had transpired, in respect of both the Society and ecology at large.

The 75th anniversary is being marked by the publication of a book compiled from the publications and archives of the Society, supported by a wide range of other published and manuscript material. In looking at the earlier history of the Society, the book provides a considerable amount of background information on trends in ecological thinking and activity.

Every reader will seek insights into the development of a particular field of ecological interest. No book can meet this need entirely. The aim of this book is to provide a context and perspective for those making more detailed investigations of the often intricate, but always fascinating, history of how ecology came to occupy its present position in the sciences in Britain.

In commenting on Pearsall's contribution to the golden jubilee symposium of 1964, J. B. Cragg wrote of how

> science is essentially an art-form in which a handful of inspired
> performers set the pace and the rest of us trudge along behind.

If this book succeeds in drawing attention to the pace-makers in ecology, and the inspiration and support they derived from the Society which they founded, it will have begun to fulfil its purpose.

It is a pleasure to record all the help given to me in the writing of this book by those who have confirmed, corrected, and extended what I might have deduced from the written record. Dr Malcolm Cherrett first drew my attention to the Society's archives, which he had played so large a part in bringing together during his period of office as Council secretary. I am grateful to the presidents of the Society, the Council and the Publications Committee, and to Dr E. Broadhead for their guidance and support, and to Dr J. P. Dempster, Dr M. D. Hooper, and other members of the Monks Wood Experimental Station. As always, I owe so much to the encouragement of Gillian, my wife.

I gratefully acknowledge the facilities extended to me in Cambridge by the University Library, Balfour Library, and Library of the Botany School. The Royal Botanic Garden, Kew, the Botany School, Cambridge, and the Tolson Memorial Museum, Huddersfield, gave permission for me to consult manuscript material in their custody. Mr A. W. H. Pearsall and Miss S. Macpherson, the Archivist of the Cumbria Record Office, Kendal, very kindly gave me access to, and permission to quote from, the personal papers of Professor W. H. Pearsall.

I thank the Botany School, Cambridge, for permission to reproduce the frontispiece and plates 3–5, 7, and 19; the Tolson Memorial Museum, Huddersfield (plate 8); the Department of Botany, University College, London (plates 9–11 and 23); the Department of Zoology, Oxford (plates 12, 15, 20–21); Professor P. W. Richards (plates 13–14); Dr J. F. Hope-Simpson (plate 16); Professor A. Burges (plates 17–18); Mrs K. Southern (plate 22); Dr N. Waloff (plate 24). The considerable assistance given me by Mr P. Barham of the Botany School, Cambridge, Mr K. Marsden of the Department of Zoology, Oxford, and Dr K. Taylor of the Department of Botany, University College, London, is gratefully acknowledged. The figures have been drawn by Mr J. Clark.

The Latin names given to plants and animals in the text are those used in the original publications.

The birth and death years of persons mentioned in the text are italicized and placed in brackets after the first major reference to that person.

Monks Wood 1987   JOHN SHEAIL

Part 1
The First Ecological Society
in the World

1.1

Beginnings

During the 19th century, increasing numbers of naturalists not only recorded what they saw, but tried to explain the patterns of distribution and abundance which they encountered. In a paper to the Literary and Philosophical Society of Newcastle upon Tyne in 1819, Nathaniel J. Winch (*1768–1838*) related 'the copious and highly diversified flora' of Northumberland, Cumberland, and Durham to the variety of exposure, situation, and soils in those counties (Winch 1825). In 1832, two important publications appeared. Hewett C. Watson (*1804–81*) published his *Outlines of the Geographical Distribution of British Plants*, giving measured altitudes for several species, and the comparative (zonal) altitude for many others (Watson 1832). The other remarkable study was by William MacGillivray (*1796–1852*), who, in his *Remarks on the Phanerogamic Vegetation of the River Dee*, sought to demonstrate the influence of altitude and topography on the distribution of Alpine plants (MacGillivray 1832a).

One of MacGillivray's heroes was Alexander von Humboldt (*1769–1859*). He had just finished editing an account of the visits made by the German naturalist and explorer to different parts of the world, during which Humboldt attempted to describe vegetation on a systematic basis (MacGillivray 1832b). In a sense, both Humboldt and MacGillivray anticipated Charles Darwin in the way they perceived the complexity of interrelationships in nature, and in how they tried to correlate the peculiarities of flora and fauna with actual environmental conditions (Thomson J. A. 1910). In his *Natural History of Dee-side and Braemar*, MacGillivray (1855) stressed how there was no simple correlation between altitude and the incidence of Alpine plants. It was as if the seeds had been profusely scattered over the district, but 'germinated only in places favourable to their development'.

Many naturalists were interested not only in plants and animals, but in other aspects of the environment. John G. Baker (*1834–1920*) gave his *Flora of North Yorkshire* the subtitle, 'Studies of its botany, geology, climate and physical geography'. Although there was some information by the 1860s on what *prevented* plants from spreading to new localities, hardly anything was known about how plants established themselves in the first place. Baker (1863) set out to remedy the deficiency by relating each species to 'certain physical conditions'. Under the heading 'station or habitat', he described how

> each species plainly has its own special power of adaptation to varied
> physical conditions, and that power is very different in different natural
> orders, different genera, and even often in different species of the same
> genus.

3

In the same way as the physiologist studied the structure and function of a plant's organs, and a systematist dealt with the diagnostic character, so the geographical botanist was concerned with the adaptive powers of plants.

Baker explained how each species had 'its own special constitution', as it related to heat and moisture levels, light and shade, and soil types. It was difficult to define the relationships precisely, especially over large tracts of land. Temperature seemed to be the most important factor, when expressed in terms of 'the sums of summer heat and the extreme minima of the colder parts of the year'. A plant was not, however, 'a mere machine, like a thermometer, but a living organism'. The way it reacted to temperature and other physical phenomena opened out 'a wide field for research and consideration'.

In 1866, a book with the title *Contributions Towards a Cybele Hibernica* was published. A second edition of 1898 provided a much more detailed account of the distribution of plants in Ireland, and was able to take account of recent advances in studying 'the soil relations of the flora'. In what may have been the first example of its kind in a British flora, the terms *calcicole* and *calcifuge* were used in a description of the influence of lime in the soil on plant distribution. The terms were taken from *Géographie Botanique—Influence du Terrain sur la Végétation*, published by Charles Contijean in Paris in 1881. The editors of the Irish flora believed that such a classification, whatever its limitations, would serve at least to draw the attention of Irish botanists to 'a neglected, though deeply interesting, branch of their science' (Colgan & Scully 1898).

In Scotland, few naturalists could rival the breadth of knowledge of Francis Buchanan White (*1842–94*). 'Neither requiring nor caring to practise his profession as a medical man', he devoted his life to a close study of 'animals and plants in their natural environment'. His *Flora of Perthshire*, published posthumously in 1898, was rated as one of the finest of its genre (White F. B. 1898). Buchanan White had always been critical of authors who had simply made lists of species. In an earlier account of the River Tay, he had written,

> we have to discover, if possible, what effect the physiographical, meteorological, and geological conditions have upon the distribution of the flora. That these conditions are potent factors is not to be denied, but in many cases they act in combination, and it is difficult to ascertain which of these several causes, acting together, is a predominant agent.

In the case of the Tay valley, White (1895) thought the chief factor was likely to be the river itself, through its ability to transport material and provide 'suitable habitats for the plants'. Unless man intervened, the river destroyed sooner or later what it had formed, only 'to repeat the construction in some other part of the river'. In an earlier study of the flora of river shingle, Buchanan White wrote of how

> I have thought that much instruction might be derived if a definite space—say fifty yards square—of a young shingle were mapped out, and a careful census taken of all the plants which sprang up in it in the course of a season.

The most conspicuous colonizers were likely to be 'the weeds of cultivation', and a host of 'more enduring species, which, if circumstances are favourable, will help to build up the future vegetation'. Whether dead or alive, the stems and branches trapped further debris in the next flood and, by degrees, the shingle grew until an island might be formed, covered with 'a characteristic flora' of herbaceous plants, shrubs, and even trees (White F. B. 1890).

It would have been strange indeed if naturalists had not been aware of the striking changes which man was bringing to the flora and fauna of the country-side. Building development often destroyed the botanist's most accessible sites. Britain was the 'Workshop of the World' and centre of the Empire. On the Tees estuary, where ironworks, wharves, and shipyards now covered much of the foreshore and mudflats, it seemed more than a coincidence that, whereas a thousand seals could be seen at one time in the 1820s, no more than three were ever seen together in the 1860s (Lofthouse 1887). Changes were also brought about by farming. In Cambridgeshire, the university professor of botany, Charles C. Babington (*1808–95*), described how the ploughing up of turf meant that 'the peculiar plants' of the chalklands had become mostly confined to small waste spots by roadsides, pits, and 'the very few banks which are too steep for the plough' (Babington 1860).

The increasing number of floras and faunas published since the mid-18th century meant that there was now a series of baselines from which changes could be gauged. Babington's *Flora of Cambridgeshire* was one of the first to compare the past and present status of species, using such sources. All kinds of reasons might be put forward to explain the changes observed. In a *Flora of the West Riding*, Frederic A. Lees (*1847–1921*) attributed the decline of the bloody cranesbill (*Geranium sanguineum*) to its being dug up, 'being so handsome'. The cat whin (*Ulex anglica*) had been eliminated from many moors by cultivation. Under the entry for *Lycopodium alpinum*, Lees (1888) wrote of how the plant was affected, even on the high moors, 'by the nameless mephitic influences of modern manufactures'.

1.2
Vegetation mapping

The writings of Baker, White, and Lees provided both models and a challenge for further studies of the local flora. Local natural history societies and field clubs played a large part in the exchange of data and experiences. Soon the younger generation of the 1890s began to provide its own insights, based on first-hand observations and a study of the growing volume of more general works, including some in German and French. Not only were the new insights promoted with great vigour, but they were addressed to an increasingly responsive audience.

1.2.1 *The botanical survey of Scotland*

The influence of this 'younger generation' of the 1890s was most readily discerned in Scotland. The contribution made by two brothers, William and Robert Smith, was outstanding. Both were graduates of University College, Dundee, where they were much influenced by the professor of biology, D'Arcy W. Thompson (*1860–1948*), and by the dynamic Patrick Geddes (*1854–1932*), who had been appointed to the J. F. White Chair of Botany in 1889. As both a theoretician and a practitioner, Geddes more than compensated for the fact that his botany department was the smallest in the United Kingdom (Boardman 1978).

William Smith (*1866–1928*) graduated in 1890. After a short spell as a science master, he became demonstrator in botany at Edinburgh University. His inclination was towards applied research, and his appointment as lecturer in agriculture for the Forfarshire County Council in 1892 gave him the opportunity to study farming practices. It also stimulated his interest in plant diseases. Having spent a year in the University of Munich, where he was awarded a PhD degree for a thesis on the deformation of trees caused by fungal diseases, he returned to the University of Edinburgh in 1894, where he took up a new post as lecturer in plant physiology. Three years later, he became lecturer in agricultural botany and assistant lecturer in botany at the Yorkshire College, Leeds (Matthews 1929, Woodhead 1929a). So far, he had shown little inclination for vegetation survey work. It was his brother, Robert, who was fast becoming its leading exponent.

In the summer of 1896, Robert Smith (*1873–1900*) began a survey of the Tay and Forth basins (Smith R. 1900a). In approaching the task, he was not only imbued with Geddes's sociological outlook on life and the biological sciences (Burnett 1964), but he had already acquired a detailed knowledge of the local flora. As a child, he had explored the Scottish east coast and Grampians. Excursions to Ireland and Norway provided further material and insights. As a part-time student, he graduated 'with distinction' in 1895, and was appointed by Geddes as demonstrator in botany. From this position, he acquired a detailed knowledge of the continental literature.

William Smith later recalled the considerable influence which Warming's book, *Plantesamfund: Grundtrak af den Okologiske Plantegeografi*, exerted on the outlook of the 'younger botanists' at that time (Smith W. G. 1924). J. Eugene B. Warming (*1841–1924*) was professor of botany in the University of Copenhagen, where he gave what was in effect the world's first course of lectures in ecology. The substance of the lectures was incorporated in his book, which became much more accessible when a German translation appeared in 1896 (Warming 1896).

The book not only distilled a lifetime's experiences, but brought together the scattered ecological work of several decades, 'presenting it under a single clear and comprehensive point of view' (Blackman F. F. & Tansley 1905). While still an undergraduate, Warming had been invited in 1863 to spend three years in Brazil, where he had every opportunity to collect and analyse plant life. His monograph (published 30 years later) on the tiny Brazilian village where he stayed, Logoa Santa, was in effect 'the first detailed ecological study of a tract of vegetation' anywhere in the world. Meanwhile, visits to west Greenland and north Finland in 1884–5 had provided him with insights into the other global extreme in plant life. A series of papers published between 1886 and 1888 laid 'the foundation of Arctic geographical botany' (Goodland 1975).

In *Plantesamfund*, a revised version of which appeared in English in 1909, Warming drew a clear distinction between floristic plant geography and 'oecological plant geography'. Whereas the former dealt with species distribution, the latter was much more concerned with 'the life forms of the species, their associations with one another, and relation to their life conditions'. Warming described how plants were adapted to a range of what he called 'ecological factors', which included light, humidity, soils, exposure, and animals (Warming 1909).

Warming recalled how the word 'formation' was first used by the German botanist August R. H. Grisebach (*1814–97*) in describing a group of plants that formed a distinct and complete feature in the landscape. Charles Darwin (*1809–82*) had focused attention on the way in which organisms struggled to survive in their environment through adaptation. From the time of the publication in 1859 of *The Origin of Species by Means of Natural Selection: or the Preservation of Favoured Races in the Struggle for Life* (Darwin 1859), there had been a stream of works describing the anatomical and physiological characteristics of plants in relation to their habitats in different parts of the world. By the end of the century, the main European centres for studying botany in terms of vegetation were Montpellier, under Charles Flahault, and Zürich under Carl Schröter.

The stimulus which Robert Smith derived from Warming's book was made abundantly clear in a paper 'On the study of plant associations', published in 1899. Warming had suggested that moisture was the main determinant of any plant association. Smith (1899) argued that any survey of vegetation should also pay close attention to:

1 The way in which each of the associations making up the vegetation of an area was distinguished by temperature, light, moisture, and food supply.

2 The particular adaptations or life-forms of the species of each association.

3 The relationship between the species, in terms of dominant social form, secondary social forms struggling for dominance, and dependent species.

4 The influence of animals and man.

For Robert Smith, the challenge was to find a way of recording these attributes while carrying out a vegetation survey over extensive areas in the minimum amount of time.

The answer began to emerge during the winter of 1896–7, which he spent at the University of Montpellier. It was Geddes's practice to encourage his most able students to spend a period under Charles Flahault (*1852–1935*), and, as a research scholar of the Franco-Scottish Society, Robert Smith accompanied Flahault on numerous surveying expeditions during that winter (Anon. 1901). Inspired by what he saw and learnt, Robert began straightaway in the summer of 1897 to devise ways of mapping the plant associations of the Pentland Hills, assisted by his brother, William (Smith W. G. 1902a).

Following Flahault's example, Robert Smith delimited his 'natural regions' on the basis of their most conspicuous species. Each species occupied so prominent a position because of its dominance, as a 'social species', over all other plants in the vegetation. In Smith's terminology,

> such a community made up of chief species, subordinate species, and
> dependent species, constitutes a Plant Association (Smith R. 1898).

Flahault recorded the presence of the most characteristic trees of an area as the means of identifying the prevailing climatic and soil conditions, and 'the presence of particular associations of subordinate species'. In the *Annales de Géographie* of July 1897, he published the first of 13 sheets intended to cover the whole of France at a scale of 1:200 000. The Perpignan sheet was chosen as the first to be published on account of its covering as many as five 'great natural regions' (Flahault 1897).

Although the Perpignan sheet gave a good impression of how the plant associations might be distinguished in Scotland, it was clear to Robert Smith that Flahault's approach would have to be modified. There had been so much deforestation in Scotland that a survey based on the chief trees would have given 'a very inadequate, if not erroneous, impression of the plant-covering of the country'. In Scotland, the main divisions would have to be based on 'the cultivated plants, the grasses, the heaths, etc.' The 'commonsense' approach adopted by Smith was based on the general physiognomy of the vegetation (mixed deciduous woods, coniferous woods, hill and Alpine pasture, heather moors, and heaths). Within these broad categories, special attention was drawn to the role of dominant species (oakwood, *Molinia* grass moor, cotton grass bog).

The first two maps of the 'Botanical survey of Scotland', together with an accompanying text, covered the Edinburgh and North Perthshire districts at a scale of half-inch to the mile (1:126 720). Taken together, Robert Smith believed they provided a representative sample of the main types of vegetation to be found at different altitudes in Scotland. Littoral, temperate, subalpine, and Alpine regions could be distinguished. The surveys of further districts would merely add to the number of minor associations dependent on factors other than altitude (Smith R. 1900a,b). In August 1900, Robert Smith died of peritonitis, at the age of only 26 years (Geddes 1900). It was his brother, William, who completed the surveys for the Forfar and Fife districts during his summer vacations (Smith W. G. 1904, 1905a).

Two factors had made it easier to carry out and publish vegetation surveys. The Ordnance Survey had at last fulfilled its goal of publishing a series of large-scale maps covering the British Isles, and recent advances in the printing industry made it possible to publish the vegetation maps in large numbers, with a range of distinctive colours (Smith W. G. 1902a). Both Robert and William Smith paid tribute to the encouragement and expert advice given to them by the publisher and printer, John G. Bartholomew (*1860–1920*). Some years earlier, his company had published a *Naturalist's Map of Scotland*, which had used colour to distinguish areas of woodland, cultivated land, moorland, and deer forest at a scale of 10 miles to the inch (1:633 600) (Harvie-Brown & Bartholomew 1893).

The greatest deterrent was likely to be the cost. The surveys for Forfar and Fife would not have been published without a special grant made by the Carnegie Trust, in recognition of Robert Smith's work. A 'fifth' map, which was 'required to complete the area', was never published.

The practical value of such maps was emphasized by two contributions to the *Scottish Geographical Magazine*. One was by Andrew J. Herbertson (*1865–1915*), who had also held the post of demonstrator in botany under Geddes and had visited Montpellier. He was on the council of the Royal Scottish Geographical Society and, for a short period, its editor (Snodgrass 1965). The other contribution was by a Belgian, Marcel Hardy, who joined Geddes from the Botanical Institute at Montpellier in 1900 (Stevenson 1978). In a paper that focused on the impoverished state of Highland agriculture and forestry, Hardy (1902) wrote of how there could be little improvement until a closer understanding was obtained of the relationships between physiographical conditions and the vegetation.

In a review of Flahault's *Annales* paper, Herbertson (1897) argued that geographers did not want the exact limits of every species of living plant charted, any more than they desired maps of every species of fossil. They needed maps of botanical associations on similar lines to those of geological formations. He continued, the geological map is a synthetic chart conveying an immense amount of information to the trained reader—each different formation telling of different structure and different fossils bespeaking different conditions of origin. The botanical map should do the same—to the student it should indicate different floral types, depending on different physical conditions.

Vegetation maps were needed so urgently that Herbertson believed it was wrong to leave their compilation to 'the spare moments of the individual worker'. 'The united efforts of the staff of a national survey' were needed.

The omens were not, however, good. Robert Smith's paper on the survey of the Edinburgh district had appeared in the same volume of the *Scottish Geographical Magazine* as the first instalment of a bathymetrical survey of the Scottish freshwater lochs (Murray & Pullar 1900). The Government had refused any financial assistance on the grounds that the work fell outside the provinces of the Ordnance Survey and Hydrographic Service. The loch survey had only gone ahead because the marine naturalist and oceanographer, Sir John Murray (*1841–1914*), together with Frederick Pullar, decided to organize it themselves. When Pullar was drowned in 1901, his father provided the funds to complete the survey (Murray & Pullar 1910, S., A.E. 1917).

The fact that there was so little prospect of Government assistance made it even more important to enlist the help of the natural history societies in carrying out the vegetation mapping. Here again, the prospects were far from encouraging. In a presidential address to the Botanical Society of Edinburgh in 1903, James W. H. Trail (*1851–1919*) read out a long list of complaints. Nothing of permanent value could be achieved until there was greater consistency and precision in the way plant records were collected and published, and greater collaboration between members of the individual societies, and between the societies themselves (Trail 1903). There were likely to be even greater problems when it came to collecting data on vegetation.

The only hope was to forge a close working relationship between the 'amateur' naturalist and the 'professional' who could provide the necessary scientific leadership (Lowe 1976). In a paper to the British Association for the Advancement of Science, setting out the aspirations of the Scottish Natural History Society, Marion Newbigin depicted the members of the natural history societies as 'the handmaiden of science', performing 'the drudgery necessary before any step in advance can be taken' (Newbigin 1901). Marion Newbigin (*1869–1934*) was herself a 'professional', holding the post of lecturer in zoology at the Medical College for Women in Edinburgh (her sex precluding her holding a full university post) (Anon. 1934).

In the event, nothing came of the various proposals. Instead, the initiative passed to England, and in particular to Yorkshire, where William Smith was by now a lecturer of what was shortly to become the University of Leeds.

1.2.2 *The botanical surveys of Yorkshire*

No one was better placed than William Smith to bring about collaboration between the 'amateur' and 'professional'. He joined the Yorkshire Naturalists' Union soon after his arrival in the county in 1897. At his instigation, the Union formed a Botanical Survey, with the aim of publishing a flora that contained not only 'floristic details', but descriptions of 'the social conditions and life of the plants'. As convener and secretary of the Survey, Smith wrote a guide, intended to help those members who wanted to take part in such surveys but who could not be expected to 'give the time necessary for the full and detailed studies which a complete study of vegetation entails' (Smith W. G. 1903a).

Although the stock of 250 reprints was soon nearly exhausted, the guide brought little in the way of tangible results. As an experiment, botanists attending the Union's field meetings in 1906 were handed Ordnance Survey maps to annotate. Whilst this guaranteed some kind of permanent record for parts of the county, the scarcity of botanists taking part in the excursions militated against the success of the scheme[1]. It soon became clear that, even in Yorkshire, the 'professional' would have to carry out the vegetation surveys himself—he could not expect the 'amateur' to do it for him.

Smith himself had begun to carry out vegetation surveys in Yorkshire in 1898, initially to test his brother's methods in another area. In 1901, he read a paper to the meeting of the British Association for the Advancement of Science, describing the botanical surveys carried out by his late brother (Smith W. G. 1902a). As Woodhead (1929b) recalled later, 'the wide outlook and the broadening of the botanical horizon' which characterized Robert's work 'came into our science as a breath of fresh air', taking botanists out of the laboratory into the field. The response to the paper was so encouraging that William Smith decided to extend the Yorkshire surveys. In collaboration with two of his senior students, he published vegetation maps at a scale of a half-inch to the mile (1 : 126 720) for a quarter of Yorkshire in the *Geographical Journal* for 1903.

Covering 1700 square miles of countryside, the maps were the first to appear for an English county. Following an introductory section by Smith, Charles Moss described the map covering the south-west of the West Riding, for which he was the principal surveyor (Smith W. G. & Moss 1903). In a second paper, William Munn Rankin (*1878–1951*) described the sheet for the north-east of the Riding, where he had carried out most of the survey work (Smith W. G. & Rankin 1903). In the *Naturalist*, John Baker congratulated the new generation of botanists on their thorough and excellent work (Baker 1903).

Charles E. Moss (*1870–1930*) spent most of his early life in the Halifax area. He nearly died in 1893, and was ordered by his doctors to spend as much time as possible out of doors. He had recently joined the Halifax Scientific Society where, at the meetings of the vigorous botanical section, he found a purpose for his long, solitary rambles over the moors of the Calder valley. Soon after he resumed his teaching career, Moss obtained a Queen's Scholarship, and

embarked upon both a teacher training and a degree course in science at the
Yorkshire College in 1895. Three of the staff at the College exerted an important influence over him. The head of the training department, J. Welton, helped
to develop his formidable logical faculties. The other two were the professor of
biology, Louis C. Miall (*1842–1921*), and William Smith (Crump 1931).

Miall's habit of constantly posing the questions 'how and why?' suited the
temperament of Moss. According to Miall (1897), it should be the aim of a
teacher not so much to impart knowledge as to excite the natural curiosity of
the pupil. When studying plants and animals, it was essential to study them on
the spot where they grew. The influence of Miall's methods can be discerned in
Moss's early papers. The first had the title, 'Why do flowers bloom in spring
and in autumn?' Moss found the answer in 'a simple application of Natural
Selection', which could be understood at first hand 'by those who have eyes to
see' (Moss 1896). Another paper commended the study of green scums for those
young naturalists who were 'tired of naming species, and wearied with text
book technicalities' (Moss 1898). A paper soon followed, comparing what he
found in the Halifax area with the 'almost continuous records' available since
the late 18th century (Moss 1900a).

Although Moss was respected for his extensive knowledge of botany and
was popular among his own generation, some of his seniors found him bump-
tious (Ramsbottom 1931). He never suffered fools gladly; his brusque manner
and plain speaking gave offence to the older school of Yorkshire botanists
(Crump 1931, Tansley 1931). In those early years, Moss may have exerted his
greatest influence as a contributor to larger works, organized by those who
recognized his talents and tolerated his off-putting manner. One of these works
was the *Flora of Halifax*, which was published in parts between 1895 and 1904
(Crump & Crossland 1904). The new flora not only had the qualities of Baker's
earlier *Flora of Yorkshire*, but was the first to describe 'the habitat or station of
a species'. As Smith (1903b) remarked, even a stranger to the county could have
gained an impression of the actual appearance of the vegetation.

Moss had taken little, if any, interest in the Scottish surveys until the summer of 1899, when one of the editors of the Halifax flora, and a close friend,
William B. Crump (*1868–1950*), gave him a copy of Robert Smith's paper on
the 'Plant associations of the Tay Basin'. Not only was Moss keen to apply the
concept of plant associations to the Halifax area, but he saw how it could sustain interest in survey work among local naturalists, now that 'the registration
of new species was pretty well exhausted' (Moss 1900b).

Through a systematic survey of the moors of south-west Yorkshire, Moss
was able to distinguish seven type-areas, each of which reflected a distinctive
pattern of historical development (Figure 1.1). The bilberry (*Vaccinium myrtil-
lus*) was dominant in the plant associations of Boulsworth Hill and other parts
where the sandstone rose to a distinct peak. The great bulk of moorland was
covered by cotton grass (*Eriophorum vaginatum* and *E. angustifolium*) (Rish-
worth type). Forming a more or less complete ring round the cotton grass moor

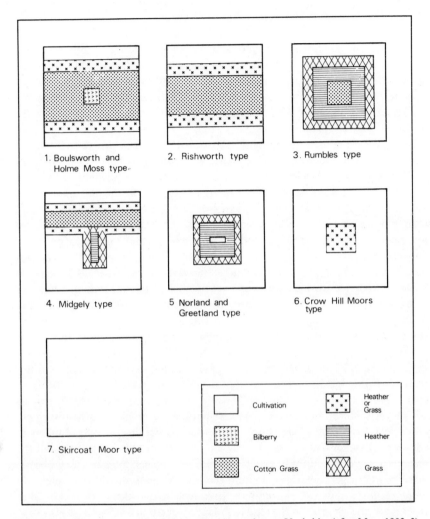

FIGURE 1.1 The seven type-areas of moorland in south-west Yorkshire (after Moss 1902–3).

was a belt dominated by heather or, where the moor edge was steep and shaley, rough grass (Rumbles type). Midgley Moor was an example of a moor that extended 'long strips or tongues far into the main valleys'. Moss (1902–3) explained how, 'in times past the spade and the plough made great inroads on the moor', and had left outlying, isolated patches now covered with grass and heather. Norland and Greetland Moor was dominated by heather. On the Crow Hill Moors of Sowerby, cultivation had left only three or four isolated patches of moor: Skircoat Moor, the last in the series of type-areas cited by Moss, had been converted entirely to cultivation.

1.2.3 'The district lying south of Dublin'

The botanical surveys of Scotland and Yorkshire, pioneered by the Smith brothers and extended by Moss and Rankin, demonstrated what could be achieved through the diligence of individual workers. Whilst natural history societies and field clubs could provide personal companionship, attentive audiences, and an outlet for publication, it was becoming clear that they were not essential to the execution of survey work.

Such conclusions were amply demonstrated in Ireland, where some of the earliest experience in vegetation mapping was gained. By far the most influential figure was Robert Lloyd Praeger (*1865–1953*), whose phenomenal vitality as a field-worker was complemented by his ability to make the maximum use of his vast store of botanical records, whether in his addresses to the Belfast and Dublin Naturalists' Field Clubs, editorship of the *Irish Naturalist*, or his own separate publications (Doogue 1986). By 1900, he had published over 120 papers, notes, and reviews on Irish floristics. Following his move from Belfast to Dublin in 1893, where he became a librarian in the National Library of Ireland, Praeger turned his attention to the whole of Ireland, making a series of field excursions to almost every county between 1896 and 1900. The publication of his *Irish Topographical Botany* in 1901 at a stroke both consolidated and extended the floristics and plant geography of Ireland (Praeger 1901a, Collins 1985).

These years of intensive field-work gave Praeger a comprehensive and unrivalled knowledge of the geography of plants. In his account of the history of Irish vegetation studies, White (1982) wrote of being unable to trace the origins of Praeger's interest in plant ecology, as opposed to floristic botany. It is, however, clear from his popular book, *Open-Air Studies in Botany*, published in 1897, that Praeger had begun to make effective use of habitats to expound the systematics and structures of plants encountered in such localities as a Connemara bog (Praeger 1897). As White reasoned, the book provided ample testimony of how far Praeger would have been receptive to the ideas contained in Robert Smith's paper, 'On the study of plant associations'.

Whatever the exact source of inspiration, Praeger had adopted the term 'plant association' by the time he came to write a series of articles in 1901 for the illustrated magazine, *Knowledge*. There he spoke of how such natural groupings were often dominated by one or more social species. In other cases, though some plants might be 'more abundant than others, the association may be a commonwealth rather than a monarchy' (Praeger 1901b). From 1903 onwards, he began a series of summer visits to the islands off the west coast of Ireland. In both his reports on Clare Island in 1903 and Achill Island in 1904 a description of the vegetation took precedence over the floristic lists (Praeger 1903, 1904). As White (1982) observed, the term 'plant association' was used quite casually in analyses of the seaboard, dunes, saltmarsh, moorland, lakes, and farmland.

Praeger attended the Glasgow meeting of the British Association in 1901, where he heard the paper given by William Smith, describing the survey work of Robert Smith and calling upon others to follow the lead (Smith W. G. 1902a). Although Praeger was very sceptical as to whether plant associations could be mapped in the way described, he decided to carry out a trial survey himself, in collaboration with George H. Pethybridge (*1871–1948*), who had recently been appointed lecturer at the Royal College of Science for Ireland, having obtained a PhD degree at Gottingen. So as to avoid any 'preconceived notions as to what the associations (if any) could, would, or should be', they decided not to read the more detailed accounts of how Robert Smith had conducted his surveys.

Between them, the two men surveyed 200 square miles of terrain between Dublin city and Loch Bray to the south, as well as the island of Lambay, the coastal area of County Dublin to the north of the city, and five square miles of the Vale of Glendalough. The surveys for the two latter areas were never published—they were probably made by Pethybridge. Work on the area south of Dublin took about 100 days, spread over the winters of 1901–5. Praeger and Pethybridge quickly discovered that it was possible to recognize and map plant associations at a fairly large scale. The field data recorded on Ordnance Survey 6-inch (1:10 560) sheets were transcribed onto extracts of four Ordnance Survey 1-inch (1:63 360) sheets, which were published as one map, together with an illustrated text of some 60 pages, in the *Proceedings of the Royal Irish Academy* in 1905.

The vegetation on the published maps was identified by its dominant species. In the moorland zone, for example, colours and a notation were used to distinguish *Eriophorum*, *Racomitrium*, and *Scirpus* moors, the *Calluna* and *Juncus* associations, and *Vaccinium* edge. At each site chosen for analysis, the species were listed and ranked in order of relative abundance, as judged by eye. The lists of each association were then tabulated and sorted, so that those occurring at all sites were listed first and those occurring in successively fewer sites ranked in order after them. The published paper concluded with detailed notes on the ecology of the littoral, agrarian, hill pasture, and moorland zones demarcated by this procedure (Pethybridge & Praeger 1905).

In a review of the paper, published in the *Irish Naturalist*, William Smith wrote of how it marked 'the opening of a new era in the investigation of the plants of Ireland' (Smith W. G. 1906). By providing so outstanding a model for all future vegetation maps, the paper also gave a much needed boost to the confidence of a recently formed British Vegetation Committee. Before looking more closely at the contribution which Praeger and Pethybridge made to that Committee, it is necessary to review the developments that had been taking place in academic botany, in both Britain and America.

1.3
Self-Conscious Ecology

Although the literature of earlier periods might contain many examples of concepts and methods which would have been described as ecological in the 20th century, it was not until the last years of the 19th century that a self-conscious and sustained approach to ecology became apparent (McIntosh 1985). The term 'Öekologie' was one of many to be promoted by Ernst Haeckel (*1834–1919*), the leading German disciple of Charles Darwin. He defined it as 'the comprehensive science of the relationship of the organism to the environment' (Haeckel 1866).

The word 'ecology' was soon taken up in its various forms. As McIntosh (1985) remarked, 'ecologists began to define ecology by doing it and recognizing that they were doing it'. Robert Smith used the term at least four times in his paper on plant associations, published in 1899. Of equal significance, biologists in general began to use the word. One of the earliest and most conspicuous occasions was the presidential address to the British Association in 1893, given by the eminent physiologist, John S. Burdon-Sanderson (*1828–1905*).

In a wide-ranging address on the origins and meaning of the term 'biology', Burdon-Sanderson (1893) identified three kinds of approach, namely the entirely experimental approach of physiology, secondly morphology, with its focus on the form and structure of species, and thirdly ecology, 'the science which concerns itself with the external relations of plants and animals to each other, and to the past and present conditions of their existence'. Reflecting perhaps on his own life-long interest in field botany (F.G. 1907), Burdon-Sanderson described ecology as 'by far the most attractive' of the branches. It was in ecology that 'those qualities of mind which especially distinguish the naturalist' found their greatest expression.

Ecology attracted not only 'good biological minds' but also the weaker students (Tansley 1951). By the turn of the century, the triviality and slovenliness of many papers gave substance to those botanists who distrusted, even if they did not openly condemn, what they perceived to be ecology. It was to combat this kind of cynicism that Arthur G. Tansley (*1871–1955*) read a paper to the Annual Meeting of the British Association in 1904, in which he sought to identify 'the problems of ecology', and to set out a fresh agenda for the future (Tansley 1904). Soon afterwards, the first ecological textbook to adopt a 'more rigorous approach' appeared in the United States. Called *Research Methods in Ecology*, it was written by Frederic E. Clements. It is the purpose of this section to look more closely at the early contributions of these two men.

Tansley's childhood interests in natural history were sustained by successive schoolmasters, and by his reading at the age of 15 of key texts on plant physiology and evolutionary theory. Having attended the intermediate science lectures at University College, London, in 1889–90, he went up to Cambridge, where he read botany, zoology, physiology, and geology. In 1893, he received the flattering invitation to become the Quain Student and assistant to F. W. Oliver, the head of the botany department at University College, London. He accepted, and was made assistant professor in 1895. He stayed until 1907 (Godwin 1957, 1977).

At first, Tansley was primarily interested in plant anatomy. His influential lectures on 'The evolution of the Filicinean vascular system' (given to the advanced botany course in the University of London) were published in the *New Phytologist*, and later issued as a reprint (Tansley 1908). They gave 'a new and welcome impulse' to that branch of botany. It was, however, Tansley's 'plasticity' which Oliver later claimed to envy. No matter what branch he took up, Tansley became its master[2]. His decision to learn German enabled him to read the German translation of what he came to call Warming's 'epoch-making book', *Plantesamfund*. As well as incorporating the essentials of Warming's book in his lecture course, Tansley tried to match the plant communities of southern England with those described by Warming (Tansley 1924).

There also appeared in 1898 'the encyclopaedic work' by Andreas F. W. Schimper (*1856–1901*), called *Pflanzen Geographie auf physiologischer Grundlage*, highlighting the relationships of plants to different soils and climate throughout the world. It was a further striking illustration of how ecology was coming to occupy a position between natural history, with its traditional botany, floristics, and plant geography, and a new kind of botany, based on experimental physical studies in the laboratory but increasingly focused on examining plants in the field (McIntosh 1985). Schimper (1903) wrote of how it was only through a close study of the conditions necessary for plant life, as revealed by experimental physiology, that ecologists could break free from the gibes of dilettantism, or of anthropomorphic trifling, which threatened to bring the subject into 'complete discredit'.

Tansley's visit to Ceylon, the Malay peninsula, and Egypt in 1900–1 gave him the opportunity to apply and assess the insights he had gained from Warming, Schimper, and other continental writers. Schimper had classified the world's vegetation into tropical, temperate, Arctic, mountainous, and aquatic, as determined by three sets of factors: heat controlled the choice of flora; climatic humidity created the vegetation; and soil picked out and blended the material supplied by the two climatic factors, and added a few details of its own.

On his return to England, Tansley was struck by the absence of any British journal suitable for publishing notes and discussion, as opposed to conventional papers setting out the results of completed research. This was one reason why he decided to finance and publish a journal of his own, called *New Phytologist*, which in its first 10 volumes included six contributions to a series, 'Sketches of vegetation at home and abroad' (Tansley & Fritsch 1905).

Not only did the new journal play an important role in fostering 'the infant growth of British ecology' (Godwin 1977, 1985a), but Tansley himself began to give the subject a greater sense of purpose and cohesion. In his lecture to the British Association in 1904 (reprinted in the third volume of the *New Phytologist*), he described how the global distribution of plant species, far from being mixed haphazardly, was arranged in geographical aggregates, which were 'the result of perfectly definite and ascertainable, though of immensely complex, causes'. Ecology was the name given to the study of these aggregates, or plant associations, and to 'the species and individuals comprising them, and their relations to one another and to their common environment'. In the same way as it had been impossible to formulate problems in comparative anatomy until the structure of organisms had been described, so it was impossible to put forward hypotheses in ecology, capable of being tested, until a body of descriptive data had been acquired. Tansley (1904) argued that it was the purpose of vegetation mapping to secure part of that information.

Tansley's agenda for ecology corresponded 'to the two stages of procedure inevitable in natural science'. The first was to characterize, enumerate, and describe the plant associations, and the second was to explain what was found. How did the plants come to live together and maintain themselves on a specific area with definite environmental conditions? Why did they exhibit distinctive morphological and physiological characteristics, and what was their functional relationship to one another and to their inorganic surroundings? Field laboratories were required to provide the facilities for observation and experimental work.

By 1904, important initiatives were under way in America. Frederic E. Clements (*1874–1945*) was one of the most outstanding pupils of Charles E. Bessey (*1845–1915*), the head of the botany department at the University of Nebraska. By the time Clements graduated in 1894, he was already actively involved in a botanical survey of Nebraska that had been initiated by Bessey's assistant and former pupil, Roscoe Pound (*1870–1964*). The main purpose of the survey was to record the vegetation of the state prior to its being ploughed up by the new settlers. It offered ample experience in field-survey methods (Tobey 1981).

A turning point for Pound and Clements may have come in 1896, when Bessey asked Pound to review a new book, *Deutschlands Pflanzengeographie*, by Oscar Drude (*1852–1933*), the director of the Royal Gardens at Dresden. The book marked the first attempt to describe in detail the vegetation of an area as large as Germany (Drude 1896). For Pound and Clements, the reading of the book may well have formalized what was already in their minds, namely the transformation of the survey of Nebraska into a phytogeographical survey (Pound 1896). They published their *Phytogeography of Nebraska* in 1897, having already used the text as the basis of a joint doctoral thesis (Pound & Clements 1900).

Ecologists studied not so much what they saw, but what their data revealed.

Pound and Clements found the existing impressionistic or qualitative methods so frustrating that they devised, by trial and error, a procedure whereby metre plots were set out in 'typical situations' across the prairie, and the number of plants of each species was counted in each. The imperceptible transition from one form of prairie to another across the plains of Nebraska could be recorded with a precision never before attained (Pound & Clements 1898a,b). Within a decade, the quadrat had become one of the most important techniques in ecology. From it there developed the permanent transect and fixed-point photography.

Soon after completing their *Phytogeography of Nebraska*, Pound and Clements went their separate ways. Roscoe Pound entered law school, eventually becoming Dean of the Harvard Law School and a famous advocate of sociological jurisprudence (Tobey 1981). Frederic Clements developed what Pound later called 'a biological philosophy' (Pound 1954). His book, *Research Methods in Ecology*, published in 1905, brought him both international fame and the offer, which he accepted, of the chairmanship of the botany department at the University of Minnesota.

The book was an important turning point both for Clements personally and for ecology at large. In seeking to identify the more important factors determining the habitat of species, Clements (1905) stressed how each factor, like the species that covered the prairies, was part of a continuum. Even soil factors could be given interval values, if their gravel/sand/clay content was continuously measured. By conceptualizing habitat factors in this way, it was possible to portray field data in the form of a graph. Eventually, it might even be possible to relate the salient features of an entire habitat in a single graphic portrait.

The book was, however, more than a discourse on ecological methods—it was an exposition in general theory, in which Clements set out to merge ecology with physiology. According to Tobey (1981), the book owed much of its success to the way in which it drew on two distinct traditions, namely the idealistic and sociological traditions. The former was represented by the writings of Humboldt, Grisebach, and Drude, who had conceived plant formations as units distinct from the plants that individually made up the vegetation. The sociological tradition was an extension of another, which perceived human society as a superorganism, or an organism in itself. Such a perception had undergone a revival during the 1890s.

Many years later, Pound recounted how he and Clements had studied closely a book called *Principles of Biology*, by the English philosopher, Herbert Spencer (*1820–1903*), 'of which we had expected great things in the days when Comtian Spencerian positivism was almost a religion to scientists' (Pound 1954). The book gave a legitimacy to the conception of human society as an organism. First published in 1874, a revised and much enlarged edition appeared in 1898–9. In compiling each part of the new edition, Spencer was guided by an acknowledged expert. On the advice of Oliver, he asked Tansley to help revise those parts dealing with plant morphology and physiology[3].

In his discussion of plant physiology, Spencer (1898–9) quoted many examples of how, 'along with the increasing multiplication of types of organisms covering the Earth's surface', there had also been a trend towards mutual dependence. Whereas in the early stages of evolution there had been both homogeneity and incoherence, the vast assemblages of living things had become both more heterogeneous and coherent. Along with progressive differentiation, there was ever-increasing integration so that, in Spencer's words, a point had been reached where

> we may recognize something like a growing life of the entire aggregate
> of organisms in addition to the lives of individual organisms—an
> exchange of service among parts enhancing the life of the whole.

It was during a visit to the Colorado Mountains in 1896 that Clements's interest in grassland ecology widened to include forest ecology. In doing so, he came to realize that, although ecologists had long been aware of habitat factors, and particularly of climate, they had overlooked the wider role of the plant as a unit of vegetation, or plant formation, which could itself be 'regarded as a sort of multiple organism'. In a section of his book on *Research Methods in Ecology*, called 'Vegetation as an organism', Clements described how, as an organic unit, a plant formation could exhibit 'activities or changes which result in development, structure and reproduction'. The changes were progressive or periodic, and might to some degree be rhythmic. As in the case of individual plants, constructive forces would prevail until the formation reached maturity, after which destructive forces became dominant.

The invasion of species and the response of others, as recorded by the quadrat method, could be perceived as stages in a succession. As well as being influenced by physical factors, the vegetation was brought about by changes within the habitat itself. These might prevent weathering, reduce run-off, bind wind-blown soils, enrich or exhaust soils, accumulate humus, or modify atmospheric factors. The trend was always towards increasing stability. Forests represented the most stable phase, and could be regarded as the end-point of all successions, wherever trees could grow. Grasslands fulfilled a similar role in prairie and plains successions, and in many Arctic-Alpine regions.

The review of *Research Methods in Ecology* which appeared in *New Phytologist* was considerably longer than any to appear in the American journals. The original intention was that the review should be written by Tansley and the Cambridge physiologist, Frederic F. Blackman (*1866–1947*). Blackman withdrew, having read the book. Although impressed by Clements's ideals and breadth of vision, he was repelled by the treatment of plant physiology[4]. It was left to Tansley to incorporate these strictures in what still purported to be a joint review. Having praised the book for its comprehensiveness and logical treatment of a difficult and little-understood subject, readers were warned against following Clements's simple hypothetical scheme for correlating factors, functions, and structure. Much more analytical work was required on the part of the specialist in physiology. Whilst it might be legitimate to portray vegetation as an organism, the term could only be used in the sense of an analogy. As quasiorganisms, both vegetation and human society itself lacked many of the essential qualities of real organisms (Blackman F. F. & Tansley 1905).

1.4

The British Vegetation Committee

In December 1904, a small committee was formed with the purpose of bringing together the very few botanists with some experience of vegetation mapping. In one sense, there was nothing remarkable in this. There were plenty of specialist bodies devoted to particular plant or animal groups. In another sense, the new committee was very different, in as much as it had within its membership academic figures who perceived the specialist objectives of the committee in a much wider context than was usual for such groups. Vegetation mapping appeared to them to be the most obvious way of giving British ecology the distinctiveness and rigour that it had previously lacked so conspicuously. It was this fusion of objectives, as represented by the varied outlook and experiences of the membership, that enabled the short history of that committee to become one of the most formative periods in the history of British ecology.

It is impossible from the evidence that survives to determine whether William Smith or Tansley made the first move in forming a committee—they may independently have come to the conclusion that one was needed. The contribution made by Smith was vital. Not only did he and his students provide an example of what could be achieved, but his own personality played a large part in forming and sustaining the committee. 'In his quiet way, a singularly attractive man', Smith combined 'the solidity and pertinacity of the best type of Lowland Scot' with a most modest nature, never desiring credit for himself. Unselfish, he was always helpful to others (Tansley 1929a).

For the Smith brothers, the surveys were only the first step in an ambitious attempt to map nations, continents, and ultimately the world (Lowe 1976). Robert Smith had intended to write a paper assessing how far his surveys had shed light on the relationships that existed between the vegetation of Scotland and that of neighbouring countries (Smith R. 1900b). For comparisons of this kind, William Smith realized that methods had to be standardized and adequate guidance given to the less experienced surveyors.

Meanwhile, Tansley had perceived the need for a greater sense of purpose and cohesion in ecology. The merits of greater liaison between workers had been brought home to him by the fact that he had tried for two years to analyse and correlate the vegetation of southern England, completely unaware of the work of Robert and William Smith (Tansley 1947a). In his lecture to the British Association meeting of 1904, Tansley called for a 'central committee for the systematic survey and mapping of the British Isles'.

On 3 December 1904 there occurred what Crump (1931) called 'the memorable meeting' convened by William Smith at his house in Leeds, when the Yorkshire workers joined Tansley in founding the Committee for the Survey and Study of British Vegetation (which was shortened to the British Vegetation Committee in 1911). In his account of the meeting, Tansley defined the purpose of the Committee as being to review the present position of vegetation surveys, co-ordinate further work, and secure uniformity of method in so far as may be desirable.

The formation and membership of the Committee highlighted how the impetus of vegetation mapping was passing out of the hands of the natural history societies. Membership was limited to those persons present at the Leeds meeting (Smith, Tansley, Moss, and T. W. Woodhead), and to those who were invited but could not attend (M. Hardy, F. J. Lewis, R. L. Praeger, G. H. Pethybridge, and W. M. Rankin). All were college lecturers, except for Praeger. Although there was scope for adding others 'engaged in the active study of British vegetation', the Committee believed it would not be 'desirable to try to obtain workers ... through natural history societies etc.'. Committee members would instead 'do what they can to obtain suitable workers to co-operate with the committeee'.

Most of the 19 meetings held by the Committee over its 10 years of life were very well attended. This was helped by the fact that many were held at the same location as, and just after, the Annual Meeting of the British Association. The cost of membership was fixed at one shilling until 1909, when it rose to half a crown. Oliver became a member at the second meeting (in London) in March 1905. Otto V. Darbishire (*1870–1934*) and W. B. Crump were elected at the fourth meeting, and Professors R. H. Yapp and F. E. Weiss (who had previously attended as 'visitors') in 1907. Three further members were elected later. Regular reports of the Committee's proceedings and activities were published in the *New Phytologist* (Smith W. G. 1905b,c).

The third meeting, held in Liverpool in November 1905, marked an important turning point in the Committee's history. By then, a six-page pamphlet had been issued, setting out for beginners 'Suggestions for beginning survey work on vegetation'. It was already clear, however, that some time must elapse before there could be any consensus on what range of scales and colour notation should be adopted. As Smith (1905b) reported, any attempt to impose uniformity would have destroyed 'the elasticity necessary in the present experimental phase of botanical survey'.

For the first time, a large part of the Liverpool meeting was devoted 'to a series of short informal papers on the investigations of members, recently published, or carried out during the season'[5]. The stimulative effect of these exchanges of information was conveyed by Praeger (1906), who described in the *Irish Naturalist* how he and Pethybridge had left the meeting, which had lasted for 13 hours over two days, deeply impressed by the way every member was 'actively and practically engaged on the work' that formed the subject of the Committee's deliberations. Many years later, he recalled how 'we discussed with many words elementary ideas respecting plant formations and associations', the mere existence of which 'some of us did not feel too sure' (Praeger 1923).

It was in Ireland that the concept of vegetation mapping had undergone its most rigorous test. At the Committee meeting in Liverpool, Praeger and Pethybridge showed examples of their 6-inch Ordnance Survey field sheets, and the final proof of the 1-inch coloured map for the area south of Dublin. They represented the first survey of the moister, western regions of the British Isles. The *Scirpus* and *Rhacomitrium* types of moorland had not previously been mapped (Smith W. G. 1905b).

The objectivity and utility of the Committee's approach to vegetation mapping were affirmed. In an introduction to their paper, published in the *Proceedings of the Royal Irish Academy*, Pethybridge and Praeger (1905) described how species were 'naturally aggregated into a number of vegetation-types or synthetic plant groups, which recur within the area wherever similar conditions of environment exist'. The species did not necessarily have any floristic relationship with one another. They were bound 'together in common comradeship or societies by similar requirements as to the necessities of existence, or even by dissimilar but mutually complementary requirements'.

For the first few years, most of the attention of the Vegetation Committee was devoted to assessing the methods and results of surveys already under way. The most extensive was that of the Scottish Highlands, carried out by Marcel Hardy. His coloured vegetation map, which was printed at the unusually small scale of 10 inches to the mile (1:633 660), first appeared in a monograph, and then in a paper published in the *Scottish Geographical Magazine*. Twelve vegetation types were distinguished on the basis of their general character and the combination of dominant and subordinate species. The general relationship of each type to geology, soils, climate, and the impact of farming and forestry was discussed (Hardy M. 1905, 1906).

As William Smith remarked later, the first surveys were essentially reconnaissance maps, taking large areas of countryside and treating them rather superficially. As each memoir accompanying the maps appeared, so greater attention was paid to the effect of environment on the patterns of vegetation discerned (Smith W. G. 1909a). This shift in emphasis was particularly striking in the work of Francis J. Lewis (*1875–1955*), a lecturer in geographical botany at Liverpool University (Balls & Brimble 1955). Over a period of three years, he surveyed 560 square miles of the Eden, Tees, Wear, and Tyne basins, recording the vegetation on Ordnance Survey 6-inch maps. His maps, published at a scale of 1 inch to the mile, in the *Geographical Journal*, revealed 'a remarkable similarity' with the plant associations recorded by Robert Smith at similar altitudes in south-east Scotland (Lewis 1904a,b).

Moss was by far the most productive member of the Vegetation Committee. He had left Yorkshire in 1901 and taught in a school at Bruton for two years. While there, he began a vegetation survey of the area covered by the Ordnance Survey sheet for Bath and Bridgwater. It was completed in the summers of

1903–4 with the help of Crump. Mapping was at a scale of 1 inch to the mile, and in some parts 6 inches to the mile. With the help of a grant from the Royal Society, the Royal Geographical Society published the vegetation map at the scale of 2 inches to the mile in 1907, as part of a memoir (Moss 1907).

By that time, Moss had carried out extensive surveys of the Peak district of Derbyshire. Anxious to abandon schoolmastering and devote himself entirely to botany, he had left Somerset in 1902 for the post of lecturer in biology at the Teachers' Training College in Manchester, where he could attend the advanced courses offered by Owen's College. As early as the Liverpool meeting of November 1905, Moss warned of how the high cost of producing coloured maps was likely to lead to delay in publishing the Peak District maps.

The Vegetation Committee decided to seek financial assistance from the Board of Agriculture. The cost of the vegetation map for the area south of Dublin had been met by the Department of Agricultural and Technical Instruction for Ireland, and the Board of Agriculture was responsible for the activities of the Ordnance Survey in England, Wales and Scotland. Tansley had interviews with William Somerville (*1860–1932*), and then with his successor at the Board, Thomas H. Middleton (*1863–1943*), both of whom had held the chair of agriculture at Cambridge. Although the Board made it clear that there was 'no fund available directly for that purpose', the Vegetation Committee found the interviews sufficiently encouraging for a subcommittee to be set up, largely at Moss's instigation, to decide on a colour scheme for future maps[6].

In a letter to the Board in August 1907, Tansley proposed that six sheets might be printed over the first three years. He accepted suggestions that further information might be given on agricultural land, but rejected the idea of adopting the scale of a half-inch to the mile. Experience had shown that whilst this scale might be adequate for illustrating the relationship of vegetation to climate, even the 1-inch scale was barely large enough for showing the edaphic relationship. Progress was slow. In late 1907, the secretary of the Board of Agriculture advised that 'the time was not favourable for approaching the Treasury', and he was still hesitant when he saw Tansley in the following May. In the autumn of 1908, a verbal offer was made to publish a map and memoir under the Board's auspices. Replies to a circular sent to members of the Vegetation Committee affirmed that Moss's map of the Peak District should be the first to be submitted for publication.

In his report of 1909, Smith (1909b) expressed the hope that 'a permanent means of publication' had been found, and that the long delay in the appearance of members' work was at an end. The optimism was misplaced. At the same meeting that Moss showed members the proofs of his maps, just received from the Ordnance Survey, Tansley announced that 'the Treasury had interposed and vetoed the publication of these Peak District maps and the memoir'. In a letter of February 1910, the Treasury informed the Board of Agriculture that there was 'no sufficient reason' why the cost of the maps or memoir should be met by public funds. Furthermore, neither the Stationery Office nor the Ordnance Survey could be made responsible for issuing the publications of a private body. The Vegetation Committee had 'no official status'[7].

Although the Board had strongly supported the Committee's application, there seemed little prospect of the decision being reconsidered. Moss doubted whether anything would be achieved unless the Government took over responsibility for the vegetation survey, and organized it on similar lines to the Geological Survey. The political situation in 1910 added to the difficulties of making further overtures and, at the Committee's meeting of December 1910, Moss reported that he would be taking steps to have the maps and memoir published through other channels. They were eventually published by the Cambridge University Press in 1913, with grants from the Royal Society and the Royal Geographical Society (Moss 1913)[8].

Although discussions were renewed in 1912 and the Committee agreed to apply to the newly appointed Development Commission for a grant, the impetus had gone. 'In view of the non-urgency of primary survey papers, and the question of a Society being formed', the Committee agreed with Tansley that 'it seemed better to postpone any definite steps'.

1.4.1 *'Work of a more detailed character'*

Although the failure to secure the necessary resources for publishing the primary surveys was a severe blow, its effect on the morale of the Vegetation Committee should not be exaggerated. The attractiveness of having a large-scale vegetation map for the whole country was as strong as ever, but its wider significance in terms of promoting ecology had already begun to wane. The main effect of the setback was to accelerate a shift of interest that was already taking place, which proved so relevant and stimulating that the Committee was ultimately forced to turn itself into a society with membership open to all.

There had always been instances of the same botanists carrying out both 'primary' surveys and more detailed types of investigation. Robert Smith had recorded the distance and direction of pine and larch seedlings from their parent trees at Tentsmuir in Fife. Most occupied the few parts that had escaped burning, and where the long heather afforded shelter (Smith R. 1900c). By the time the Vegetation Committee was founded, field-work had raised an ever-increasing range of questions. Why did plants group themselves as shown on the maps? Why was one species dominant in one part, and another elsewhere on ground that seemed to be similar? Enough had been mapped and published to make it clear that 'mere observation by the eye could not answer these questions' (Praeger 1923).

As William Smith concluded, in one of his reports on the Committee, 'work of a more detailed character was required' (Smith W. G. 1905a). For Tansley, the growing concern with causal mechanisms had the effect of bringing the interests of Committee members even closer to his own. Never concerned with

survey and classification for their own sake, he had, under the considerable influence of F. F. Blackman at Cambridge, come to see his own contribution
more and more in terms of what Godwin (1958) called 'physiological ecology'.

It was another founder-member of the Vegetation Committee who
pioneered the trend towards more intensive studies. Although Thomas W.
Woodhead (*1863–1940*) had carried out a survey of 66 square miles of the
countryside south of Huddersfield, most of his attention was focused on a small
number of woods within the area, plotting their species composition on Ordnance Survey 25-inch (1:2500) maps. Woodhead's interest in natural history
had first been aroused by classes held at the Mechanics Institute in Huddersfield. He joined the local naturalists' society, and then the Yorkshire Naturalists' Union, and began to teach botany and zoology at the Institute (which
became the Huddersfield Technical College). After two years, Woodhead decided to leave his office job in the local woollen company to become a full-time
teacher. He held the post of head of botany for the next 30 years (Woodhead
1923, Pearsall 1940a).

In what was probably the first intensive study of its kind in Britain, Woodhead drew attention to the role of soils in influencing the distribution of species.
In those parts of Birks Wood where a few inches of humus overlay the loams, he
found that the main components of the undergrowth occupied distinct zones in
the soil profile (Figure 1.2a). Far from the species being in severe competition
with one another, the soil requirements, modes of life, periods of vegetative
growth, and times of flowering and fruiting were each very different.

First in papers published in the *Naturalist* (Woodhead 1904), and then in a
paper read to the Linnean Society in December 1904, Woodhead used the term
'complementary association' to describe the kind of relationship where, insofar
as any plant was affected by competition, it came from other plants of the same
species. A very different situation prevailed in those parts of Birks Wood where
stiff clays or light sandy soils were encountered by the soil auger. Here, the bluebell was virtually eliminated by competition. The few bulbs present were tightly
packed between the roots of small trees, yellow dead nettles, dogs mercury, and
arum (Figure 1.2b).

Between the reading of his paper to the Linnean Society and its eventual
publication in 1906, Woodhead spent a year in Zürich, studying under Carl
Schröter (*1855–1939*), who was then completing his *Das Pflanzenleben der
Alpen* (Schröter 1904–8). Never before had the habits, structure, and adaptations of Alpine plant life been described so closely (Woodhead 1908, Tansley
1939a). The influence of Woodhead's year in Switzerland may be gauged from
the way he revised and extended his paper in order to illustrate how the more
'plastic' species among the xerophytes and mesophytes were the most likely to
invade one another's territory. They could adapt themselves so much more
easily to different environments (Woodhead 1906). The revised paper obtained
for its author the award of doctor of philosophy in the University of Zürich
(Tansley 1906a).

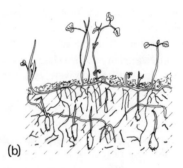

FIGURE 1.2 The influence of soil profiles on species distribution in Birks Wood, Yorkshire (after Woodhead 1904).

It was not many years before 'the personal associations formed in Committee' began to find tangible expression in print (Tansley 1947a). In 1909, a document was drafted in Cambridge, largely by Moss and Tansley, designed to bring order to the use of terminology. Ever since the International Congress of Geographers in 1899, there had been attempts at successive botanical congresses to reach agreement on a standard terminology—none had succeeded.

Moss used his considerable grasp of the published literature to argue that there was plenty of evidence in Britain, where the climate was so uniform, to support a hierarchy of three fundamental units of vegetation, based largely on soil type. The term 'plant formation' should be used to describe the whole of the vegetation occurring in a definite and essentially uniform habitat. A 'plant association' characterized minor differences within the otherwise uniform habitat, and the third term, 'plant society', corresponded with even less fundamental differences in the habitat. Some idea of the areal extent of each of the units was conveyed by the fact that plant formations could be recorded on quarter-inch (1 : 253 440) maps, whereas 1-inch maps were needed to show plant associations. Plant societies and the smaller examples of associations could only be plotted on 6-inch maps (Moss 1910, 1913).

The draft, which formed the basis of a paper on the 'Fundamental units of vegetation', published in the *New Phytologist*, was endorsed by the Vegetation Committee, and forwarded to the International Congress of Botanists at Brussels. It soon became apparent that neither this nor any other document would attract the consensus required. In their capacity as reporters to the Congress, Flahault and Schröter criticized the hierarchy for being based on too much hypothesis and subjectivity. Not only did climate exert a greater influence than Moss implied, but the relationship between plant communities and soil types in Britain did not necessarily hold for other countries and regions (Smith 1912a)[9].

Moss had drawn heavily on his own field surveys to illustrate how the different units of the hierarchy related to one another, and had 'a life history' of their own. He quoted examples from the Somerset Moors and Peak District to illustrate two kinds of succession. The lowland peat moors of Somerset probably began their history as associations of aquatic or marsh plants. Intermediate associations of *Myrica*, *Tetralix*, and *Molinia* were still found. The final stage was represented by a closed association of *Calluna* (Moss 1907). Whatever the phase of development, the formation remained the same 'organism'. Instances of retrogression in the Peak District could be found where streams had cut into the cotton grass moor, causing the hitherto uniform cover of peat to be broken into peat hags until, eventually, the underlying rock was laid bare and colonized by siliceous grassland (Moss 1913).

No scheme of botanical survey could ignore man as a biological agent (Smith W. G. 1909a). In his survey of the Pentland Hills, Robert Smith had recorded the way in which the cutting of drains could lead to the replacement of rushes or *Sphagnum* with dry pasture and heath. In this and other ways, the human impact grew stronger every year, so that a vegetation map soon became out of date, and acquired 'a new interest from the historical point of view' (Smith R. 1900a). The Liverpool meeting of the Vegetation Committee in 1905 discussed the feasibility of compiling maps of 'reconstructed' vegetation (Praeger 1906).

Moss's survey of the Bath and Bridgwater district of Somerset was the first to contain so much cultivated land. He found that the only clue to the character of the primitive vegetation was often the species occupying old pastures, or growing on the stone walls that divided the arable fields. On the peat moors, most of the native associations had been destroyed in the course of converting the moors to farmland. The only exceptions were where the turf had been cut for fuel. Here there were instances of history repeating itself. The wet hollows left by the turfcutters were colonized by the species characteristic of the primitive bog, forming 'substituted plant associations' (Moss 1907).

The effects of past land use and management had an obvious bearing on the choice of criteria in any classificatory system. In the first of what was intended to be a series of papers categorizing the principal types of vegetation in Britain, Moss, Rankin, and Tansley (1910) began by assessing how far any natural woodland had survived in a country that had been so intensively cultivated and thickly populated. From the evidence of the primary surveys already completed, they concluded that most woodlands were neither virgin nor plantations *de novo*. Even where more or less planted, the constancy with which the same association of trees occurred on similar types of soil over extensive, and often widely separated, tracts of countryside suggested that they were the lineal descendants of the primitive forest cover. The authors attributed the survival of so much seminatural woodland to the 'innate conservatism of the English landowner, as well as the backward state of forestry practice in the country'.

At the meeting of the Vegetation Committee in late 1907, Tansley put forward a tentative classification of English oakwoods, drawing on the earlier surveys of Scotland, Yorkshire, and Somerset. These seemed to suggest a primary division between the woods of *Quercus sessiliflora* in the drier parts, and those of *Q. pendunculata* in the moister parts[10]. Further investigations indicated that whilst some differences in the woodland pattern could be attributed to the effects of altitude and the east–west transition in climate, the decisive factor was usually differences in soil. Very wet soils encouraged an alder–willow series to develop. In other parts, it was the lime content of the soil that determined which of two types might prevail. An oak and birch series was typical of siliceous soils, and a beech and ash series was more frequent on lime-rich soils (Figure 1.3). Each of these soils could be subdivided on the basis of water and humus content.

The comparatively advanced state of woodland ecology, and the more rigorous approach being adopted in survey work, were demonstrated by a paper given to the Vegetation Committee in December 1910 by R. Stephen Adamson (*1885– 1965*), a graduate student of the Botany School in Cambridge, describing 'the physical coefficients of woodland plant societies'. At Gamlingay Wood, near Cambridge, Adamson found a close relationship between light and soil water content and a 'fairly definite range' in the occurrence of plants in the ground flora. 'By using the degree of these factors as ordinates', it was possible to construct a chart, showing 'a distribution pattern for each species which was later found to coincide in a striking way with the actual distribution in the wood' (Adamson 1912).

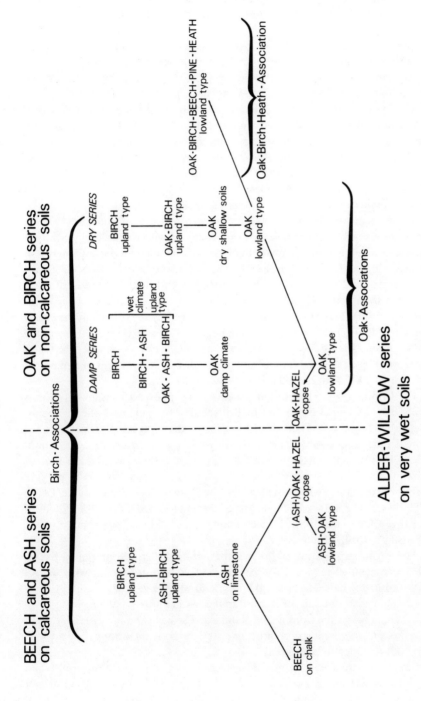

FIGURE 1.3 The influence of soils on woodland types (after Moss *et al.* 1910).

1.4.2 *The International Phytogeographical Excursion*

In seeking to define the different types of vegetation encountered in Britain, Committee members were able to draw not only on their discussions indoors, but also on the excursions that frequently followed the meetings (Smith W. G. 1907, 1909b). One of the first was to Crockham Hill in Kent in March 1907, as the guests of Tansley. They visited the oak with beech coppice, and stands of pure beech, on the Lower Greensand, and then moved to the higher ground, covered by stunted oaks, heath, and pine heath. Descending to the Weald, members saw 'good examples of nearly pure oak with a varied undergrowth of shrubby plants, almost in primitive condition'. The Committee minutes recorded how 'the excursion formed a noteworthy feature of this meeting and it was thoroughly appreciated'.

These excursions began to take on an even wider significance in 1908. In July of that year, Tansley took part in a tour of Switzerland, led by Schröter and organized in conjunction with the International Geographical Congress in Geneva. He was further impressed by the considerable benefits to be gained from excursions led by those with considerable knowledge of the local flora, and from the interchange of comments made by the remainder of the party (Tansley 1947a). These impressions were reinforced by the success of the excursion which Praeger led to the low-lying bogs and hills of Connemara, and the Carboniferous Limestone region of the Clare–Galway border, after the Committee meeting at the Dublin meeting of the British Association in September 1908 (Lewis 1908, Praeger 1908).

The Committee meeting at Dublin gave full support to Tansley's proposals for an international excursion of the British Isles. In the course of correspondence with Schröter and Flahault, it was decided that August 1911 would be the optimal time, and a subcommittee of Tansley, Lewis, Moss, Praeger, Smith, and Weiss was appointed to make arrangements. Because of difficulties of accommodation in the more remote areas, it was decided to invite only 12 foreign participants, who would be led by Tansley and G. Claridge Druce (*1850–1932*). The author of three county floras and secretary of the Botanical Exchange Club (Allen 1986), Druce's knowledge of British floristic botany was, in Tansley's word, 'unsurpassed'.

The 'business programme' for the International Phytogeographical Excursion recommended a 'warm tourist suit and waterproof cloak, with flannel shirts and strong nailed thick soled boots'[11]. It warned of how

> the British standard of comfort in the smaller hotels and inns is very
> much below the Continental standard. The beds, though generally
> clean, are not always comfortable, and the cooking of food frequently
> leaves much to be desired.

In the event, there were 11 guests. They were F. E. Clements (Minneapolis) and H. C. Cowles (Chicago) with their wives, O. Drude (Dresden), P. Graebner (Berlin), C. A. M. Lindman (Stockholm), J. Massart (Brussels), C. H. Ostenfeld

FIGURE 1.4 The route of the International Phytogeographical Excursion, 1911.

(Copenhagen), and C. Schröter and L. A. Rübel from Zürich. Of the four others invited, Warming, Flahault, and C. A. Weber accepted, but had to drop out at the last moment. Other Committee members joined for part of the month-long excursion, which ended at the venue of the meeting of the British Association in Portsmouth (Figure 1.4) (Tansley 1911a).

In Tansley's words, the Excursion 'was, by all available testimony, an outstanding success' (Tansley 1947). Cowles regarded its chief result as being

to internationalize for all time the subject of plant geography, and to
divest it of the provincialism which has hitherto too greatly
characterized it.

In his contribution to a series of impressions and reflections, published in the
New Phytologist, Clements (1912) wrote of how such an excursion afforded the
chance 'of scrutinizing one's own concepts'. For him, 'the problem of the moor,
with its scientific, economic and practical aspects', was the outstanding problem
for British vegetation studies. There was no more alluring a field for study than
the correlation of peat deposits and their successions with those of the present
day.

The most permanent bequest to British ecology was the decision of the
Excursion's subcommittee to produce 'a book having the character of a guide
to British plant associations, which would however be wider in its scope than a
guide for this excursion alone'[12]. With Tansley as editor, seven members of the
Vegetation Committee contributed sections to what became the first attempt at
'a scientific description of British vegetation'. Called *Types of British Vegetation*, Tansley (1912a) later recalled how 'the definite article was advisedly
omitted in naming the book in order to avoid any suggestion that the work included *all* existing types, a suggestion which would have been misleading'.
Dedicated to Warming and Flahault, advanced copies of the book, published
by the Cambridge University Press, were available at the start of the Excursion
(Tansley 1911b). Well illustrated and handy in size, the book was sold for as
little as six shillings. This barely covered production and advertising costs, and
the whole edition had to be sold as quickly as possible. Copies soon became a
second-hand rarity.

Tansley acknowledged 'a special indebtedness' to Moss for providing the
theoretical basis for classifying the types of vegetation. In a review, published in
the *Botanical Gazette*, Cowles (1912) commented on how all plant geographers
seemed to believe in the reality of the terms 'formation' and 'association', and
that the formation included various units or associations, but both the book
and the discussions held during the Excursion underlined how British plant
geographers applied the terms very differently from anyone else. In Cowles's
view, the British concept of the formation, as representing the aggregate of
plant associations as a successional series in a given habitat, was the most work-
able yet proposed. Although so much out of harmony with other proposals, it
had to be admitted that those who opposed the British concepts had signally
failed to agree among themselves on alternatives.

The publication of the book provided an opportunity not only to distil all
that had been published on the 14 formations, and their plant associations and
societies, which had so far been distinguished, but to report on work that had
previously been described only to meetings of the Vegetation Committee. One
such case was the survey carried out by Marietta Pallis (*1881–1963*) on the
Norfolk Broads (Stearn 1985). At the Liverpool meeting of the Committee in
1909 she had illustrated, with the help of lantern slides and diagrams, the trans-

ition of vegetation from open water, through an association of plants with submerged leaves and another with floating leaves, to open or closed swamp. The opportunity given to her and others to contribute to *Types of British Vegetation* brought their work before a much wider audience than would otherwise have been possible.

1.5
The British Ecological Society

In his review of *Types of British Vegetation*, the American botanist, H. C. Cowles, recalled how the study of vegetation had played until recently a very small part in British botany. Now, through the work of the British Vegetation Committee and the publication of this collaborative study, the British had not only 'caught up with their American and continental brethren, but in organization at least, they have forged ahead' (Cowles 1912). However true this might be in an international context, there were increasing doubts as to the effectiveness of the Vegetation Committee in promoting vegetation studies in Britain.

In a sense, the problems grew out of the success of the Committee. It was the custom to invite the staff of the botany department in which the committee was holding its meeting to join the field excursions. It was not long before the Committee came under pressure to widen its membership so as to involve others in all its discussions and activities (Salisbury 1964). How could this be achieved 'without lessening efficiency'? As Smith (1905b) emphasized, in his report on the Liverpool meeting of 1905, it was only by keeping numbers low that every member could exert a direct influence over the committee's affairs.

Tansley's early discussions with the Board of Agriculture gave impetus to discussions. The Government was more likely to make grants to a formal society than to 'a small informal committee with no regular constitution'. There were, however, considerable drawbacks to affiliating with an existing society. Moss suggested turning the Committee into a society 'to include all persons interested in ecological botany', including 'university botanists not publishing on ecological work', as well as teachers and other naturalists. Moss estimated that up to a hundred people would join. In return for their subscription, they would expect 'a small publication including general summaries, and free reprints of papers published'.

The minutes of the winter meeting of 1907 simply record that, after some general discussion, Moss proposed 'that the subject be deferred at present'[13]. Smith (1909b) later reported that 'a considerable majority of members' believed the Committee was 'the best instrument for advancing the subject'. There were 'few serious workers not already represented directly or indirectly on the Committee'. Tansley (1947a) later recalled how 'some of us were reluctant to dissolve the Committee, which had done very good work and had been a real source of inspiration for its members, for the sake of a society which might not be so successful'.

The Committee made, however, what William Smith called, one fundamental change. A category of associate membership was instituted, which was intended to include

> research students beginning ecological work and other botanists engaged in research within the scope of the Committee's work, or those directly interested in and useful to the Committee.

Smith (1912a) described the category as 'a matriculation stage', and of the 15 associates eventually admitted over the period 1910–12 Marietta Pallis and C. B. Crampton were among the first. Later associates included E. J. Salisbury and R. G. Stapledon (whose inclusion was to be highly significant in view of the inspiration ecology derived from advances in grassland husbandry).

It was not only external pressures that caused the Committee to reappraise its status; the rapport between members weakened as careers developed in different directions. Having completed their 'model' study of the area south of Dublin, Pethybridge and Praeger separated. Pethybridge turned to plant pathology (M., A.E. 1956), and Praeger focused once again on floristics (White J. 1982). For Woodhead, ecology led to studies of man himself, local history, and industry. Among his many civic responsibilities, he played a prominent part in founding the Tolson Memorial Museum near Huddersfield, intended to show how closely man was related to his surroundings (Pearsall 1940a). Munn Rankin became the principal of the new technical college at Bournemouth in 1911, from where he moved to a similar post at the new college in Burnley, Lancashire, in 1917. The posts left him with little time for field-work (Anon. 1950–1)[14].

The first member of the Committee to leave the country was Hardy, who in 1906 sailed for Mexico, where he organized one of the few successful land colonization and reclamation schemes in Latin America (Stevenson 1978). From that time onwards, he served as a kind of roving geographical consultant, an experience which enabled him to write two pioneer textbooks on plant geography, describing 'the conditions in which plants flourish, and their distribution in the great geographical divisions of the world' (Hardy M. 1913, 1920).

The greatest loss to the Vegetation Committee was Moss. During his years at Manchester, he became increasingly anxious to obtain an academic and purely botanical appointment. To his great surprise, Albert C. Seward (1863–1914), the professor of botany in Cambridge, offered him the post of Curator of the University Herbarium in 1907. Although it meant some financial sacrifice, Moss accepted (Ramsbottom 1931). With a valuable herbarium to hand, and particularly after the rebuff from the Treasury over plans to publish the vegetation surveys, Moss turned increasingly to floristic work. The Cambridge University Press took up his proposal to publish a new, critical British flora. The first volume appeared in 1913 (Crump 1931, Allen 1986).

In a review of Moss's vegetation survey of the Peak District, eventually published in 1913, Tansley remarked on how British ecology was 'very much the poorer through the withdrawal of the author's logical and incisive mind from vegetation to the domain of floristic and taxonomic studies' (Tansley 1913a). The outbreak of war soon cast doubt over the future of the Cambridge flora, and this, together with what Tansley called domestic worries, caused Moss to leave Cambridge in 1917 for the newly created post of professor of botany in the South African School of Mines and Technology (later the University of Witwatersrand), where he straightaway applied himself to the study of the flora of the Transvaal and the establishment of an herbarium. He died unexpectedly in 1930, having almost completed the first critical survey of the flora.

Changes in circumstances also affected William Smith and Tansley. Smith left Leeds in 1908 to take up a post as lecturer in biology at the Edinburgh and East of Scotland College of Agriculture, where he remained for 18 years, working mainly on the agricultural improvement of hill pastures (Woodhead 1929). In correspondence with Tansley, Smith wrote of how 'ecological work (at Leeds) did not promise to afford that income which I considered I needed'. Edinburgh was, however, geographically much more remote, and Smith, confronted with the demands of his new post, was tempted to drop out of the Vegetation Committee. He remained, largely because he knew that if he did resign the secretaryship would pass to Cambridge, and 'the North-men would not be so loyal'[15].

During 1912, and the discussions which led eventually to the formation of the British Ecological Society, Tansley seriously considered applying for the new chair of botany in Sydney. He had moved from University College, London, to Cambridge in 1907, where he took Seward's former post as lecturer in botany. The prospects for teaching ecology seemed bright. He took every opportunity to escape from the lecture room and laboratory by joining Moss in leading students on day and half-day excursions. The advanced class went to the North Downs and Weald in 1911, and to the Cotswolds, Forest of Dean and Malvern country in 1912 (Tansley 1947a). There was, however, little support or inspiration to be gained from other members of the botany department.

There survive, with Tansley's draft application for the Sydney post, two testimonials, written by Oliver and F. F. Blackman, and a letter from William Smith, who paid tribute to Tansley's qualities of leadership. He wrote,

> you have always been open to reason when matters came under
> discussion, and I don't know anyone who has so much the confidence of
> all concerned. There has always been too that enthusiasm not too much
> cooled by caution which has led us into doing what luke-warmness and
> over-caution would not have done.

If Tansley left the country, Smith foresaw two possibilities, apathy might overcome ecology because there 'would be no definite person pushing its claims'; alternatively, ecology 'would fall into the hands of dogmatic unteachable people who would take things with a high hand and make heretics of any who did not agree with them'. In Smith's words, they would 'exploit the subject for popularity only and will prostitute it unscrupulously for their own ends'. The inevitable outcome would be the segregation of the subject into various units.

Even now, great care had to be exercised in keeping together 'the northern hill-men and the southerners'[16].

Whatever his reasons for considering and rejecting the Sydney post, Tansley remained in England, and soon emerged as the leading figure in setting up the new Ecological Society (Godwin 1977). The question of forming a Society had been discussed informally on several occasions and, at a meeting in May 1912, Tansley put forward a fresh set of suggestions. He recalled how the Vegetation Committee had achieved a great deal in a short space of time. Members could not possibly have acquired so much knowledge without the frequent meetings and the stimulus of united effort. The Committee had played an important part in compiling and publishing papers and books, and in organizing the International Phytogeographical Excursion. In a sense, however, the Excursion and the publication of *Types of British Vegetation* had marked the close of 'the initial stage of the committee's career'. There was evidence that 'the stream of new ideas that directed and informed the first labours' was being exhausted, and of an increasing tendency towards specialization, leading in turn to a greater diffusion of studies[17].

The Committee was already taking on the character of a very small society. As Tansley pointed out, there had been a steady addition of members and associates since 1908. A much smaller proportion of members attended meetings than in the earlier years. By turning this small body into a strong society, it should become easier to attract the funds necessary for the publication of primary survey memoirs and maps. The need for a journal was more pressing than ever. As ecological research moved away from the more geographical aspects, the journals of the Royal Geographical Society and its Scottish counterpart became less and less appropriate as publishing outlets. None of the existing journals could guarantee space for ecological papers.

The Vegetation Committee agreed that there were good grounds for establishing 'a British Ecological Society', but everything hinged on whether the membership would be large enough to give the Society 'a position of authority' and funds to support a journal. The minimal requirement was seen to be about 100 members, each contributing a guinea a year. A circular was sent to all members of the Committee, asking them to seek the views of everyone likely to be interested. The response was so encouraging that, at its meeting in December 1912, the Committee resolved unanimously 'to create a Society to take its place and carry on its work' (Tansley 1913b). Some intimation of the doubts that might have been expressed was given in a note, published anonymously in the *Naturalist*, which cast doubt on 'the wisdom of adding another to the long list of Scientific Societies'. There was a danger of members withdrawing 'support from older institutions'. It would have been better to add an ecological section to the Linnean Society or another existing body (Anon. 1913).

At its two-day meeting in December 1912, the committee defined the object of the new Society as 'to promote and foster the study of Ecology in its widest sense'. Membership would be open to anyone subscribing to that purpose and paying a subscription (which was raised from a guinea to 25 shillings in 1919). Associates would pay 12/6d. A motion tabled by F. E. Weiss and C. B. Crampton, enabling membership to be restricted in the event of 'a majority of non-active workers' posing a 'possible danger to the interests of Ecology', was withdrawn, following assurances that 'such a contingency was controllable'. The principal benefit to members would be the free receipt of the quarterly journal, which was to be called the *Journal of Ecology: the organ of the British Ecological Society*[18].

Much time was taken up with debate over the rules of the Society, with Smith and Woodhead critical of their over-elaboration—misgivings that proved to be well-justified (Salisbury 1964). A resolution tabled by Tansley and R. H. Yapp was carried, whereby the Society was to be governed by a Council consisting of 12 members, including a president and two vice-presidents. The Council would, in the first instance, be made up of members from the Vegetation Committee. Two members would retire annually. In order to ensure a continuous infusion of new blood, they would not be eligible for re-election immediately. The president would hold office for two years, and deliver an address. Members of Council would be elected by members of the Society at the annual meeting. There would be at least one other meeting each year.

By adopting the all-embracing title of the 'British Ecological Society', everything possible had been done to attract a large and diverse membership, and thereby the financial support needed to launch the journal. Within a month or so of the December meeting, the Committee had become sufficiently confident as to discuss the possibility of producing a larger journal of much wider scope than originally contemplated. Despite fears that British work might be 'smothered', it was agreed that the new journal should include contributions from all parts of the world. Not only would this attract a far wider foreign circulation, but it would help to 'counteract insular views'. Since the British were the first in the field, it seemed, in Tansley's words, 'a pity to miss the opportunity of publishing the first comprehensive journal which would at once take its place among the standard botanical journals of the world'[19].

On 12 April 1913, the Society's inaugural meeting was held in the botany department at University College, London, attended by about 47 members, including 15 members and associates of the Vegetation Committee. Approval was given to the founding of the new Society and dissolution of the Vegetation Committee. Tansley became the first president, William Smith was made the first honorary life member. An underwriter and keen amateur naturalist, Hugh Boyd Watt (*1858–1941*) was added to the Council in the role of treasurer (Shanks 1943). The first issue of the *Journal of Ecology* was printed in time for the meeting, and was widely circulated as a way of attracting publicity for the new Society.

Two key appointments were those of secretary and editor. At the instigation of Smith and Tansley, Francis Cavers (*1876–1936*) was appointed to both. As Smith confided in a letter of July 1912, the choice of a suitable secretary had been 'a source of considerable trouble'. Cavers was probably 'better than any of our younger associates'. Among a succession of posts, he had been an assistant lecturer at Leeds between 1901 and 1904. In Smith's words, Cavers was 'a tremendous absorber of literature and when he gets on to research he goes at it hard'. He had published a 'monumental' series of papers on 'The inter-relations of the Bryophyta' in the *New Phytologist* (Cavers 1910)[20].

The appointment of Cavers as secretary and editor proved a wise one. As Tansley wrote later, he was 'an extremely hard, untiring worker' showing, as editor, 'conspicuous powers of masterly condensation and clear exposition'. By 1915, however, Cavers had decided to abandon one career for another. He trained as a medical doctor, practised for 10 years in north London, and then retired to lead a busy life editing two journals on cancer research (Tansley 1936, Shaw 1969). Following his resignation from the British Ecological Society in 1916, E. J. Salisbury became honorary secretary, and Tansley took on the role of editor.

The British Ecological Society was the first ecological society in the world— the Ecological Society of America was formed two years later, and the first issue of its journal, *Ecology*, did not appear until 1920 (Shelford 1938). The significance of the British headstart, and the transition from a Vegetation Committee to an Ecological Society, should not, however, be exaggerated. The numbers of those actively involved in the Society's affairs remained very small. Membership only just exceeded 100 in 1917. It was only through library subscriptions to the journal, and the most rigid economies in secretarial expenses, that the Society survived. With membership so sparse and scattered, attendances at meetings and field excursions were low. It meant, however, that those who attended could actively participate in all that was being said and displayed (Salisbury 1964).

Just over a year after the Society was founded, war broke out. Although it was resolved to carry on 'business as usual', a reduction in the size of the journal seemed inevitable in view of the resignations already received and the inevitable loss of some foreign subscriptions. In common with other journals, it was agreed that copies of the *Journal of Ecology* lost in torpedoed ships should be replaced free of charge to subscribers. Not only did the war bring to an end overseas excursions, but it ruptured the often close links with German scholars. It meant an inevitable dislocation of plans and careers. Tansley was leading a party from the Cambridge Botany School to Provence and the fringes of the Mediterranean Alps when news of the Sarajevo assassinations broke in June 1914. Tansley (1912b) had earlier described the remarkable transition between the typical Mediterranean vegetation of the coast and that of the Alps (all within a distance of 30 miles). Within two years, two of the youngest and brightest members of that party had been killed fighting in the war.

Of all those who perished, the loss of Albert S. Marsh (*1892–1916*) was particularly severe. His scholastic and university career had been an unbroken success (Price 1916). In a paper in the *Journal of Ecology*, he provided the first example of a detailed comparison between topography and plant communities, based on a contemporary survey and old Ordnance Survey maps of the salt marsh and sand dunes of Holme-next-the Sea in Norfolk (Marsh 1915). The survey, completed in the long vacation of 1913, was used by the Cambridge botany department to measure subsequent changes during the inter-war years (Wadham 1920, Conway 1933).

Marsh's interests were wide-ranging. At Tansley's suggestion he took over an experiment to see how two closely-allied species of the bedstraw, *Galium*, reacted when grown in competition under controlled conditions. Whilst both could live on calcareous soils and peat, competition between the two species meant that a different dominant characterized each soil type. Some 50 years later, the work was to be quoted as an outstanding example of pioneer research in the field of interspecific competition (Harper 1964). It was Tansley who brought the trial to a conclusion and published the results (Tansley 1917). Marsh had been shot through the heart by a sniper's bullet in the trenches of Armentières in 1916 (Tansley 1916a).

Part 2
The First Quarter-Century, 1913–1938

2.1
Disillusionment in 'a Goldfield of Unquestionable Richness'

In a sense, the founding of a national society to represent and promote the needs of ecology was an extraordinary pretentious thing to do. Despite the appeal of ecology to the younger generation and many without professional posts or qualifications, there were few signs as yet of ecology being accorded the kind of recognition which Tansley had envisaged in his paper, 'The problems of ecology', given to the British Association meeting of 1904. Most botanists still had to be convinced that the study of 'plants for their own sakes as living beings in their natural surroundings' deserved to be placed at the forefront of the botanical sciences.

In his presidential address, the first to be given to the British Ecological Society, Tansley (1914) blamed the virtual neglect of ecology in the universities on 'the natural conservatism of the adherents of the older and well-established branches' of botany, as well as 'some tincture of jealousy' over the way in which younger workers were being attracted to 'a newer area of development'. It was, however, one thing to diagnose a problem, and another to have the means of resolving it. Tansley likened the predicament of the ecologist to that of the miner 'standing over a goldfield of unquestionable richness, but as yet tested only here and there'. In his words, 'we scarcely know where to begin, and how to locate the richest bodies of ore'. Ecological techniques were still so rudimentary, and training so inadequate, that it was hardly surprising that ecologists met with

> the scepticism, if not the jeers, of those who have other mines to work, and are not likely to be specially anxious that ours should be too remunerative an undertaking.

While Tansley might be the acknowledged leader of ecological thought and progress in Britain (a leadership given tangible form in his roles as founder-president of the British Ecological Society and, later, an outstanding editor of the *Journal of Ecology*), his kingdom remained extremely small. Matters were not helped by the disruption and disenchantment of his own career. University teaching came almost to a halt during the war. Not eligible for military service, Tansley had no alternative but to accept a more or less routine clerical post in the Ministry of Munitions in London (Godwin 1958). As he remarked, in a letter of September 1915, the work was at times 'quite interesting but I feel the 8 hours a day 6 days a week, which is natural considering the freedom I have been used to'. The routine left little time for writing and editing[1].

Tansley returned to Cambridge in 1918, where Godwin (1977) later remembered being shocked to find 'so eminent a man so poorly housed' in a sort of chicken-coop partitioned from the elementary laboratory. In Godwin's words, it was an age when university departments were under strong paternalistic control. Any proposal for improvement or reform was likely to be rebutted publically with a firm, even acrid, reply, and most notably from Seward, the head of the Botany School at Cambridge, himself a very successful and popular lecturer (Godwin 1985b).

In such circumstances, it was hardly surprising that Tansley sought intellectual stimulation elsewhere. He had long been interested in psychology. E. Pickworth Farrow later recalled how his own interest in the subject was kindled in 1912, when Tansley recommended that he and other students in the class should read a proof copy of Bernard Hart's book, *The Psychology of Insanity* (Farrow 1942). Tansley developed a considerable interest in the work of Sigmund Freud (*1856–1939*), some time before it became widely known in Britain (Tansley 1940a).

From printed sources alone, Tansley mastered the themes being developed by Freud, and appraised them in the light of the findings of biologists. He published a book in 1920, called *The New Psychology and its Relation to Life*, which was reprinted many times, and translated into Swedish and German (Tansley 1920a). Public interest was so great, and requests for help so pressing, that other publications followed. In the summer of 1922, Tansley visited Freud. He resigned from his university post at Cambridge a year later. Drawing on his private income, he moved with his family to Vienna, where he spent several months studying under Freud.

2.1.1 *'Botanical Bolshevism'*

Tansley's departure from the Cambridge Botany School followed several years during which he played a prominent role in campaigning for reforms in education. He had been elected to the Royal Society in 1915, and took a leading part in various initiatives to promote scientific research, both pure and applied, between 1917 and 1919 (Godwin 1977). As thoughts turned to post-war reconstruction and its implications for every level of teaching and research, Tansley saw both the opportunity and need to press the claims of ecology.

Using as his pretext a letter to him as editor of *New Phytologist* on the subject of teaching in ecology, Tansley published what soon came to be known as an 'encyclical'. It was signed by F. F. Blackman and his brother, Vernon H. Blackman (*1872–1967*), Frederick W. Keeble (*1870–1952*) (who was director of Wisley Gardens), Oliver and Tansley himself. The memorandum, which appeared in the *New Phytologist* of December 1917, called for a fundamental reappraisal of elementary education as a prerequisite for rescuing botany from its dangerous predicament.

The authors of the 'encyclical' complained of how the teaching of botany in schools and university departments had changed little since the time when evolution was first accepted. In its present form, the study of morphology held out no attraction for 'the type of mind that wants a deeper insight into the working of plant processes or into the part which plants play or can be made to play in the economy of the world'. By giving so much pre-eminence to morphology, plant physiology had been relegated to a subordinate and almost separate status. All too frequently, it was taught as a dreary catalogue of results from inconclusive experiments (Blackman F. F. *et al.* 1917).

It was no longer enough to tinker with the proportion of time devoted to the teaching of plant physiology. What was very desirable before the war had become 'an imperative necessity'. If botanists were to do their share in the struggle to re-establish normal life, elementary teaching had to be based not only on a knowledge of the structural life history of plants but on the physiology of individual plants. It should lead to a study of plants of different ecological types and a consideration of such concepts as competition and the social life of plants. Such a range of topics would have a direct bearing on practical life, providing the scientific basis for farming and forestry, the utilization of waste lands, coastal-protection works, and the whole spectrum of activities where man exploited plant life.

The memorandum succeeded in its immediate objective of stimulating comment. Under the heading of 'Botanical Bolshevism', Frederick O. Bower (*1855–1948*), the Regius Professor of Botany at Glasgow, attacked 'the authors of the manifesto' for seeking their utopia by subordinating something which they admitted to be good in itself. That was the spirit which had ruined Russia and endangered the future of civilization. Bower contended that the arguments of the signatories were based on the completely false premise that there was an antithesis between physiology and morphology. A more constructive approach would have been to seek greater collaboration between the different branches of the science (Bower 1918).

Among those to respond to Bower was M. M. Cheveley Rayner, a lecturer in botany at University College, Reading, whose recent research had demonstrated the dependence of *Calluna vulgaris* on a mycorrhiza (Rayner 1915, Guillebaud 1949). She argued that, just as Bolshevism was a reaction 'against intolerable conditions in the body politic', so botanical Bolshevism had arisen from a situation where grave disorders called for desperate remedies. The effects of the prevailing morphological bias had been to perpetuate 'the common attitude that Science is something essentially different and apart from the facts of everyday life'. By giving greater prominence to the newer 'phases of botanical knowledge', students would come to recognize 'the close interdependence of plant and animal life' (Rayner 1918).

Another contributor to the debate was Marie C. C. Stopes (*1880–1958*), a palaeobotanist in the botany department at University College, London. She contended that, before tackling the bigger and more complex aspects of botany, students had to be given 'an accurate mental picture' of the organs and tissues of plant life. Anatomy, microscopic studies, and morphology formed the 'permanently stable bricks and mortar' required by the physiologist. Without them, his structures and edifices would be very flimsy indeed (Stopes 1918).

Whatever the wisdom of introducing students to the rudiments of ecology, Woodhead drew attention to the fact that no explicit use had been made of botanists during the war (Woodhead 1919). Tansley brought the debate to an end with 'a postscript', stressing how the overriding purpose had been to urge a shift in the main centre of interest of elementary teaching—a change from 'interest in the lines of descent to an interest in the plant in all its manifest aspects as a living organism' (Tansley 1919).

A major change had come about—largely as a result of the recent writings of several contributors to the debate. Bower himself had written a new book on the *Botany of the Living Plant* (Bower 1919). Woodhead had published a textbook on *The Study of Plants. An Introduction to Botany and Plant Ecology*, and, in 1920, there appeared a textbook on plant biology, written by Cheveley Rayner and her husband, William Neilson Jones (*1883–1974*) (Woodhead 1915, Jones & Rayner 1920). F. E. Fritsch and E. J. Salisbury had also embarked on what was to become a long series of collaborative publications for schools and colleges (Fritsch & Salisbury 1914).

No longer was it possible to reproach British botany for the extent to which students depended for their textbooks on translations of German works. In 1923, there appeared a further textbook from Tansley, called *Practical Plant Ecology. A Guide for Beginners in Field Study of Plant Communities* (Tansley 1921–1923). It probably did more than anything else to hasten the introduction of ecology to schools. As Godwin (1977) remarked later, the lateral attack, as waged through the textbook, proved far more effective than any direct assault on 'the citadels of traditional biological teaching'.

2.1.2 Patrons of ecology

In their struggle for recognition, the pioneers of ecology drew heavily on the support of those botanists whose reputations for teaching and research had already been established, and who were known to be sympathetic towards the infant ecology. Three men in particular stand out for their support, namely F. E. Weiss, F. E. Fritsch, and, most important of all, F. W. Oliver.

The enormous influence exerted by Oliver on his own and succeeding generations can be discerned even in the careers of Weiss and Fritsch. The research interests of Frederick E. Weiss (*1865–1953*) spanned both botany and zoology. He went up to University College, London, in 1884, intending to read botany. There, he fell under the influence of E. Ray Lankester (*1847–1929*), and graduated in zoology in 1888. It was not until he received an offer from Oliver to become the first Quain Student in the botany department that Weiss turned again to botany. Three years later, he was appointed professor of botany at Owen's College, Manchester, where he remained until 1930, frequently acting as pro-vice chancellor, and as vice-chancellor between 1913 and 1915. In his studies of both living and fossil plants, Weiss tried to relate their structures to the environments in which they grew (Weiss 1925). He was a member of the British Vegetation Committee, which frequently met in his department, and was elected president of the British Ecological Society in 1923 (Thomas 1953).

Felix E. Fritsch (*1879–1954*) became professor of botany in the East London College in 1911. He had previously spent nine years as a member of Oliver's department at University College and, like him, he had become highly critical of those who sought to assess the significance of the various factors determining plant life solely on the basis of experiments carried out in the laboratory. There seemed little point in discovering the effect of diverse reagents if the significance of the response under natural conditions remained unknown. It was much more useful to start by making detailed and frequent observations, and then to test the validity of the inferences made in the field by way of experiments.

It was a point of view which Fritsch promoted, both in his research and in the textbooks written in collaboration with E. J. Salisbury. His interest in aquatic systems had developed in an earlier academic post in Munich. As an 'observer' at the second meeting of the British Vegetation Committee in 1905, he outlined a project to record the periodicity of algae in ponds, and asked for the help of Committee members in forwarding monthly samples[2]. In collaboration with M. Florence Rich (*1865–1939*), he made monthly observations over five years on a pond at Abbots Leigh, near Bristol. Salisbury took monthly samples over four years from a pond near Harpenden in Hertfordshire (Salisbury 1954). At first the investigations were solely concerned with recording the effects of seasonal and less regular changes of light, temperature, and other factors on algae (Fritsch 1906), but, as sample after sample was examined, it became clear that regular monitoring could provide important insights into reproduction and interspecies relationships (Fritsch & Rich 1907, 1909).

His series of textbooks evolved out of a publisher's request to Salisbury, who was by then a member of his department at the East London College, to write an elementary textbook. Salisbury later recalled how, having discussed the contents of each chapter, he would extemporize while Fritsch typed the manuscript, oblivious to the maroons and air raids in the London streets around them (Salisbury 1964). As the preface explained, the text was unusual in providing a chapter on soils in relation to plants, and 'a somewhat detailed account of vegetation as a whole'. The book was revised and reprinted many times, and became the first of five textbooks from the same authors, which, taken together, were to play a large part in stimulating interest in 'the consideration of plants as living wholes, rather than as assemblages of organs, tissues and cells' (Fritsch & Salisbury 1914, Clapham 1980).

It was immensely important for ecology that so eminent a palaeobotanist as Francis W. Oliver (*1864–1951*) should advocate the closer study of plants in their actual habitat, and that he should have published so many research papers developing that approach to botany. While at Cambridge, Oliver spent two long vacations in Germany, where his interest in plant geography was aroused by A. F. W. Schimper. He succeeded his father as head of the botany department at University College, London, in 1888 (Oliver 1927, Thomas 1953). His field parties, made up of members of his department as well as distinguished botanists from other universities, played an important part in preparing the ground for the introduction of ecological studies into university curricula (Salisbury 1964).

It all began in 1903, when Oliver and Tansley led a group of advanced students round the Broads and coastline of east Norfolk, spending three days studying the marsh and water associations of Higham Sound in detail (Anon. 1903). The next year, it was decided to devote a fortnight to studying a more accessible area, where the vegetation was composed of comparatively few species. This aim was to study the relationship of plants to physiographic conditions, and to measure the speed with which changes took place (Anon. 1904). Oliver was already familiar with the Channel Islands and nearby French coast, and he chose the estuary of a small stream, the Bouche d'Erquy, 25 miles west of St Malo. The expedition, made up of 27 members, proved an 'unequivocable success', and was followed by three further annual expeditions (Hill 1909).

In the early years, the expeditions were regarded 'by some as a bold and risky innovation'. Crowds and friends of the students would assemble on the platform at Waterloo to see the St Malo boat–train depart (Oliver 1927). As well as providing a valuable training for the students, the expeditions were an opportunity to study 'the various fundamental problems' related to vegetation and its mapping (Tansley 1905). On their first visit, Oliver and Tansley (1904) experimented with a technique which they called 'the method of squares'. It was soon abandoned for the quadrat method pioneered by Pound and Clements. First on the heather-covered 'chert' diggings in Surrey[3], and then on the expeditions to the Bouche d'Erquy in 1906–7, they found it particularly useful for recording changes in plant associations from year to year (Oliver 1906, 1907).

In the opinion of E. J. Salisbury, it was these kinds of investigation that did more than anything else to save 'British ecology from becoming dominated by a more static description of vegetation units' (Salisbury 1952). While convalescing from an attack of pleurisy in Norfolk, Oliver discovered the great potential of Blakeney Point for research and teaching purposes. Visiting parties could be accommodated in local hotels, tents, and a field station which he had built. A fortnight's visit of July 1913 was organized into six sections, namely general survey, floristics, faunistics, *Suaeda fruticosa* and shingle stabilization, and soil and chemical sections[4].

Despite the fact that shingle beaches made up an estimated 300 miles of coastline in England and Wales, very little interest had been taken in their structure and development. In a paper of 1912, Oliver put forward a tentative classification, drawing particular attention to the effects of differences in exposure and mobility, and the nature of the soils, on the plant life of the beaches (Oliver 1912). In order to discover more about these functional relationships, Oliver and Salisbury (1913) published a paper in the first volume of the *Journal of Ecology*, describing how *Suaeda fruticosa* established itself on the shingle banks, and how, over the course of time, its rooting system might slow down or arrest the landward movement of the banks.

Another species to excite Oliver's interest was *Spartina townsendii*, which had, over a 20-year period, spread to almost every inlet and bay of Poole Harbour in Dorset. Photographs taken of the same locality at different points in time showed how rapidly the small, scattered patches of the hybrid could spread and fuse together to form a continuous meadow (Oliver 1925). Despite the obvious implications for coastal management, details of the plant's life history were largely unknown. As Oliver (1920) remarked, whether the plant was regarded 'as a botanical phenomenon, a weed which seriously threatens navigation, or a gift of providence capable of being put to a variety of uses', there was an urgent need for closer investigation.

In his own contribution to the debate in the *New Phytologist* on the 'encyclical' of 1917, Oliver (1919) pointed out how botanists should involve themselves much more closely with matters of public concern. There was a pressing need to develop the applied side of their science, building up links with industry. His own interest in the practical implications of vegetation studies began in 1908, when he gave evidence to the Royal Commission on Coastal Erosion. He described how

> by the study of the properties and capacities of sand grasses, etc., man has learnt how to tame the sand dune. It has no terrors left; its fixation has become a fine art.

Using the data, photographs and charts acquired during the expeditions to the Bouche d'Erquy, Oliver demonstrated how a similar approach might be developed for manipulating rates of tidal siltation (Royal Commission 1909).

In his presidential address to the British Ecological Society in December 1916, Oliver spoke of how ecologists could 'do much to establish their subject in general esteem', if they applied themselves to what might be called 'profitable channels', especially at a time when the greater exploitation and improvement of land had become a national necessity. Turning to studies of maritime vegetation, he spoke of how he was continually astonished at the extent to which data from both autecology and synecology could be applied for economic ends (Oliver 1917). In a book called the *Tidal Lands*, written in collaboration with an engineer, he went into much greater detail as to how vegetation might be used for such purposes as coastal protection and flood protection (Carey & Oliver 1918).

2.2
The Ecological Constituency

It is never easy to know when to form a new society. It is obviously pointless to do so until there are well-defined objectives and large numbers of people wanting to join. On the other hand, if delayed too long, there is a danger of at least some of the objectives being taken up by another body—perhaps an existing society—which would then have a vested interest in preventing the new society being formed. In all these respects, the founding of the British Ecological Society was an outstanding achievement. Not only were there no existing bodies claiming to speak for ecology, but no rivals emerged to challenge the British Ecological Society as the only national body, which spoke for the whole of British ecology.

Who joined the Society? Nearly everyone whose name appeared on the first membership list to be published in 1917 considered himself to be a botanist. The fascination of plant life had arisen through botany, and it was through botany that opportunities for survey, research and publication, and full-time employment, had come. The founding of the British Ecological Society was never intended to sever these bonds. Whilst members might rebel against the way in which botany was taught or studied, their commitment to, and indeed their dependence upon, botany increased. Ecology continued to draw much of its inspiration from botany.

The complex way in which botany stimulated thinking in ecology, and in turn came to benefit from the adoption of an ecological approach, can be illustrated by reference to the fierce controversy that waged over species differentiation, and the 'age and area' hypothesis, during the early years of the Society. The principal figure in the controversy was John C. Willis (*1868–1958*).

In the course of compiling his *Manual and Dictionary of the Flowering Plants and Ferns*, published in 1897, Willis had become increasingly sceptical of the widely accepted notion that species were arranged in a kind of casual assortment as a result of natural selection operating in response to local needs and conditions. Following his return to Cambridge in 1915, he published a series of papers in the *Annals of Botany*, describing how it was age, rather than natural selection, that decided how great an area should be occupied by any given species. Taking a group of species in a region, free from major barriers to migration, it was inevitably found that the commonest species (in terms of area occupied) were the longer-established species. Endemic species tended to be rare and much younger (Willis J. C. 1916, 1922, Turrill 1958).

No one denied that age could play a part in determining the range of a species. Willis was criticized, however, for taking so little account of the extreme complexity of such issues as mutation and endemism. While the age–area hypothesis might help to explain the distribution of species which had not yet reached their potential limits, it took no account of the many cases where species had died out, or survived as relicts of a much wider distribution. Critics drew attention to the wide range of other factors, including human intervention, which could affect the geographical extent of a species. No matter how uniform a region might appear to be, there were likely to be many barriers to the spread of a species (Sinnott 1917).

Perhaps the greatest contribution of the hypothesis was to stimulate interest in plant geography generally—an interest that was soon reflected in the pages of the *Journal of Ecology*. One of the first contributors was Harry Godwin (*1901–85*), whose interest in ecology was first aroused by the discovery of the works of Clements and Tansley in the library of the Cambridge Botany School. He had heard of the age–area hypothesis from Willis, in the course of a series of lectures, and, while still an undergraduate, Godwin set out to survey the flora of a series of ponds of different ages, in order to see whether differences in their floristic richness could be correlated with the lengths of time over which colonization had taken place. As his paper demonstrated, there was a high degree of correlation (Godwin 1923, 1985b).

Of much more immediate significance to ecology were the findings of James R. Matthews (*1889–1978*), a botanist whose research interests covered a remarkably wide spectrum (Gimingham 1979). In 1914, Matthews had published a study of the successional stages of marsh and swamp vegetation around White Loch Moss in Perthshire (Matthews 1914). He quickly realized that if Willis's hypothesis could be applied to the British Isles, an extremely useful method would have been discovered for resolving many of the controversies as to the origins and distribution of the British flora. In a paper published in the *Annals of Botany* in 1922, Matthews sought to apply the age–area hypothesis to the flora of Perthshire, which was believed to be entirely post-glacial in origin. He was able to draw a distinction between the 133 Arctic–Alpine species, which were essentially relicts confined to the higher hills, and the remaining 605 species, belonging to a lowland or temperate flora, the spread of which seemed to correspond closely with Willis's hypothesis (Matthews 1922).

From this initial investigation, Matthews went on to investigate much more closely the distribution of species. In 1934, he left Reading, where he had been professor of botany for five years, to become the Regius Professor of Botany at Aberdeen. In the same year, he was elected president of the British Ecological Society. In his presidential address, he described the 11 geographical elements which made up the British flora. According to Gimingham (1979), few presidential addresses have ever received such wide acclaim, securing a permanent and important place in phytogeographical literature. The address emphasized how many of the uncertainties in plant distribution were essentially ecological problems, to be investigated as much by the ecologist as by the plant geographer (Matthews 1937).

2.2.1 *The contribution of agricultural science*

In the same way as the farming of Nebraska coloured the thinking of Frederic Clements and his concepts of plant succession, contemporary developments in agricultural practice and research had a considerable bearing on plant ecology in Britain. It is the purpose of this section to look more closely at the contribution made by botanists working in agricultural research.

The years of the British Vegetation Committee corresponded with a time when the influence of the Rothamsted Experimental Station began to extend beyond the field of agricultural science. The publication of the book *The Soil* by A. Daniel Hall (*1864–1942*), the director of Rothamsted, stimulated a general interest in edaphic factors (Hall 1903). There followed the classic study on the agriculture of the soils of Kent, Surrey, and Sussex by Hall and E. John Russell (*1872–1965*) (Hall & Russell 1911). Many years later, Salisbury (1964) recalled how these publications, and a tour of the famous grass plots at Rothamsted, made him realize that 'many aspects of agricultural science were, in fact, the ecology of an environment, albeit largely artificial'. The first field excursion which Salisbury arranged as secretary of the British Ecological Society was to Rothamsted[5].

Many papers were published in the *New Phytologist*, and later in the *Journal of Ecology*, demonstrating the need to look more closely at the relationship of weed species to soil conditions and different crops. In a series of pot experiments, Winifred E. Brenchley (*1883–1953*) found that competition between crops and the weed flora was largely determined by the timing and duration of competition for food, water, and light between plants, irrespective of the species involved (Brenchley 1920).

Ecologists were particularly interested in grassland practices. Advances in that sphere of husbandry reflected the farmer's ability to manipulate not only single-species crops but plant communities (Sheail 1986). A paper in the *Journal of Ecology* in 1915 used the detailed records kept of the experimental plots at Rothamsted to trace the influence of such factors as soil water content on the way in which formerly arable land reverted to 'natural' conditions (Brenchley & Adams 1915). The role of rainfall in determining the persistence of individual plants and species in a sward was highlighted by R. George Stapledon (*1882–1960*), who used the exceptional drought of 1911 to monitor 'the progressive relation between climatic conditions and the personnel of the several pastures' he had already begun to study within and around the grounds of the Royal Agricultural College at Cirencester (Stapledon 1911, Russell 1961).

During trials on grass and clover mixtures at the Cockle Park Experimental Station, the great superiority of wild white clover over the commercially available strains was discovered (Pawson 1960). In his observations at Cirencester,

Stapledon (1913) found that it was the unsown, indigenous plants which survived in the largest numbers, which he attributed to their greater 'inherent vitality'. Soon after his appointment in 1912 as an Adviser in Agricultural Botany,
based at Aberystwyth, Stapledon began to collaborate with Thomas J. Jenkin
(*1885–1965*) in identifying ways of promoting the development of indigenous
plants in established swards. A succession of posts from 1913 onwards as agricultural organizer for Brecon and Radnor, and then as Adviser in Agricultural
Botany at the University of North Wales, Bangor, had given Jenkin opportunities to carry out censuses on a large number of grasslands (Jenkin 1919, Thomas
1966). He had independently noticed the superiority of self-sown indigenous
Festuca duriuscula over commercial mixtures. Early trials indicated that the
most effective way of encouraging such plants was to 'disturb' the pasture constantly through 'reasonably heavy grazing', so as to create plenty of opportunities for colonization (Stapledon & Jenkin 1916).

In order to identify which of the treatments studied in the trial plots might
be most applicable to particular tracts of farmland, a systematic method was
required for classifying grasslands. Stapledon made full sense of the insights he
gained from a year spent at Cambridge (1909–10). Although ostensibly taking a
diploma in agriculture, he used every opportunity to gatecrash lectures given by
Moss, and to join the ecological excursions led by Moss and Tansley (Waller
1962). He was made an associate member of the British Vegetation Committee
and, in 1913, completed a vegetation map at the scale of 6 inches to the mile (1 :
10 560) of over 100 000 acres of north Cardiganshire (Davies 1936). Stapledon
was struck by not only the variety of communities recorded, but the extent to
which such surveys could provide the means 'of determining rapidly and with
reasonable accuracy the practical possibilities of any particular habitat' for
farming (Stapledon 1933a).

Under Stapledon's direction, the mapping of the grasslands of Wales proceeded at a scale of 1 inch to the mile (1 : 63 360), and often 6 inches to the mile,
based on the premise that each grassland type was made up of 'a complex community of a number of different species representing a number of different
growth forms and physiological relationships towards the habitat'. Stapledon
(1914) found that the most objective way of recording 'the percentage contribution of each species to the total herbage' was to analyse as many samples of each
type as possible. The results were eventually published at the 'quarter-inch' (1 :
253 440) scale in 1936. In the words of the accompanying text, the plant community was 'the visible expression of all the factors of environment, climate,
edaphic and biotic'. The plant served as a yardstick 'by which the sum effect of
those factors is expressed' (Davies 1936).

The grassland survey provided an outstanding opportunity to build up a
large collection of seed and plants, removed *in toto*, from characteristic habitats. The idea itself was not new. In an analysis of the herbage of old grasslands
carried out in the 1880s, William Fream (*1854–1906*) had taken turves from 25
rich, old pastures in England and Ireland, and grown them side by side in their

original soils (Fream 1888). Stapledon's appointment first as director of the
official Seed Testing Station of the wartime Food Production Department, and
then, in 1919, of the newly founded Welsh Plant Breeding Station at Aberystwyth, provided him with unprecedented facilities for raising the plants from
various sources under controlled conditions. The results made him even more
sceptical of textbook accounts of averages; too little attention had been paid to
the qualities of variation and uniqueness.

By recording systematically all the types of a series found in a characteristic
habitat, and by working out their numerical relationships, Stapledon (1928,
1933b) was able to demonstrate by reference to the cockfoot (*Dactylis glomerata*) and other species how the differences in the scope for cross-fertilization and
the power to establish a perennial habit largely decided what kind of ecogroup
of plants would dominate the sward. So far such investigations had only been
possible where the immediate agricultural benefits warranted the cost of the research facilities provided. Stapledon hoped that ecologists would soon be able
'to obtain sufficient financial backing to found for themselves a station un-
hampered by agricultural or other economic calls upon its methods of enquiry'.
Only then would it be possible 'to elucidate with reasonable assurance the
fundamental problems connected with the distribution, acclimatization and
evolution of plants'.

2.2.2 *The contribution of geological science*

Stimulation of another kind was provided by the geologist. Collaboration with
geologists not only served to emphasize the transient nature of plant communities in a dynamic environment, but brought the work of ecologists before a
much wider audience. The potential benefits of working together were most
strikingly evident in the various studies made of highland Britain. The upland
moors were not only the first areas to be mapped ecologically, but, as William
Smith remarked, the distribution of heather vegetation provided outstanding
opportunities for studying 'the exquisite adaptations to environment' (Smith
W. G. 1902b).

A critical source of evidence in unravelling the history of moorland and
other types of vegetation were the subfossil remains preserved in peat and other
kinds of organic sediments. The first person to recognize the significance of
these remains was an officer of the Geological Survey, Clement Reid (*1853–
1916*), while surveying the Cromer Forest Bed of north-east Norfolk in 1876.
From that time onwards, he kept a record of any remains he found in the 'newer
Tertiary deposits' (M., J.E. & N., E.T. 1919). In a book on *The Origin of the
British Flora*, published in 1899, Reid gave details of the stratigraphic position
and geological age of all the known plant-bearing deposits of post-Miocene age,
giving details of the plant species and distribution of each (Reid 1899).

F. J. Lewis began to investigate the post-glacial history of the moorlands
during his primary surveys of north-east England. As he explained to the

British Vegetation Committee at the Liverpool meeting of 1905, the deeper mosses probably dated back to the later phases of glaciation, and the differences in the composition of the horizons were likely to reflect shifts in climate during the intervening period. If it were possible to correlate the horizons found in different parts of Britain with those described by Scandinavian workers, a much better understanding of present-day vegetation would emerge (Smith 1905b, Lewis 1906a, 1907a).

Over the next few years, Lewis investigated the peat profiles of about 40 different locations in Scotland, England, and Iceland, setting out the results in a series of papers read to the Royal Society of Edinburgh. Not only was there evidence of several changes in vegetation since the last Ice Age, but the fact that many of the sequences discerned in the different areas could be correlated with one another suggested that the changes had been brought about by climatic, rather than edaphic, factors (Lewis 1906b, 1907b, 1910).

These various studies of subfossil remains marked 'a new era for the student of plant succession' (Woodhead 1929b). In the south Pennines, it became clear that the woodlands reached their climax during a climatic optimum as long ago as 3–4000 BC. Thereafter, soil leaching and the water content of the soil increased greatly as a result of a wetter climate. Archaeologists found numerous artefacts of the Neolithic and Bronze Ages lying *in situ* in peat. The Roman road at Blackstone Edge ran over peat 22 inches in depth (Woodhead 1924). As the peat accumulated, the woodland cover degenerated, and moorland plants became more conspicuous.

It would be misleading, however, to suggest that the only contribution geologists made to ecology was to introduce longer chronologies for vegetation development. To do so would be to overlook the contribution of Cecil B. Crampton (*1871–1920*), who joined the Geological Survey in 1900 and played a leading part in discovering one of 'the outstanding treasures of Scottish geology'—the hornfels sediments which had escaped the Moine–Schist metamorphism of the northern Highlands (Bailey 1952). As Tansley recalled many years later, Crampton could also be credited with having 'attacked a side of field ecology too much neglected' by British workers at that time (Tansley 1948).

An early intimation of Crampton's interest in the relationship of plants to their environment can be found in a paper on the use of fossils in deducing how coal was formed. Quoting from Schimper, he used the evidence of competition between dominant species in the plant formations of the present day to demonstrate how natural selection must have determined the distribution and character of the vegetation making up the coal measures (Crampton 1906).

In the course of surveying the geology of Caithness, Crampton recorded details of the vegetation, both on maps and in his field notebooks[6]. At first, he intended to incorporate them as an appendix to the published memoir (Crampton & Carruthers 1914), but the head of the Geological Survey for Scotland concluded they would take up too much room. On the advice of Bower, the manuscript was sent to the British Vegetation Committee 'for criticism and suggestions as to mode of publication'. Tansley, Moss, Oliver, and Smith were unanimous in agreeing that it represented 'an important contribution to descriptive ecology', and that the Committee should itself 'adopt' the memoir for publication, provided that a grant could be obtained. Through the good offices of the head of the Geological Survey in Scotland, a grant was made by the Carnegie Research Fund (Crampton 1911).

In his accounts of the vegetation of Caithness, and of the Ben Armine district of Sutherland, surveyed in 1910–12, Crampton demonstrated how the typical moorland plant associations could be correlated with (what were geologically) comparatively stable land surfaces. In his terminology, they were stable formations. The other plant associations occupied surfaces that were undergoing perpetual change as a result of geological and physiographical forces. These migratory or neogenic formations were often found on the sea coast, flood plain or mountainside, where the habitat was constantly 'on the move' in terms of detail, although the vegetation as a whole maintained its general nature. The boundary between the two formations was also always changing (Crampton & MacGregor 1913).

It was not until Crampton met members of the Vegetation Committee in 1911 that he became aware of the publications of Henry C. Cowles (*1869–1939*), the professor of botany in the University of Chicago (Tansley 1940b). Cowles had also been trained as a geologist. It was through John M. Coulter (*1851– 1928*) of the botany department that Cowles came to read Warming's pioneer work (Cooper 1935). Between 1896 and 1898, Cowles and his students sought to classify the vegetation of parts of Michigan on the basis of differences in the water content of the soil. They found little correlation unless close account was taken of the role of dynamic forces at work in the environment (Cowles 1899).

In his studies of the dune systems of Lake Michigan, and of the vegetation of the Chicago area generally, Cowles (1901) came to realize that changes in plant societies were largely determined by physiographic factors, both in the past and at the present day. They did not occur haphazardly—there was a progressive tendency towards a definite end. The denudation of the uplands led to deposition in the lowlands. Plant life was affected wherever hills were eroded, lakes became infilled, and coastal plains extended. A definite succession of plant groups could be discerned. The changes were cumulative, with each locality comprising plants that were relics of an older stage as well as species typical of prevailing conditions.

When Crampton came to revise his paper on 'The geological relations of stable and migratory plant formations', which he had first read to the British Vegetation Committee, he made detailed reference not only to Cowles's earlier papers, but to another, published in 1911, on 'The causes of vegetation cycles'. Crampton (1912) regarded it as a great step forward in appreciating the differences in the origins of plant communities on the more stable parts of the Earth's crust, and those affected 'by the topographic successions'.

Cowles (1911) discerned three cycles of succession, namely regional successions (which were the outcome of secular changes in climate), topographical successions, and thirdly biotic successions, due to plants and animals. Crampton believed the situation was even more complicated. In a paper read to the Royal Physical Society of Edinburgh in 1913, he classified habitats into three categories. First, there were natural habitats, which could be further subdivided into stable and migratory habitats. Secondly, there were artificial habitats, comprising fields, quarries, and other types of made-ground. The third category, which Crampton called 'altered' habitats, was the outcome of more subtle forms of human intervention, and included the grazing of farm animals, timber felling, and smoke pollution (Crampton 1913).

Among contemporary botanists, it was William Smith who gave Crampton the greatest encouragement. His confidence in Crampton's perception of stable and migratory associations was enhanced during the International Phytogeographical Excursion of 1911, when, on a visit to Ben Lawers, Schröter and Rübel found several examples of a snow-flush association, commonly found in the Alps. It was an outstanding illustration of a migratory association (Smith W. G. 1912b). In a letter to Tansley in July 1912, William Smith wrote of how he was increasingly impressed by the practical value of ecology. He found 'Crampton's ways of looking at things immensely useful in tackling grasslands and even arable land'[7]. At the meeting of the British Association in 1912, Smith and Crampton gave a joint paper to the agricultural section, demonstrating how grasslands could be categorized according to geological and topographical features, as well as to climate and land management practices (Smith W. G. & Crampton 1914).

To William Smith fell the task of completing unfinished work. Crampton's career was tragically cut short by mental illness, causing him to resign from the Geological Survey in 1914. He died in 1920. Through Smith's encouragement, his approach to vegetation studies was taken up by Donald Macpherson (*1886–1917*), whose research in the Moorfoot Hills demonstrated how plant associations dominated by *Nardus stricta* were largely confined to areas where the peat hags had become exposed to water, wind, and biotic agents. Smith and Macpherson were both on the staff of the Edinburgh and East of Scotland College of Agriculture, and it was Smith who published the results of the survey in the *Journal of Ecology*. Macpherson died in a French military hospital in 1917 (Smith W. G. 1918).

2.3
The Plant Community and the Ecosystem

Having illustrated a range of formative influences on ecological thinking, it is time to look at the different ways in which the leading proponents of the ecological approach sought to communicate their ideas and information to fellow ecologists, and to the scientific community at large. What use was made of the meetings and publications of the British Ecological Society?

For Tansley, there seemed little prospect of ecology achieving a more prominent position until ecologists agreed on some of the more fundamental concepts of their science. In his presidential address to the British Ecological Society in 1914, Tansley regretted the lack of progress that had been made over the previous 10 years in reaching a consensus on such questions as the classification of vegetation. Far from the situation improving, phytogeographers, particularly in America and on the continent, seemed to be erecting ever more elaborate hypotheses and complex hierarchies (Whittaker R. H. 1962, Shimwell 1971).

Some dissension was only to be expected. As Tansley (1914) remarked, concepts were 'creations of the human mind which we impose on the facts of nature'. Differences of viewpoint often reflected the circumstances in which the respective observations were made. As W. H. Pearsall pointed out, in a presidential address to the Yorkshire Naturalists' Union in 1937, there was considerable evidence that climate determined the soil types of the great continents, and it was only in areas of recent geological disturbance, where erosion and deposition had produced large tracts of *new* soils, that marked variations in soil type, and therefore of vegetation, were likely to be encountered. Yorkshire was one of these areas (Pearsall 1938).

By the time the British Ecological Society was founded, ecology had become fragmented into an exceptionally large number of schools. Those in continental Europe tended to focus on the statistical recording of the composition and structure of vegetation. The American approach was based on the doctrine of development and succession. Colinvaux (1973) quoted instances of ecologists in that country being 'so busy looking for (what proved to be) a non-existent climax that they forgot to record what was actually growing there'. British workers were also greatly influenced by successional theory, but, according to Godwin (1930), they concentrated much more than their American counterparts on the causal mechanisms by which succession took place.

It was the object of the International Phytogeographical Excursions to help break down the geographical isolation of workers, and to expose the wide and deep differences of training and outlook to wider scrutiny. Tansley was among the 10 Europeans who took part in the second International Phytogeographical Excursion, which was led by Cowles and Clements in 1913. In his report for the *New Phytologist*, Tansley (1913–24) wrote of how

> in the vast field of ecology America has secured a commanding position and from the energy and spirit with which the subject is being pursued by very numerous workers and in its most varied aspects, there can be little doubt that her present pre-eminence in this branch of biology— one of the most promising of all modern developments—will be maintained.

During 1913–14, Clements travelled extensively in the western half of the United States, and became more and more convinced that his concept of the plant formation as an organism could be applied universally. He began to write a new book, which described how each formation had a life history of its own, and could be traced through successive stages or 'seres' to a climax community of adult organism. In a letter to Tansley in May 1914, Clements wrote of how 'I am getting more and more enthusiastic about it the longer I work it over and try it on'. He hoped the manuscript would arrive early enough for Tansley to draw on the text in preparing a second edition of his *Types of British Vegetation* (which in the event never appeared)[8].

Clements believed that if Tansley's support could be secured for his developmental concepts, they were likely to be put into instant and general commission. As well as being the most outstanding European ecologist, Tansley stood 'in the midst of a thoroughly well organised situation' in Britain. He was, so to speak, 'the managing director', capable of asserting a controlling, or at least a guiding, influence. He was uniquely qualified to assess how far the hypotheses formulated in 'the enormous stretches of fairly uniform untouched climaxes' of America could be applied to the variously modified forms of vegetation in Europe.

For Tansley, the strength of the Clementsian system lay in its philosophical sweep and comprehensiveness. As he remarked both at the time and later, Clement's book, *Plant Succession. An Analysis of the Development of Vegetation*, was the most important to appear since the 'foundation' works of Warming and Schimper. Although the importance of succession had long been recognized, there was still a tendency 'to treat plant communities too statically' (Tansley 1916b). Clements demonstrated how the laws of succession could be applied not only to all the plant communities in the world, but, perhaps even more significantly, to all plant communities since the beginning of time (Cooper 1926).

The main weaknesses of the new book arose from a rigidity and tendency to argue *a priori* from the proposition that the plant formation was

> an organism whose existence is determined solely by climate, and that all other plant communities, whatever their degree of individuality and permanence, must be interpreted in terms of the climatic climax or formation.

Tansley (1947b) believed the only way to build a more secure theoretical structure was to adopt a more empirical approach. Theory had to be built up piece by piece, as knowledge of vegetation and habitat factors increased.

The greatest point of controversy was Clement's perception of the plant formation as an organism. As early as the Annual Meeting of the British Ecological Society in 1917, Tansley challenged the view that every form of succession could be regarded as 'an organic unity'. A more considered response came in his paper, 'The classification of vegetation and the concept of development', published in the *Journal of Ecology* in 1920. In that paper, Tansley denied that any aggregation of plants could possess *all* the characteristics normally associated with an organism, namely the ability to arise, grow, mature, and die in the same way as an individual plant or animal. One obvious difficulty was what Tansley called the psychical relationship, namely the evidence of co-operation between organisms, whether deliberate or instinctive. As Tansley (1920b) pointed out, highly integrated plant communities depended instead on a balance of competition between individuals of the same species and different species.

An early critic of Clements's perception of vegetation was Henry A. Gleason (*1882–1975*), who moved from the University of Michigan to take up the post of head curator of the New York Botanic Garden in 1919 (McIntosh 1975). His interest in plant succession had first been aroused by Cowles's paper on physiographic ecology in 1901. Over the next 20 years, he became increasingly sceptical of Clements's concept of an irreversible trend leading to a climax. In Illinois, there was abundant evidence of forests succeeding prairie, and of prairie following forest. For Gleason (1922), the key to understanding succession was migration. The vegetation of any region was the outcome of

> repeated migrations of diverse floristic elements, arriving in the region
> from various directions, persisting there for various lengths of time, and
> finally retreating under the pressure of environmental changes which
> made their position no longer tenable.

Everywhere, there was a 'mingling of species of originally different history', and 'the persistence of isolated relics of early migrations'.

Far from perceiving plant associations as organisms in the Clementsian sense, with a series of functions distinct from those of the individual plant, Gleason became certain that their population was made up of whatever species happened to be in the vicinity and could withstand the prevailing environmental conditions. His misgivings as to how far ecological theory had outdistanced field observations came to a head in 1926, when he wrote a paper on 'The individualistic concept of the plant association', demonstrating how 'every species of plant is a law unto itself'. Its distribution in time and space reflected the species' own 'individual peculiarities of migration and environmental requirements' (Gleason 1926).

Tansley was present at the International Conference on Plant Science at Ithaca, when George E. Nicholls of the Osborn Botanical Laboratory at Yale both 'pulverized' and 'ridiculed' Gleason's paper (Duggar 1929). Although Gleason received some support from taxonomists, his views were clearly anathema to the ecologists present. In his own words, he was condemned to spend the next 10 years as 'an ecological outlaw' (Gleason 1953).

How did Tansley react? In his own paper to the Ithaca conference, Tansley emphasized that his support for the analogy of vegetation as an organism or quasiorganism was conditional on its not being pushed too far, or being made the pretext for 'illegitimate deductions'. The concept of plant successions could be of scientific value only if there was coherence in the sequence of succession to enable comparisons to be made and laws to be formulated. These 'uniformities' had to arise from the fact that the vegetation, like all forms of life, was dynamic, and that the vegetation was progressing towards a climax, or a state of equilibrium.

If the ecologist began by considering 'the existing communities as members of a developmental series', distinguishing those which were relatively permanent from those which were transitory, and 'tracing out the actual sequences and climaxes', he was at once brought into contact with the actual processes at work in the environment, and guided in his research towards the species and habitat factors of particular significance. Although most successions were affected by a combination of internal and external factors, Tansley (1929b) argued that it was useful to draw a distinction between autogenic succession, in which the successive changes were brought about by the action of the plants themselves on the habitat, and allogenic succession, in which changes were instigated by external factors.

In his Ithaca paper, Tansley reminded his audience of how such concepts as that of the plant community and climax community were merely devices for enabling the ecologist to discover connections which he might otherwise overlook. They helped to penetrate more deeply 'the web of natural causation'. Until something better was available, such crude analogies as the depiction of the plant community as an organism helped to crystallize thought and distinguish detail. As an heuristic device, they seemed an appropriate aid to the new field of investigation, where basic phenomena were only just being recognized.

By the 1930s, sufficient progress had been made for ecologists to want something less shallow (Tobey 1981). For Tansley, the point of revolt came when a three-part paper published in the *Journal of Ecology* appeared to push Clements's theories to their logical limit. The author was John Phillips, whose previous appointments to the Forestry Department in South Africa, and then to assist in the control of tsetse fly in Tanganyika, provided ample opportunities to develop the Clementsian approach. He spent several months in America, working with Clements, and succeeded Charles Moss as professor of botany at Witwatersrand in 1931 (Phillips 1954).

In his three-part paper, Phillips (1934–5) examined the terms 'succession', 'development', and 'climax', and concluded by putting forward the concepts of a complex organism or biotic community. In Tansley's words, it was 'the pure milk of the Clementsian word'. It was as if Clements was the major prophet, and Phillips had become the chief apostle. In an editorial footnote to the third part, Tansley acknowledged that some readers might think the subject matter too philosophical for an ecological journal. He had, however, decided to publish it *in extenso* because of the way it 'rounded off' the author's theory of vegetation, and would assist readers in comprehending it.

Tansley's response came in the form of a paper, entitled 'The use and abuse of vegetational concepts and terms', in which he attacked Phillips on almost every point. He could hardly have found a more devastating way of doing so. The paper was written in response to an invitation from the Ecological Society of America for a contribution to the volume of the journal, *Ecology*, intended to commemorate the retirement of Henry Cowles as chairman of the botany department at Chicago in 1934.

Tansley had visited the scene of Cowles' early research during the International Phytogeographical Excursion of 1913. He showed 'numerous fine lantern slides' of the sand dune systems of Lake Michigan to the first Annual Meeting of the British Ecological Society in December of that year. At the Ithaca conference of 1926, Tansley had used Cowles's research in order to illustrate concepts in succession. Now, in the commemorative volume, Tansley made every use of the opportunity to draw on hindsight in conferring on Cowles the kind of praise he had previously reserved so conspicuously for Clements (Tansley 1935).

Having acclaimed Cowles as 'the great pioneer of the subject', whose early work had demonstrated both the reality and universality of vegetation succession, Tansley went on to attack the followers of Clements, and by implication Clements himself, for their abuse of both the concepts and terminology of plant ecology. Whereas the interest and enthusiasm arising from Cowles's work had led 'to such great results', Tansley asked rhetorically, in another part of his paper, what researches had been stimulated or assisted by the Clementsian concept of the complex organism, or biotic community.

Tansley had long been dismissive of the 'monoclimax' doctrine, whereby there was only one kind of climax in any climatic region. The evidence from studies of the heath and fen in Britain suggested that not all communities would become woodland. Phillips explained away these types of climax by describing them as stages leading to *the* climatic climax. For Tansley, this sounded too much like a case of making the facts fit the theory. A much more scientific approach would be to recognize, describe, and study all the relationships of actually existing vegetation, and then to see how far they conformed to a general hypothesis that had been provisionally adopted.

In his paper to the Ithaca conference, Tansley introduced the much wider concept of the ecological system, in which climate, soils, plants, and animals would function as part of a system, each with a functional relationship to the other. Together with other biotic and abiotic elements, a plant community could be regarded in a very real sense as part of an ecosystem, an example of the vast number of physical systems of the universe, which ranged from the universe itself to the atom. Within each, there was a certain autonomy in organization and a tendency to move towards an equilibrium. The vegetation climax represented the highest form of integration, and the nearest approach to perfect dynamic equilibrium that could be attained under the system developed in the given conditions and with the available components (Tansley 1935).

In putting forward the concept of the ecological system, Tansley drew widely on the philosophies of physics and mathematics, and particularly on the book *The Universe of Science*, by Hyman Levy (*1889–1975*), the professor of mathematics at Imperial College, London (Levy 1932). The human mind perceived only 'events' or interrelated phenomena. The act of perception was primarily an act of 'isolation', in which the mind seized on a group of events out of the flowing stream of events. These 'isolates' became 'wholes' in the sense that they could be discerned from the larger environment, and that their components were sufficiently integrated to enable them to withstand a certain amount of buffeting from changing systems and the environment at large. All systems were evolving. In Tansley's words, there was 'a kind of natural selection of incipient systems, and those which can attain the most stable equilibrium survive the longest' (Tansley 1935).

2.4
'The British Islands and their Vegetation'

The British Ecological Society celebrated its quarter-century in 1938 by electing Tansley to a second term of office. He used the occasion of his second presidential address to review progress over the previous 25 years, and to demonstrate how the adoption of the concept of the ecosystem would confer a unifying sense of purpose on ecology (Tansley 1939b).

Within a few months, Tansley provided ample evidence of how this unifying concept might be applied. It came in the form of a new book, *The British Islands and their Vegetation*. It had originally been Tansley's intention to edit a new and larger edition of *Types of British Vegetation*, in which those who had carried out detailed studies would be given the opportunity to deal more fully with their special fields. In the event, it proved impossible 'to define limits acceptable to the possible contributors, and in a subject growing as rapidly as it then was, uniformity of treatment would have been impossible'. Tansley accordingly decided to shoulder 'the Herculean task of integration' himself. It was a task for which he was eminently well suited[9].

It was never part of Tansley's temperament to serve as 'the artisan concerned with laying the bricks and doing the joinery'. As Godwin (1977) recalled, he was neither a gifted taxonomist nor an experimentalist. His *métier* was in planning the buildings; he was the architect. Through his lectures and papers, Tansley responded to, and in many senses inspired, the more confident stance adopted by ecologists during the inter-war period. He himself had accepted, with grateful alacrity, the invitation to apply for the Sherardian Chair of Botany in Oxford in 1927, which he occupied until his retirement as emeritus professor 10 years later.

This new-found confidence was reflected in the preface to the new volume, *The British Islands and their Vegetation*. Tansley (1939c) wrote of how, in 1911, 'we wrote practically all we knew and a good deal that we guessed'. Now, in 1938, it was possible to draw on the work of a new generation of ecologists, with its 'deeper knowledge and better training'. Pearsall later recalled how one of Tansley's favourite techniques was to assemble a small group of persons with special expertise, or to use the occasion of a field excursion of the British Ecological Society, 'to thrash out in the field some sort of uniformity of approach'[10]. The outcome of this 'field-work' was 'an ecological event of the first magnitude'.

In a review for the *Journal of Ecology*, Pearsall (1940b) wrote of how, although written on broadly similar lines to *Types*, the new book possessed 'a breadth, a unity and detail which were lacking in the older work'. The modest octavo volume, designed for the pocket, had been replaced by a tome in the larger format of the *Journal of Ecology*, extending to nearly a thousand pages and illustrated by 162 plates with 418 photographs, as well as 179 text figures (Stamp 1940). For the writing of the book, Tansley was awarded the Gold Medal of the Linnean Society. The book was reissued as two volumes in 1949, when a slighter and more popular version was published under the title, *Britain's Green Mantle* (Tansley 1949).

As Tansley himself emphasized, *The British Islands and their Vegetation* was essentially 'a compilation of abstracts from the published literature'. As such the volume drew attention to the enormous advances that had been made in three spheres of British ecology, namely the identification and characterization of different vegetation types, secondly the appreciation of the significance of the biotic factor, and thirdly the reconstruction of the vegetation of the past. The remainder of this section will illustrate how these advances came about.

2.4.1 Studies of the hydrosere

Progress was particularly striking in respect of the hydrosere, where an outstanding contribution was made by William H. Pearsall (*1891–1964*). Pearsall's family moved to a village near the Lake District when he was six years old. As he later recalled, his father (who was also William H. Pearsall) was 'a good amateur botanist and ecologist', who exchanged specimens with Claridge Druce. Together, father and son devised a three-pronged, weighted dredge, which they used to drag up aquatic specimens from the bottom of lakes, with one of them rowing and the other using the dredger and making notes[11].

Pearsall left his father's school in 1905, and after four years at Ulverston Grammar School entered the Victoria University of Manchester, where he changed from chemistry to botany after one year. The head of the botany department was Weiss, who had himself published an account of the submerged vegetation of Windermere (Weiss 1909a). Pearsall obtained a first-class honours degree in botany in 1913, with a distinction in his subsidiary subject, chemistry. During the following year, he completed the university diploma course in teaching.

With the assistance of his father, Pearsall now embarked on a systematic survey of all the lakes in the area, with the aim of relating the distribution and abundance of plant life to the physical features of the shoreline and the transparency and chemical composition of the water. A grant of £20 from the Royal Society helped to defray the cost of analysing the water samples which his father collected on a monthly basis. Both were engaged on the detailed mapping of the comparatively small and little-known Esthwaite Water when war broke out in the summer of 1914. They repeated the survey in August 1915 (Clapham 1971). In a paper published in the *Journal of Ecology* in 1917–18, Pearsall recounted how the aquatic and fen successions 'were so complete, that a detailed survey would prove an admirable starting point for a description of the plant communities of the other lakes, and of the relationships underlying their description' (Pearsall 1917 and 1918).

Pearsall volunteered for the Royal Artillery and suffered permanent damage to his hearing during active service in France. Four days after the Armistice, he wrote of his relief at leaving 'the drab monotony of the modern battlefield, where a few broken bricks mark the site of a once flourishing village'. With the active encouragement of Weiss and Tansley, he applied for a variety of posts, including one at Rothamsted; he secured an assistant lectureship in botany at Leeds, becoming a lecturer a year later, and a reader in 1922 (Clapham 1971)[12].

For Pearsall and his generation, it was more than a question of resuming their pre-war research. The rough impact of war service had broken their academic mould. There was a much more individualistic outlook. In Pearsall's own words, the outcome was 'a great outburst of new ways of tackling biological problems' (Pearsall 1959)[13]. As the authors of the 'encyclical' in the *New Phytologist* had hoped, the new generation was much more attracted to the experimental, ecological, and genetic aspects of the subject.

In a further paper of 1918, Pearsall put forward an alternative approach to the prevailing system for classifying lakes. Rather than follow the continental classificatory system, whereby Esthwaite Water would be characterized as eutrophic, and Ennerdale and Wastwater as oligotrophic lakes, Pearsall perceived the 11 largest lakes in the Lake District (discounting the impounded Thirlmere) as forming a series, each one occupying a different place somewhere between the extremes of Esthwaite and Ennerdale/Wastwater. The floating-leaved, reed swamp, and fen communities encountered in the different lakes were not so much separate entities, but well-marked phases of a long succession, starting from bare stones and finishing in moor (Pearsall 1918, Macan 1970).

A distinctive feature of both this paper and another that looked more generally at the aquatic vegetation of the Lake District was the importance which Pearsall accorded to edaphic factors. As C. H. Mortimer recalled later, Pearsall 'looked at lakes through the eyes of a plant ecologist, and saw them as stages in an evolutionary succession'. He discovered parallel correlations between geology, soil processes, water chemistry, and the plant and animal life. In the Cumbrian climate there were two types of plant succession, depending respectively on the presence or absence of silting by inorganic material relatively rich in basic ions (Pearsall 1920).

Drawing on the many thousands of observations made with 'the unsparing assistance of his father' in the Lake District, Pearsall published a paper in the *Proceedings of the Royal Society* in 1921, which considered the aquatic vegetation in relation to the extent, depth, and age of the lake basins. Because all shared the same post-glacial origin, and contained rocks of similar character, any differences in vegetation could be attributed to variations in the rates of erosion and sedimentation of the rock basins, induced by inequalities in the durability of the underlying rocks. By distinguishing rocky from relatively silty lakes, as revealed by their vegetation, a contrast could be drawn between the primitive and more highly evolved lakes of the Lake District (Pearsall 1921).

2.4.2 The biotic effect

In writing his book, *The British Islands and their Vegetation*, Tansley was able to draw not only on an increasing knowledge of edaphic factors, but also on the growing awareness of the role of biotic forces in influencing the course of vegetation succession. As Stapledon commented, it was not until ecologists began to take account of the impact of grazing animals on plant life that ecology began to enter 'a mature stage'. It was impossible to understand the community as a whole, and its separate problems, without close reference to 'the interactions of animals with plants, and of animals and plants with all the other factors of habitat and environment'. In Stapledon's words, all this was true 'to a unique degree of grasslands and all that pertains to grasslands' (Stapledon 1933a).

One of the earliest intimations of the significance of the animal–plant relationship occurred at a meeting of the British Vegetation Committee in December 1907, when Weiss recounted how he had taken up a suggestion of Tansley to look more closely at the way *Ulex* appeared to invade pure heather moorland in a strikingly rectangular pattern. Observations on the North York Moors revealed that the bushes appeared to follow the course of old trackways, the ruts of which could still be seen under the heather. Weiss remembered seeing a monograph on European myrmecochorus plants, published in Sweden in 1906, calling attention to the role of ants as 'an ecological factor of no small importance' in dispersing seed. Experiments conducted by Weiss indicated that ants would carry food and nesting materials along such pathways, including the seed of the *Ulex* (Weiss 1908, 1909b)[14].

One of the first detailed studies of the impact of animal life on plant communities was carried out by E. Pickworth Farrow (*1891–1956*). At the suggestion of Adamson, and under Tansley's supervision, Farrow made a study of the breck heaths of Breckland. In his book, *Plant Life of East Anglian Heaths, being Observational and Experimental Studies of the Vegetation of Breckland* (which was based largely on papers previously published in the *Journal of Ecology*), Farrow (1925) acknowledged the early encouragement given him by Moss, and the many insights he had gained from Oliver at Blakeney Point.

In the course of studying the *Calluna* and grass heaths of Breckland, Farrow (1916) became increasingly sceptical of the widely held assertion that their distribution was determined by differences in the lime content of the soil. He noticed how, on approaching a *Calluna* association, one first encountered isolated *Calluna* plants of a small height, and then more numerous plants of greater height until the association itself was reached. Many of the dwarf plants were smothered by *Cladonia*. Farrow was puzzled by the character of this transitional zone until he noticed that it bore a striking similarity to what occurred on a much smaller scale around the rabbit burrows situated in the *Calluna*

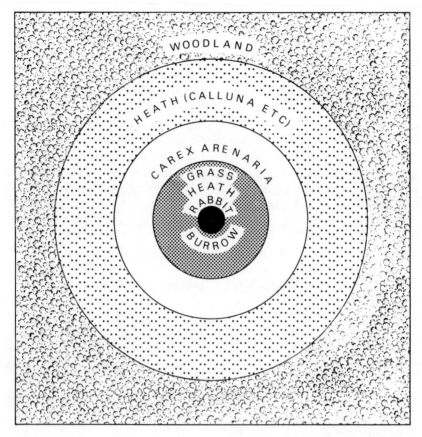

FIGURE 2.1 The zonation of vegetation around a rabbit burrow in Breckland (after Farrow 1917).

(Figure 2.1). The transitional zone, with its smooth, rounded bushes, corresponded to those parts where the *Calluna* heath was degenerating to grass heath as a result of rabbit attack.

Farrow used fixed-point photography and predetermined transects, as well as rabbit exclosures, to assess the speed and character of this form of degeneration. In his words, the result was 'a very striking and beautiful example of a dynamic biotic succession' (Farrow 1917a). The minute-book of the British Ecological Society recorded how Farrow led a field meeting of 15 members to the Breckland in 1917 to see 'the various features of the area' that had been 'described in the *Journal of Ecology*'[15]. Farrow (1917b) argued that much greater account should be taken of biotic factors in determining the effects of climate and soils on the distribution and character of associations.

Meanwhile, Tansley had initiated a long-term study of the course of succession within the Ditcham Park estate on the Hampshire–Sussex border. As a means of assessing how far biotic pressures, namely the very high rabbit populations, were responsible for maintaining the balance between shrub land and open grass, two rabbit exclosures of 820 and 390 square metres were maintained over the period 1908–20. In a paper to the Annual Meeting of the British Ecological Society in December 1921, Tansley described how a length of woodland had been included in the exclosures, and how a comparison of the charts made in 1909, 1914, and 1920 made it possible to trace the spread of woody plants from the edge[16]. By 1920, the density of the shrubs colonizing the grassland had more than doubled. *Calluna vulgaris* had established itself, and four healthy beech plants were recorded for the first time. Tansley (1922a) believed that if the exclosures had remained, the loose scrub would have given way to ashwood and, perhaps in half a century, to beechwood.

In a general study of the Hampshire–Sussex border, Tansley and Adamson (1925) drew attention to two contrasting types of change that were taking place on the chalk grasslands traditionally close-cropped by sheep. Where the rabbit population was very high, the turf had been grazed down below 2·5 cm, and, in extreme cases, was destroyed, leaving only those few species which the rabbit found generally unpalatable. Where rabbits were less abundant the effects of the post-war trend from sheep to cattle grazing were more evident. The turf became deeper, and there was a tendency for the grassland to be invaded by *Potentilla erecta* and *Calluna vulgaris*, species generally associated with acidic conditions. Their presence might reflect trends towards deeper soils, caused among other factors by the accumulation of humus above the chalk.

Studies of the biotic impact on woodlands were started by Alexander S. Watt (*1892–1985*). Having obtained an Arts degree and a BSc in agriculture at Aberdeen in the same year, 1913, Watt decided to focus on forestry. He applied for a Carnegie Scholarship to spend a year in Germany, with a view to returning to Aberdeen to assist in teaching a new degree course in forestry. He obtained the scholarship, but war intervened, and he moved instead to Cambridge where, under Tansley's supervision, he investigated the regeneration of British woodlands. He was appointed in 1915 to the new post of lecturer in forest botany and forest ecology at Aberdeen, where he stayed until 1916, when he was called up for war service in the Royal Engineers. Watt survived the Battle of the Somme, but was so badly gassed in the later stages of the war that he was released early from the army in 1918. Before returning to Aberdeen, he spent a term in Cambridge so as to fulfil the regulations for a BA 'by research' degree (Gimingham 1986).

From his studies of the fate of the acorn crop, and the disappearance of beech mast and seedlings, Watt (1919, 1923) discovered that grazing was often so intensive as to prevent regeneration taking place, except where dense scrub afforded some protection from the larger herbivores. This phenomenon was most graphically demonstrated at Kingley Vale in the South Downs. In a paper given to the Annual Meeting of the British Ecological Society in January 1925, Watt described how the yew-woods represented a temporary, albeit drawn out,

phase in a succession from open grassland to yew-wood to ash/oak, and ultimately to beech. The woods were in effect 'migratory'. No succession could take
place, however, where grazing was so intense that it prevented the yew seedlings
establishing themselves on the nearby grasslands, and the ash/oak saplings developing where gaps appeared in the canopy of the older yew trees (Watt 1926).

In correspondence with Tansley in 1921, Watt wrote of how 'I should feel
most content doing whole time research work but opportunities for this I am
afraid are strictly limited or non-existent, when account has also to be taken of
adequate remuneration'[17]. His studies of the fate of crops of seed soon widened
into an analysis of the structure and development of beech communities. In
1924, Watt was awarded one of the first Cambridge PhD degrees for research
on the beechwoods of the western South Downs, carried out during the summer
vacations. This was later extended to include examples from the Chilterns and
the subspontaneous beechwoods of Aberdeenshire (Watt 1925, 1931, 1934). In
1919, Watt was appointed the Gurney Lecturer in Forestry at Cambridge.

Meanwhile, the study of biotic factors in fenland vegetation was opening up
new perspectives on another major habitat. Once again, it was Tansley who
acted as the mentor. In 1921, he organized a bicycle excursion for the newly
formed Botany School Ecology Club to the National Trust's nature reserve at
Wicken Fen, near Cambridge. One of those to take part was Harry Godwin. As
Godwin later recalled, 'we learned to identify the remarkable wealth of fen
plants', and saw at first hand the way in which the areas of free-floating aquatics
were being invaded by reed swamp and pure sedge. As surface levels rose, the
sedge was replaced by bushes, and a fen carr ultimately gave way to damp oakwood (Godwin 1978).

It was Godwin who soon provided evidence of the impact of livestock grazing and the cutting of vegetation on primary succession. With Tansley's active
encouragement, and while still a research student of F. F. Blackman (studying
the metabolism of cherry laurel leaves), Godwin had begun to record the vegetation of marked localities on the reserve. Once his thesis on plant physiology
was completed, and he had been appointed a junior demonstrator in the Botany
School in Cambridge, it became much easier to devote time to studying the successional changes taking place (Godwin & Tansley 1929).

Until then, it had been commonplace to portray breaks in succession as sub-
climaxes, when the normal course of succession was arrested for as long as the
repressive factor was applied. In two papers, published together in the *Journal
of Ecology* of 1929, Godwin (who was by now a lecturer) challenged such an
interpretation, and put forward the alternative concept of a 'deflected' or
'specialized' succession.

As he explained, the primary succession at Wicken Fen was not so much
arrested as deflected by such activities as livestock grazing and the harvesting of
sedge. Where the fen continued to dry out, and the sedge was cut at four-yearly
intervals, the absence of competition from *Cladium* and bush growth made it
possible for the tussock-forming grass, *Molinia caerulea*, to flourish, forming a

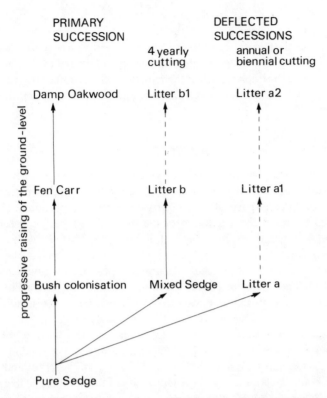

FIGURE 2.2 Primary and deflected successions at Wicken Fen, Cambridgeshire (after Godwin 1929).

Cladio–Molinietum. The transition to a pure litter community of *Molinia* was accelerated where the sedge was harvested more frequently (Figure 2.2). The outcome was a chequered pattern in the vegetation, representing the various stages in primary succession and in the two deflected successions (Godwin 1929a,b).

The interest aroused by the types of plant succession at Wicken Fen encouraged Godwin to extend his investigations to the Norfolk Broads, and to investigate reports of acid-loving plants growing on some of the peat lands. Attention was focused on Calthorpe Broad. By systematically measuring the surface level and soil acidity, it became clear that species composition changed markedly as the peat level rose and became more remote from the Broad. The plants associated with alkaline, open water and fen gave way to acidophilous species (Godwin & Turner 1933).

Not only did this further evidence of the transitory nature of plant communities make the ecologist keener to build up some kind of historical perspective, but his increasing knowledge of such phenomena as deflected successions and the trend towards soil acidity made him better qualified to interpret the evidence surviving from the past. By the time Godwin and another Cambridge botanist, John S. Turner, gave their joint paper on Calthorpe Broad to the Annual Meeting of the British Ecological Society in January 1933, they had begun to recognize the implications of the remains of Scots pine, a tree often associated with the natural transition to acid bog, within the peat deposits of the East Anglian fens.

2.4.3 *Pollen and peat stratigraphy*

In writing his book, *The British Islands and their Vegetation*, Tansley was able to demonstrate not only the role of the ecologist in building up a picture of the vegetation of different parts of the country, but the enormous potential for discovering more about the vegetation of the past. The scope for historical research had been greatly extended by E. J. Lennart von Post (*1884–1951*) of the Swedish Geological Survey, who had developed and applied a technique for analysing the pollen found in successive layers of peat so as to give a consistent picture of what had happened to the vegetation since the end of glaciation.

One of the first British botanists to draw attention to the significance of pollen diagrams in unravelling 'the postarctic history of the forests of north-western Europe' was Salisbury who, in the course of a regular series of notices in *Science Progress* (Salisbury 1922a), summarized 'an extensive and fascinating paper' by O. Gunnar E. Erdtman (*1897–1973*), a pupil of von Post and a member of the botanical laboratory of the University of Stockholm (Hedberg 1973). It was to Salisbury that Erdtman turned for advice as to where to publish a paper drawing the attention of British botanists to the results of visits made by Erdtman to the Highlands and Islands of Scotland in July 1922, when, with the help of officers of the Geological Survey, he had collected over 800 peat samples. The paper was published by the Linnean Society, and was followed some years later by another, on the post-glacial history of British forests, in the *Journal of Ecology* (Erdtman 1924, 1929)[17].

Detailed insights into the vegetation of the different interglacials began to emerge from studies of the large numbers of interglacial freshwater deposits found in Jutland and north-west Germany. The tentative scheme of pollen zones for the last interglacial was taken up by botanists and geologists alike (Jessen & Milthers 1928, West 1981). Despite differences of detail, a parallelism could be detected in the development of vegetation throughout north-west Europe. As von Post told the International Botanical Conference at Cambridge in 1930, the broadly synchronous nature of the series of pollen zones found across the continent suggested that the overriding effect of climatic change since the last ice age had been to cause the vegetation belts of Europe to move steadily northwards. This had been accompanied by corresponding shifts in altitudinal range, and movements towards and away from the Atlantic coast (Brooks & Chipp 1931).

It was Tansley who encouraged first Margaret, and then Harry, Godwin to take up the study of pollen analysis. Albert Seward made laboratory space and equipment readily available. Seward had himself recently published a book on the history of plant life, which concluded by calling for more study of the Quaternary Period (Seward 1931, Thomas 1941). Analysis of peat from deep and extensive excavations at Wiggenhall St Germans, where a new sluice and pumping station were being installed, showed that the fenland peats of East Anglia were full of pollen (Godwin & Godwin 1933). A review paper on the 'Problems and potentialities of pollen analysis' attracted so much publicity for the technique that it became the first of a long series of papers to be published in the *New Phytologist* (Godwin 1934, 1985b).

The need for close collaboration with archaeologists was obvious. During 1932, a Fenland Research Committee was formed, made up of both professionals and amateurs, with Seward as president and an archaeologist as secretary (Godwin 1978). Drawing on a wide range of geological, archaeological, historical, and botanical data from the excavations at Shippea Hill and from other parts of the East Anglian Fenlands, a chronology was pieced together, within which the individual deposits could be relatively dated for the first time (Godwin et al. 1935, Godwin & Clifford 1938, Godwin 1940a).

In March 1935, the Royal Society organized a meeting on the origin and development of the British flora, under the chairmanship of Seward. At the meeting, Godwin described how, by analysing the tree pollen preserved in the Fenland peat profiles, it had been possible not only to trace changes in the dominant tree species but to speculate on the composition and structure of the associated vegetation and climatic conditions. In the light of what had been achieved, he suggested that an examination of the stratigraphy and associated plant remains of bog systems would repay close study (Anon. 1935).

The opportunity to investigate such systems arose later in the year, as a result of a chance meeting between Godwin and G. F. (Frank) Mitchell, a recent graduate of Trinity College, Dublin. Godwin had been invited to act as chauffeur on a tour of Ireland, which Tansley (who was preparing to write *The British Islands and their Vegetation*) had arranged with Hugo Osvald (*1892– 1970*), a leading authority on Swedish raised bogs. In the course of the tour, they met Knud Jessen (*1884–1971*), whose pioneer work on the interglacial deposits of Jutland and north-west Germany had aroused so much interest. At Praeger's instigation, the Royal Irish Academy had invited him to investigate the glacial and post-glacial deposits of Ireland. Mitchell was deputed to act as his 'courier and brain picker' (Godwin 1973, Mitchell 1976).

Godwin and Mitchell realized that the most effective way of consolidating and extending what they had seen and learnt was to collaborate in studying one of the comparatively few raised bogs covered with natural vegetation, little affected by cutting and burning. They chose for their purpose a complex of three raised bogs near Tregaron in the Teifi valley of Cardiganshire. Having successfully petitioned the Royal Society for a Swedish-made peat borer, a fortnight's investigations took place in July 1936. Another summer excursion was mounted by Godwin in 1937.

A paper on the stratigraphy and morphology of the bog system (Godwin & Mitchell 1938) was followed by another in the *Journal of Ecology*, describing the surface ecology. The fact that the centre was occupied by Scirpetum and Molinietum, and only the wetter, peripheral areas by Sphagnetum was taken to indicate a trend towards a drier, more continental climate—a hypothesis that seemed to be supported by the discovery of a discontinuity between the surface vegetation and the underlying peat, which was largely composed of *Sphagnum imbricatum*, a species now absent from the living cover of the bog (Godwin & Conway 1939).

It was an announcement in the *Manchester Guardian* of the resumption of archaeological excavations at the Iron Age lake village of Meare in the Somerset Levels that caused Godwin to investigate what proved to be 'the western homologue' of the East Anglian fens. During 1936–7, he investigated not only the village site but the remnants of the raised bogs. From peat stratigraphy and pollen analyses, it became clear that the massive timber causeway across Meare Heath (and three other trackways) had been built during the transition from the Sub-Boreal to the Sub-Atlantic phases, when the higher rainfall resulted in the raised bog surfaces, on which the causeways were built, becoming flooded with calcareous water. The flooding was 'at once the reason for the building, preservation and contemporaneity of the trackways' (Godwin 1941, Clapham & Godwin 1948).

When Godwin came to assess the contribution of pollen analysis to studies of forest history, at a meeting of the Linnean Society in March 1940, he was able to cite evidence from a wide range of localities. Drawing on insights gained from studying vegetation successions and regional climatic controls at the present day, Godwin outlined how the pollen zonation established for the East Anglian fenland might be related to those reported for other parts of the country (Godwin 1940b, 1946).

In his presidential address to the British Ecological Society in 1943, Godwin reviewed the implications of what had been discovered for future investigations of the coastal peat beds around Britain and the North Sea. It was clear from pollen analyses that the formation of the present-day submerged peat beds had continued throughout the post-glacial period. This provided not only an independent and sensitive means of dating the changes that had occurred in land/sea levels, but it was possible that these dateable deposits might contain evidence of vegetation successions that were no longer available for study in the 'living' vegetation of today (Godwin 1943).

2.4.4 *The life conditions of plants*

As Salisbury wrote in a review of *The British Islands and their Vegetation*, no one had done more than Tansley to promote the study of 'phytogeography' in

Britain. The new book was a fitting culmination of his 20 years as the editor of the *Journal of Ecology* (he retired in 1937), outlining what had been achieved in descriptive ecology, and indicating the gaps that remained. In Salisbury's words, future progress was likely to lie in the direction of causal ecology, namely 'the study of the biological relations of species, and the applications of the principles of physiology and the factors of competition' to the elucidation of problems affecting plant and animal life under natural and seminatural conditions (Salisbury 1939).

Tansley himself was well aware of the need for research in this direction. In his presidential address to the British Ecological Society in 1939, he spoke of how a deeper understanding of 'whole plant communities' could only come from research on

> the autecology of the dominants, the conditions under which dominance is established, and the behaviour of dominant species in competition within their geographical range but outside their range of dominance (Tansley 1939a).

There were already numerous examples of how this major field of enquiry might be extended.

It was during the International Phytogeographical Excursion of 1911 that the British participants first learned of 'the system of life forms', developed by the Danish botanist, Christen C. Raunkiaer (*1860–1938*), who shortly afterwards succeeded Warming as professor of botany at Copenhagen. William Smith was so impressed by what he heard that he summarized the contents of Raunkiaer's publications in the first issue of the *Journal of Ecology* (Smith W. G. 1913). In his search for a standard by which different communities could be compared, Raunkiaer used as his criterion the degree to which the plant was adapted to withstand the most critical or rigorous season of the year, in other words the nature and extent of protection afforded by the dormant perennating buds or shoot-apices. From this, Raunkiaer developed the concept of the 'biological spectrum', by which different climatic regions could be characterized by different biological spectra, namely the proportion of different life-forms in their vegetation (Raunkiaer 1934).

Among those stimulated to look more closely at the life-forms of plants, and their relationship to the environment, was Richard H. Yapp (*1871–1928*). Despite considerable obstacles, Yapp had succeeded through evening classes in achieving his ambition of going to university. He graduated in botany from Cambridge in 1895, and took part in the university expedition to the Malay States in 1900. There, he developed an interest in the effects of the environment on individual species, which he developed on his return to England (Weiss 1929).

In the course of his studies on the National Trust's reserve at Wicken Fen (which he continued even after his appointment to the chair of botany at Aberystwyth), Yapp provided detailed descriptions of the close relationship that existed between differences in soil moisture and the spatial distribution of the

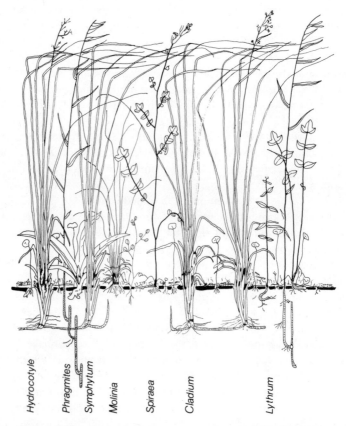

FIGURE 2.3 Stratification of both shoots and roots in sedge vegetation at Wicken Fen, Cambridgeshire (after Yapp 1909).

fen species. His most original contribution was, however, to discover the extent to which the stratified structure of the individual stands of vegetation influenced evaporation and temperature (Figure 2.3). Yapp realized that the structural peculiarities of the vegetation might play as significant a part as the morphological and anatomical modification of leaves and other transpiring organs in reducing the adverse effects of excessive transpiration (Yapp 1908, 1909).

Yapp took as the theme of his presidential address to the British Ecological Society in 1921 the concept of the 'habitat'. This was usually taken to imply 'a uniformity of life-conditions' in the sense of there being a 'prevalence of a particular complex of factors' in operation throughout the habitat. His own studies of Wicken Fen and elsewhere suggested that this degree of uniformity was only likely to occur in the corresponding parts of the habitat. The stratification and layering effect of the vegetation might be so marked that the smaller, subordinate species might live in a climate that was very different from that of the larger, dominant species. In a sense, the organs of the various growth forms and species reacted not only with their habitat but also, in a wide variety of ways and degrees, to one another (Yapp 1922).

One of the most outstanding figures to emerge in the study of the life conditions of plants was Edward J. Salisbury (*1886–1978*). As a student at University College, London, he impressed Oliver so much with his field knowledge of plants that he joined the expeditions to the Bouche d'Erquy while still an undergraduate. Upon graduation, he became the Quain Student, and assisted Oliver in detailed studies of Blakeney Point. Between 1914 and 1918, he held the post of senior lecturer in Fritsch's department at the East London College (Clapham 1980).

It was in his home county of Hertfordshire, around Harpenden, that Salisbury acquired his knowledge of plants and began to develop his own research interests. He joined the Hertfordshire Natural History Society in 1908, took an active part in its affairs, and gave a lecture, while still a postgraduate, showing how competition between furze and bracken on Harpenden Common was affected by the increasing incidence of fire and a falling water-table. The bracken was displacing the furze most rapidly on the higher ground (Salisbury 1912, Sheail 1982a).

Through his studies in Hertfordshire, Salisbury became one of the country's leading workers in woodland ecology (Clapham 1980). He was able to correct the widespread impression that scrub generally represented a retrogressive phase of woodland. In a paper on 'The ecology of scrub' in Hertfordshire, Salisbury (1917) cited examples of where scrub had developed from arable and heathland, and represented a progressive, rather than a retrogressive, succession. As Salisbury (1916, 1918) demonstrated, even the intensively managed oak–hornbeam woods of the county could offer important ecological insights. Coppicing brought about an 'ebb and flow' in the two major components, namely the 'shade flora' and the 'marginal flora', highlighting the kind of dynamic interrelationships that existed between the components of any woodland community (Salisbury 1923).

Salisbury returned to University College in 1918, and became a reader in 1924. He succeeded Oliver as the Quain Professor of botany in 1929. Increasingly, he focused on the vegetation/soil relationship. In a lecture to the London Intercollegiate Botany Society in 1920, he described how the presence or absence of species on different soil types was likely to reflect a much greater range of factors than ecologists had suspected. Even where the soil itself was responsible, the reason might be attributable to any one of many factors. Calcareous and non-calcareous soils, for example, differed in many ways; there were differences in the concentration not only of calcium, but also of hydrogen ions in the soil solution (Salisbury 1920).

Whatever the functional relationships between soils and plant life, they were unlikely to be static. There was a tendency over time for all soils to become more acid, with consequent changes in the character of vegetation. Experience, first in Hertfordshire and later elsewhere, indicated to Salisbury (1921) that old woodland and meadow sites were particularly suitable for revealing the effects of progressive leaching over long periods. In a collaborative study of the Durmast oakwoods (*Quercetum sessiliflora*) near Malvern in Worcestershire, Salisbury and Tansley (1921) put forward a chronology whereby the Malvern limestones were once dominated by ash, with an accompanying calcareous flora. As leaching became more pronounced, and humus accumulated, conditions became more suitable for the Durmast oak, with its predilection for acid soils, so that, by the present day, the ash remained abundant only on the steeper slopes where fresh soil was constantly exposed and the build-up of humus was retarded.

Studies made along tracts of coastline provided striking confirmation of a general trend towards a higher organic content, and diminution of available salts (and particularly calcium salts) in naturally undisturbed soils. At Blakeney Point, there was a change from appreciably alkaline conditions in the embryo dunes to marked acidity in the later phases (Salisbury 1922b). In the case of the dune systems of Southport, Lancashire, it was possible to assign an approximate age to each part, using maps dating back to the early 17th century. Whether rich or poor in calcium initially, the average carbonate content fell at first very rapidly. Over a longer time period, the average content fell by a fairly uniform rate of 0.6–0.8% per annum, with the bulk of the calcium probably derived from decaying organic matter (Salisbury 1925).

The title of Salisbury's presidential address to the British Ecological Society in 1929 was 'The biological equipment of species in relation to competition'. How far did competition determine the character of vegetation? Salisbury recalled how, in earlier culture trials, he had found that many plants failed to flourish, not so much because of some inimical factor in the soil, but because of their inability to withstand the competition of other species when growing in that soil type (Salisbury 1920). Such observations posed the question, 'How far was dominance the consequence of selective depression or of selective stimulus?' Whichever was the case, it was determined by the relative vigour of the species making up the vegetation.

An understanding of how far one species could compete with another could only come from a greater knowledge of the 'biological equipment' of the species involved. How far were the species able to cope with differences in radiant energy induced by the height and shading of other species? How far did rooting systems affect one another? As Salisbury (1929) reminded his audience, there were few precise data on plant propagation and dispersal, longevity, and the scope of vegetative multiplication. No one had ever studied the reproductive capacities of even the chief constituents of a community, either in Britain or abroad.

Comparatively few autecological studies had been published by the *Journal of Ecology*, despite the apparent head start given by the Annual Meeting of

December 1914, when papers were read by Lilian Baker on the wood sage (*Teucrium scorodonia*) and T. A. Jefferies (a former student of Thomas Woodhead) on the purple moor grass (*Molinia caerulea*). Based on a detailed study of a tract of moorland near Huddersfield, Jefferies described how this plant was one of the sturdiest species in the vegetation. This was due to its compact tussock habit, adaptation to rapid transpiration and assimilation, abundant provision for absorption and food storage, and development of woody and strengthening tissues. The most important limiting factor in the species' distribution was the availability of fresh water (Jefferies T. A. 1915).

It was Salisbury who played the key part in encouraging the British Ecological Society to embark on the publication of 'an ecological flora of Britain'. After a lengthy discussion at the Annual Meeting of 1928, a subcommittee was appointed, made up of Tansley, Oliver, Woodhead, and Ramsbottom, with Salisbury as convener. This recommended that the work should be issued separately from the *Journal*, either as a whole or in parts, under the editorship of Salisbury. It was hoped the proposed flora would generate greater interest in autecological studies, and highlight the many lacunae that remained to be filled (Salisbury 1928). In the event, nothing was achieved for another 10 years.

Meanwhile another, and much more successful, venture was initiated under the aegis of the British Ecological Society. The minutes of the Annual Meeting of 1927 record how 'Dr A. W. Hill opened a discussion on the subject of transplantation experiments in which Dr Salisbury, Professor Oliver, Professor Yapp, Professor Weiss and Mr Ramsbottom took part'. One of the country's leading plant taxonomists, Arthur W. Hill (*1875–1941*) had been appointed the director of the Royal Botanic Gardens at Kew in 1922 (Brooks 1942). Hill drew attention to the need for carefully devised experiments to determine the effects of different soils on selected plant species. It was agreed that the Council of the Society should form a committee in consultation with Hill to initiate experiments and to raise the necessary funds[18].

In taking this initiative, Hill may have been prompted by William B. Turrill (*1890–1961*), who joined Kew as a temporary technical assistant in 1908, and retired as the keeper of the herbarium and library 49 years later. Turrill returned from war service convinced that the only way to achieve a greater understanding of species and their groupings into genera was to extend the 'purely herbarium approach to plant systematics based mainly on morphology to one in which anatomical, ecological, cytological, genetical and chemical factors were taken into full consideration'. From 1924 onwards, he began to collaborate with Eric M. Marsden-Jones (*1887–1960*) in a series of trials both at Kew and in the latter's garden at Potterne, Wiltshire, where an extensive collection of herbaceous species from known wild localities had already been established (Hubbard 1971, Turrill 1961).

The British Ecological Society formed a Transplant Subcommittee. Turrill and Marsden-Jones joined Oliver, Salisbury, and Tansley (chairman) on the Subcommittee, and were mainly responsible for compiling the periodic reports which appeared in the *Journal of Ecology*. By growing uniform genetic material from several species on different types of soil, the aim was to discover the extent to which edaphic factors induced differences in plasticity, persistence and mortality, and phenology (Marsden-Jones & Turrill 1930, 1933, Salisbury 1963). The trials continued until 1940 (when Marsden-Jones had to leave Potterne), and provided a basis for detailed studies of the knapweeds (*Centaurea*) and bladder campions (*Silene*), published later by the Ray Society (Marsden-Jones & Turrill 1945)[19]. In 1950, Turrill became the first professional taxonomist to be elected president of the British Ecological Society (Turrill 1951).

PLATE 2 William G. Smith.

PLATE 1 Robert Smith.

Mr.Pethybridge   Mr.Parkin   Mr.Arber   Mr.Boodle   Mr.Hill   Mr.Burdon
Mr.Smith Mr.Woodhead   Mr.Lang   Prof.Yapp
Prof.Goethart Mr.Gwynne–Vaughan Mr.Darbishire Mr.Wager Prof.Johnson Mr.Waddell Prof.Buller Mr.Bentley
    Mr.Groom   Mr.Druce   Prof.Phillips Mr.Rendle Mr.West   Prof.Ewart
Dr.F.F.Blackman   Mr.Keeble Dr.Pierce   Mrs Scott
        Miss Robertson M de Candolle Prof.Hillhouse   Prof.Büsgen
Miss Fraser Mr.Kidston Mdme.Zacharias Miss Dawson Miss Pertz Prof.Errera Prof.Vines Prof.Ward Dr.Stapf Dr.Gardiner Prof.Green Dr.Scott Miss Stopes Mdlle.Bertrand
Miss Ewing Mr.Ewing Prof.Zacharias Prof.Eriksson Prof.Klebs Miss Sargent Prof.Engler Mr.Darwin Sir W.Dyer Prof.Reinke Dr.Lotsy Prof.De Candolle Prof.Chodat Prof.Bertrand
Mr.V.H.Blackman Mr.Tansley Miss Matthaei Dr.Fritsch Miss Dale Prof.Weiss Mr.Miyake Prof.Oliver Miss Thomas Miss Gibson Miss Ford   Prof.Fujii

PLATE 3 Section K, Botany, of the British Association meeting at Cambridge, in 1904, during which Tansley read his paper, 'The problems of ecology'.

Mr.Wilmott Miss Hume Mrs Arber
Miss Ridgway Miss Powell Mr.Gwynne-Vaughan
Prof.Osborne Mrs.Osborne Miss Lorraine Smith Miss E.Thomas
Miss Pearce Mr.Davie Miss Lockhart Mr.Ewing Mr.A.M.Smith Mr.Compton
Mr.Richardson Mr.Lee Prof.Priestly Mr.Laidlaw Mrs Tansley Prof.Lewis Dr.Arber
Mr.Stiles Mr.Adamson Miss Wigglesworth Mr.Cavers Miss Halket
Miss Dale Mr.Maugham Miss West Mr.West Mr.Cheeseman
Mr.Delahunt Mr.Crump Dr.Lawson Mr.Highley Miss Chambers Mr.Ewing Dr.Fraser Prof.West Mr.Thomas Miss Baker.
Mr.T.G.Hill Miss Welsford Mr.Brooks Mrs.Reid Prof.Oliver Prof.Seward
Mr.Lechmere Mr.Williams Mrs Brooks Mr.Clement Reid Mrs Clements Prof.Bottomley Dr.Moss Prof.Yapp
Mr.Wager Dr.Benson Prof.Massart Mr.Tansley Mrs Weiss Prof.Lindman Dr.Ostenfeld Dr.Rendle
Mrs.Cowles Prof.Grabner Prof.Clements Prof.Trail Prof.Drude Prof.Weiss Prof.Schröter Prof.Phillips Dr.Rübel Dr.Scott Prof.Cowles Dr.Green,
Dr.Gluck (president)

PLATE 4 British Association meeting at Portsmouth in 1911, at the conclusion of the first International Phytogeographical Excursion.

R.C.McLean T.S.Prankerd E.P.Farrow M.K.Dagget I.M.Allen A.C.Halket H.H.Thomas E.J.Salisbury W.N.Jones
J.Lomax W.E.Hiley Mrs McLean R.R.Gates N.Bancroft A.I.Davey E.M.Delf E.M.Berridge M.C.Rayner A.G.Tansley
E.M.Rankin F.Brooks E.W.Thomas A.B.Rendle E.R.Saunders Sir Daniel Morris A.Remer A.C.Seward Lorraine Smith F.W.Oliver

PLATE 5 British Association meeting at Bournemouth in 1919.

PLATE 7 R. H. Yapp.

PLATE 6 C. E. Moss.

PLATE 8 T. W. Woodhead.

PLATE 10 F. W. Oliver beside the hut at Blakeney Point, July 1928.

PLATE 9 F. W. Oliver with women students at Blakeney Point, about 1913.

PLATE 11 E. J. Salisbury at Blakeney Point in the mid-1920s.

PLATE 12 C. S. Elton in Bagley Wood, Oxford, 1926–7.

PLATE 13 Oxford University Expedition to British Guiana, 1929. Front row (left to right), L. Slater, M. Nicholson, Major R. W. G. Hingston (leader), B. Nicholson and O. W. Richards. Back row, P. W. Richards, M. Creswell, J. E. Duffield, T. A. W. Davis, S. T. C. Livingstone Learmonth and N. Y. Sandwith.

PLATE 14 Oxford University Expedition to British Guiana, 1929. The living and working accommodation of Paul Richards.

PLATE 15 V. S. Summerhayes beside one of the point quadrat cages used in studies of the grazing impact of voles, July 1935.

PLATE 16 W. B. Turrill on the British Ecological Society summer excursion, Aberystwyth, July 1939.

PLATE 17 Field course in the Chiltern beechwoods in 1938. Left to right, A. S. Watt, J. Turner, G. Metcalfe. Foreground, M. H. Clifford and Verona Conway.

PLATE 18 Botany School, Cambridge, Cairngorms Expedition, 1938. Front row (left to right), G. Metcalfe, Joyce Ingram, P. Richards, E. F. Warburg, A. S. Watt, J. B. Jones. Middle row, Delia Simpson, Mary Robertson, Betty Brown, M. Ingram, B. Wilkinson. Back row, A. Burges, Joan Valentine, B. Robertson, E. W. Jones, F. Dainton and D. Valentine.

PLATE 19 H. Godwin.

PLATE 20 D. Lack, with his wife, Elizabeth, 1950.

PLATE 21 Ecology field course in Wytham Wood, Oxford, 1953. Left to right, C. S. Elton, G. Blane, H. N. Southern, A. Macfadyen, M. Todorovic, D. Duckhouse, M. Boyd, T. Myres, T. Bagenall, J. Lock and R. B. Freeman.

PLATE 22 H. N. Southern in Wytham Wood, February 1955.

PLATE 23 W. H. Pearsall at Blakeney Point in June 1955. Left to right, R. S. Clymo, R. Knowles and W. H. Pearsall.

PLATE 24 O. W. Richards, with J. P. Dempster, September 1963.

2.5
Animal Ecology

In a sense, the 'replacement' of the British Vegetation Committee by the British Ecological Society gave a very misleading impression of what was happening to British ecology. The attention of members remained firmly fixed on plant life. Even when studies were made of animal life, their prime purpose was to help explain how the distribution and character of plant communities had evolved. It was not until 1927 that the first book on animal ecology was published in Britain. Written by Charles S. Elton, it appeared in a series of textbooks on animal biology, edited by Julian Huxley. Simply called *Animal Ecology*, the book played a major part in establishing the other 'half' of ecology in Britain.

As in the case of plant ecology, it is difficult to draw a dividing line between the Victorian naturalists and the first practitioners of animal ecology. In their account of the history of ecology in Scotland, Gimingham *et al.* (1983) noted how the authors of the *Vertebrate Fauna of the Moray Basin* clearly perceived the impact of habitat destruction and the spread of alien animal species on the distribution of native vertebrates (Harvie-Brown & Buckley 1895). As an assistant in the natural history department of the Royal Scottish Museum, James Ritchie (*1882–1958*) had ready access to such sources when he came to compile his book on *The Influence of Man on the Animal Life in Scotland*, published in 1920. In it he chronicled the decline and, in some cases extinction, of species as a result of forest destruction, marsh drainage, and the killing of predators for game preservation (Ritchie 1920).

In one of his papers, Crampton drew attention to a study, published in the *Naturalist*, relating the geographical distribution of freshwater mollusca to the plant associations recorded in the South Lonsdale district of Lancashire, as part of a wider survey of the Morecambe Bay lowland peat bogs carried out by Munn Rankin (Crampton 1913, Rankin 1910). As the authors of the article pointed out, mollusca were even more 'spot-bound' than many plants, which were able to disperse their seed over long distances (Kendall C. E. Y. *et al.* 1909).

The main author of the study, the Reverend C. E. Y. Kendall, moved to Northamptonshire, where he was able to make frequent visits to localities within a five-mile radius of Oundle. Among the 95 species of mollusca identified, it was possible to distinguish characteristic, dominant, and associated species for each type of habitat. As he remarked to the Liverpool Biological Society in 1929, it was clear that the doctrine of the 'web of life' applied as much to mollusca as any other group. If botanists, lepidopterists, coleopterists, and others worked together in the study of a locality, their combined observations 'would produce a perfect picture of the life, animal and vegetable, in their locality'. If the observations made for one locality were compared with those of others, it might be possible 'to arrive at conclusions of real value' (Kendall C. E. Y. 1921–2, 1929).

In drawing attention to this 'large new field open to the naturalist', Kendall was mapping out a course for animal ecology which had been followed by plant ecologists for the previous 30 years. In tracing the progress made, close reference has to be made (as in the case of plant ecology) to the deaths and dislocation brought about by the First World War. Some intimation of the losses is conveyed by the papers published by Sydney E. Brock (*1883–1918*), a tenant farmer from Linlithgowshire, who died of his wounds in a military hospital on Armistice Day, 1918 (Evans W. 1919).

Both in a paper published in *British Birds* in 1914 and in a manuscript discovered after his death, Brock perceived the aims of the bird ecologist to be twofold. One was to identify and assess the character of the present-day links between bird-life and the environment. The distribution of a species was far from accidental. Each was part of a larger bird association, dependent on a particular plant formation and its accompanying invertebrate fauna. Nine types of bird association could be distinguished in Scotland. They included those of the moorland, woodland, coastline, and sea (Brock 1914, 1921).

The second objective was to reconstruct 'the original aspect of the avifauna', tracing the faunal changes induced by the direct and indirect influence of man. Brock explained how the origins of the environmental bias, which provided the 'immediate link attaching the species to its present habitat', were to be sought in 'the past environmental relations of the race'. They were literally or metaphorically 'the expression in an individual of the sum of racial experience'. Their persistence reflected not only 'the endurance of ancestral experience', but also the relative constancy or uniformity of 'the racial environment' inherited by each generation of birds.

As Brock observed, it was likely that botany would soon be joined by zoology in having a large and enthusiastic body of workers engaged in ecological research. The fact that the relation of birds to the environment was exceptionally intricate should make the subject even more attractive for study. It was also likely that this complexity would cause animal ecology to follow the lead of plant ecology in drawing heavily on the guidance and input of academic figures, both in Britain and abroad.

Although there is no evidence of Brock being aware of it, the first example of that kind of inspiration appeared in 1913, in the form of a *Guide to the Study of Animal Ecology*, written by Charles C. Adams (*1873–1955*), a member of the zoology department at the University of Illinois. Adams described ecology as 'a science with its facts out of all proportion to their organization or integration'. His textbook was the result of a 10-year search for some kind of consistent and satisfactory working plan for handling the almost bewildering number of facts of ecological significance (Adams 1913).

There also appeared in 1913 the first book to look in detail at a particular animal community. It had the title, *Animal Communities in Temperate America as Illustrated in the Chicago Region. A Study in Animal Ecology*. The author was Victor E. Shelford (*1877–1968*), a member of the zoology department of the University of Chicago. A former student of Cowles, he had studied for his doctoral research the relationship of tiger-beetles to the sandy areas around Lake Michigan. From this and later studies, Shelford (1913) concluded that it was possible to define the animal communities of different habitats by means of the physiological reactions of those 'indicator' species which were confined to a narrow range of habitat.

The publications of Adams and Shelford were soon brought to the attention of British workers. Long abstracts of both appeared in the *Journal of Ecology*, in which Shelford also wrote a paper, demonstrating how knowledge of animal life would contribute much to the deeper understanding of ecological principles and problems. Animal succession was just as marked as that of plants. The fact that it was promoted by the same factors suggested that there was a generally close agreement between the two types of community (Shelford 1915).

Shelford's book caught the attention of Julian S. Huxley (*1887–1975*), who graduated in 1909 and served briefly as a lecturer in zoology at Balliol College, before leaving Oxford to become a research associate at the newly founded Rice Institute at Houston, Texas. Huxley was almost the only senior British zoologist to consider ecology and behavioural studies important during the 1920s and early '30s (Thorpe 1975). Although his early research was in comparative physiology and the study of Protozoa, he soon realized that his real love, scientific bird-watching, was likely to produce the more important results. His first major paper described the courtship display of the great crested grebe (*Podiceps cristatus*) (Huxley 1914, 1970). After war service, he returned to Oxford as a fellow of New College and senior demonstrator in the department of zoology and comparative anatomy. There he encountered 'an enthusiastic group of undergraduates and began to show to full extent his capacity to stimulate the possessors of eager minds anxious to forget the war and devote themselves to learning' (Baker 1976).

While still in his first year at New College, Charles Elton was given a copy of Shelford's book by Huxley. Largely through the encouragement of his oldest brother, Elton was already an experienced field naturalist. His surveys of a series of flooded marl pits near Liverpool, and various ponds and pools in the Lake District and around Oxford, had dispelled any notion of the water bodies representing some kind of simple system, or closed microcosm (Elton & Miller 1954).

During his second year at Oxford (1921), Elton was invited by Huxley to join a university expedition to Spitsbergen. The original intention had been to study only the bird life, but, as the idea took shape, the party of ornithologists was expanded to include geologists, botanists, and zoologists. Although nominally Huxley's assistant, Elton was given virtually a free hand to pursue his own interests, namely an ecological survey of animal life. No better destination could have been chosen. Ecological detail was reduced to the minimum—there were so few species present that the 'ecological web' could be seen as a whole (Anon. 1925)[20].

Whilst it would be wrong to exaggerate the parallels between the development of botany and zoology, the sense of frustration that drove a small number of botanists to publish their 'encyclical' in the *New Phytologist* was shared by a similarly small number of zoologists. The writings of Charles Darwin had brought about a similar situation as in botany—zoologists responded to the challenge by drifting into museums, where they spent the next 50 years describing and classifying the living world, or they retired to their laboratories, where they made detailed studies in comparative anatomy and descriptive embryology. Not only did much of the work prove to be speculative, but the scope for speculation had been largely exhausted by the turn of the century (Hardy A. C.1968).

In both his criticisms and response to the prevailing situation, Elton emerged as the leading 'zoological Bolshevik', attacking those who saw salvation simply in terms of increasing the amount of weight given to animal physiology. Elton argued that whilst laboratory work should constitute over half the training of a zoologist, it was deplorable that so little provision was made for field training (Elton 1924a). In his own choice of research on living animals, as studied in the field, and his decision to pursue it from Oxford (rather than through a career in an agricultural or fisheries establishment), Elton was able to 'blaze a pioneering trail in the new domain of terrestrial animal ecology' in the company of those he most wanted to influence.

For Elton (1927), ecology was a new name for a very old subject. It meant 'scientific natural history'. In a series of lectures on 'the future of animal ecology', given in the University of London in 1929, he argued that the problem was not so much a lack of data as the way in which the subject had been split into 'innumerable tiny subjects'. The 'scattered state of ecology' could be compared to that of

> an active worm that has been chopped into little bits, each admirably
> brisk, but leading a rather exclusive and lonely existence and not
> combining to get anywhere in particular.

This disintegrating tendency was very largely due to the absence of any proper working hypothesis which would cut right across the divisions of specialists.

Elton (1930) believed that such co-ordinating principles could be found in the study of animal evolution. They could be encapsulated in terms of the food cycle, size, niche, and pyramid of numbers, the significance of which could only be measured through detailed investigations in the field. It is the purpose of the following sections to review the circumstances leading up to the writing of Elton's volume on animal ecology in 1927, and the decision taken shortly afterwards by the British Ecological Society to launch a new journal, the *Journal of Animal Ecology*.

2.5.1 *Animals and the food cycle*

The expedition to Spitsbergen in 1921 gave rise to three papers in the *Journal of Ecology*. John Walton, a Cambridge botanist, described the succession of vegetation on the intertidal flats and raised shingle beaches (Walton 1922). There followed, a year later, a paper on the plant and animal communities of Spitsbergen and Bear Island, and a further long paper in 1928. These were written by Charles Elton and Victor S. Summerhayes (*1879–1974*), who had graduated in 1920 from the botany department at University College, London, and was awarded the Quain Studentship (Brenan 1975).

Elton had taken a copy of Shelford's book to Spitsbergen. The inadequacy of Shelford's method of distinguishing different animal communities, on the basis of their physiological response to the habitat, soon became apparent. There were so few species confined to a single habitat or plant association that they played a very small part in the life of the community. Most individuals of a species were found at different times in a variety of plant communities. Summerhayes and Elton (1923) found it much easier to identify animal communities in terms of their food relationships. From their observations on Bear Island, they illustrated the kinds of food chain and food–host relationships likely to be encountered in terms of what they called 'a nitrogen cycle' (Figure 2.4).

Elton acted as chief scientist on further expeditions to Spitsbergen in 1923 and 1924 (Anon. 1929). These gave him the opportunity to visit other parts, and to look more closely at the numerous habitats of the highly diversified terrain. In a paper given to the Annual Meeting of the British Ecological Society in 1928, Elton described how there was a gradation of both plants and animals, from the simple to the complex according to the degree of climatic amelioration. On the eastern side of Spitsbergen, with its low temperatures and fog, no more than nine species of animals could be found among the few lichens and mosses, whereas over 50 species of animals were recorded in the Cassiope heaths of the western part (Summerhayes & Elton 1928)[21].

Back in Oxford, Elton tried to carry out a similarly comprehensive survey of the succession of communities colonizing the bare mud dredged from a section of the Oxford Canal. He failed, largely because of the difficulties of identifying the different species 'except through the direct labours of specialists'. One of those to give considerable help was Owain W. Richards (*1901–84*), another of Huxley's students, and the Christopher Welch Scholar and Senior Hulme Scholar between 1924 and 1927. Elton wrote of how many of the ideas discussed in his book, *Animal Ecology*, were discovered with Richards and had benefitted from his extensive knowledge of insects.

Richards believed it was much more important to extend his knowledge of the British insect fauna than to concentrate on some narrow PhD topic. His years of intensive collecting contributed many tens of thousands of specimens to museum collections. In the course of visiting the research site of Bagley

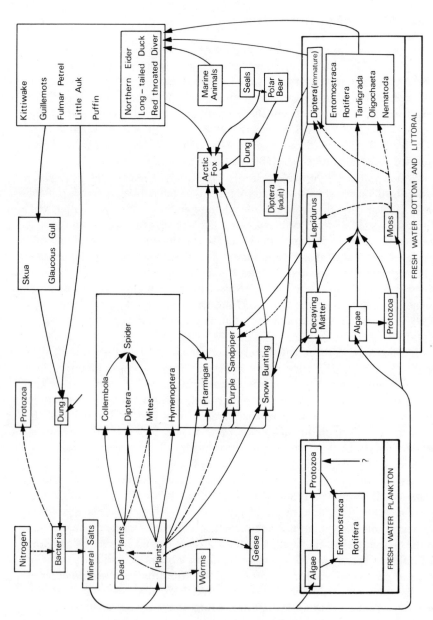

FIGURE 2.4 Food relationships on Bear Island (after Summerhayes & Elton 1923).

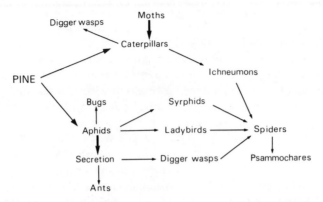

FIGURE 2.5 Food cycle on young pine at Oxshott Heath, Surrey (after Richards 1926).

Wood near Oxford over the period 1922–27, Richards kept a detailed record of the invertebrate life occupying the rotten fence posts of the site. As he pointed out in a paper published in the *Journal of Ecology*, the results provided a striking example of the difficulties of classifying species which might be present for different reasons, belong to different food chains, and associate with quite different insects in other habitats (Richards O. W. 1930).

The abortive attempt to study the excavated muds of the Oxford Canal was followed by a much more complete investigation of the relationship between the animal life and the plants colonizing the clear-felled areas of Oxshott Heath in Surrey. Summerhayes had recognized the potential of the area for studies of successional change in 1920, and Richards was quick to see the advantages of collecting from a community being studied simultaneously by plant ecologists (Summerhayes & Williams 1926). In a separate paper, published in the same issue of the *Journal of Ecology*, he used 29 detailed tables to illustrate how the animals collected from each habitat tended to form communities attached to particular plants as a result of being part of a chain of food relationships (Figure 2.5). So dependent were the animal communities on the dominant plant species that when new dominants arrived, and the food chain was replaced by another, a rapid and complete change in fauna occurred (Richards 1926).

The value of food chains as a means of expressing the relationship of organisms to one another was further demonstrated by a quantitative study of the fauna of stream beds in the West Riding of Yorkshire published in the *Journal of Ecology* in 1929. Analyses of the density and percentage composition of the species sampled showed the extent to which the conditions favourable for the growth of unicellular and filamentous algae, namely a stable substratum, also facilitated the development of a considerable insect fauna. The authors of the study followed Elton in using the term 'niche' to describe the role or mode of life of those organisms which converted, for example, vegetable matter into animal food. Chironomidae, *Ephemerella* and Naididae were the 'cattle' which grazed the extensive algal pastures and themselves became the nourishment for a wide variety of other invertebrates, which in turn fell prey to carnivorous vertebrates (Percival & Whitehead 1929).

The scope and relevance of research on food chains was perhaps most strikingly illustrated by research in the marine environment. In 1924, the Ministry of Agriculture and Fisheries published the results of a study of the food and feeding habits of the North Sea herring, carried out by Alister C. Hardy (*1896–1985*), an assistant naturalist who was appointed to the Ministry's Fisheries Laboratory at Lowestoft in 1921 (Wynne-Edwards 1987). As Hardy (1924) explained, a thorough knowledge of the relationship of the herring to other organisms in the environment was required if the habits and migrations of the fish were to be understood. On the basis of specimens taken from 60 different locations, and a quantitative analysis of microscopic stomach contents, Hardy attempted to convey the extremely complex relationship of the plankton community to the different stages in the life cycle of the herring.

2.5.2 *The regulation of animal numbers*

Meanwhile, there was increasing interest in the way in which animal populations were regulated. Elton described such studies as 'the chief scientific goal of pure ecology' in the sense that it was only possible to understand the mechanisms adopted 'after most of the other phases of ecological research had been covered in a fairly extensive way' (Elton 1930). As with his research on food cycles, Elton's involvement can be traced from his undergraduate days, and particularly his reading of three books.

Two days before his final schools in zoology, Elton read the newly-published study of population problems by Alexander M. Carr-Saunders (*1886–1966*). Carr-Saunders had returned from the war to take up an appointment as demonstrator in the department of zoology and comparative anatomy; he was in charge of marine biology on the Spitsbergen expedition of 1921. The book was the first to demonstrate how it was unusual for human societies to allow numbers to rise until they pressed against the means of subsistence. Carr-Saunders (1922) used the evidence of anthropologists and prehistorians to suggest that it was those societies with 'good health, a leisured poise, and a fine physique' which were the most likely to limit their numbers to a size that could make the fullest use of the resources available (Phelps-Brown 1967). Although the volume dealt with human populations, Elton found it 'full of exciting ideas of a general nature', of considerable relevance to the study of animal populations.

On his return from Spitsbergen in 1923, the boat called at Tromso, where Elton spent his last £3 on a copy of Robert Collett's book on Norwegian mammals. This was how Elton first learned of the phenomenon whereby lemmings appeared to migrate every three or four years. The copy of the book is now in

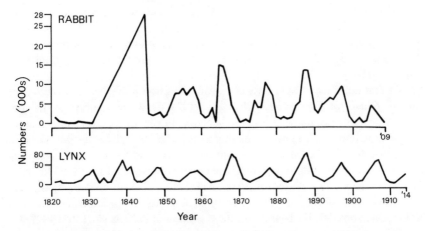

FIGURE 2.6 Periodic fluctuations of the rabbit and lynx in Canada, according to the fur returns of the Hudson's Bay Company (after Hewitt 1921).

the Elton Library at Oxford. Interleaved in the pages dealing with the lemming is a page-by-page translation into English, written by Elton (Collett 1911–12)[22].

It was Huxley who drew Elton's attention to a third book, a study of the conservation of wildlife in Canada. It was written by C. Gordon Hewitt (*1885–1920*), who died of pneumonia shortly after completing the manuscript. Hewitt left his post as lecturer in economic zoology at Manchester University in 1909 to take up the appointment of Dominion Entomologist. He was greatly influenced by the early American conservationist, William T. Hornaday (*1854–1937*), and played a major role in drafting legislation for the protection of game and wildlife (Scott 1920). For Elton, the most exciting part of the posthumously published book was the chapter on the periodic fluctuations in the populations of fur-bearing animals (Hewitt 1921).

The timing and magnitude of these fluctuations were highlighted in the fur returns of the Hudson's Bay Company, covering the period 1821–1914 (Figure 2.6). The company had posts throughout the country, and particularly the Canadian north. Hewitt believed that if a careful and intensive study were made of the reasons for these fluctuations, it should be possible for the industry to predict the years when the fur-bearing animals would be abundant.

It had been commonly assumed that there was a harmony—a balance of nature—in natural animal communities, where any tendency on the part of a species to increase in population would be cancelled out by some controlling factor. As Elton (1930) observed, the maintenance of such a balance seemed the most logical way for natural selection to provide the best possible conditions for species. The fact of the matter was that no such balance existed. Herbert Spencer had said as much in his book, *First Principles*, published in 1862, but its significance had passed unnoticed. After a species had been greatly thinned by its enemies or a lack of food, the surviving members were much more favour-ably placed. Food was relatively plentiful, and predators had diminished for want of prey. Such conditions encouraged a rise in numbers to a point where food again became relatively scarce, and predators were more numerous (Spencer 1862).

At first unaware of Spencer's observations, Elton published an essay in the *British Journal of Experimental Biology* in 1924, drawing attention to the con-cept of unstable populations and the benefits of obtaining a closer understand-ing of how and when fluctuations might occur. Such knowledge would be of value not only to the fur industry, but to farmers and foresters in forecasting when pest populations might reach plague proportions. Elton (1924b) con-sidered three possible causes for these oscillations—climate, epidemics, and ani-mal food. Whichever was the more important, the fluctuations in numbers could have considerable bearing on the evolution of the species (Elton 1929a).

Elton adopted two approaches to studying population cycles. One was to map and follow trends on a wide geographical scale. In this, he was greatly helped by George Binney, who had organized the three Spitsbergen expedi-tions, and was now on the staff of the Hudson's Bay Company. Having been appointed the biological consultant to the company, Elton was able to keep a running record of fur returns by distributing annual questionnaires. Through diligent searches of the company records, he traced, for example, the cyclic be-haviour of the lynx population back to 1736. At a conference on wildlife cycles, convened in Labrador, Elton put forward evidence for a 10-year cycle in wild-life throughout North America. The fact that the cycle did not correspond with any recognized climatic period suggested that 'a combination of more than one astronomical factor' might be involved (Elton 1933a).

The other approach was to carry out detailed studies of wildlife populations close at hand, and for that purpose the site of Bagley Wood, near Oxford, was chosen. Between 1925 and 1928, John R. Baker (*1900–84*), Edmund B. Ford, and Elton collaborated in a study of the wood-mice and bank-vole populations. First Denys Kempson, and then A. Douglas Middleton (*1904–87*), assisted in the field-work, extending over 600 trapping nights during which 2000 mice and voles, and several hundred shrews, were taken. The first systematic investi-gation of its kind, it confirmed that the populations rose to peaks of abundance and then crashed. The reason for such cycles remained, however, a mystery. Despite close studies of the parasites, health, and general ecology of the animals, there was no definite evidence as to the cause of the cyclic change (Elton *et al.* 1931).

In order to survive, a species population had to be large enough to avoid being eliminated by some chance catastrophe, and yet small enough to prevent overcrowding. Elton (1930) argued that the key to achieving this necessary de-gree of elasticity was through migration. Indeed, migration was so essential and commonplace that it could be regarded as 'a real ecological law among animals'. This was most clearly demonstrated in the case of bird migration.

A considerable body of data had been amassed on bird migration, largely as a result of bird-marking schemes initiated simultaneously, but independently, by A. Landsborough Thomson (*1890–1977*) and Harry F. Witherby (*1873 1943*), in conjunction with his journal *British Birds*. Landsborough Thomson started his scheme while still reading natural history at Aberdeen in 1909 (Nicholson 1977). After the war, he took up an administrative post with the newly formed Medical Research Council and resumed his studies of what he called, in a book of 1926, 'one of the most remarkable examples of animal behaviour'. Few other countries were so well placed for the study of bird migration (Thomson 1926).

It took time to realize that almost every group of motile animals exhibited some form of migratory impulse. The first scientist to draw attention to the phenomenon among invertebrates was Carrington B. Williams (*1889–1981*). It was during his first year as an agricultural entomologist in the Colonial Service that he saw his first indisputable example of butterfly migration, taking place over several days along a 10-mile front in British Guiana. Despite this and later spectacles, he regarded such movements as a local curiosity until he took up an appointment with the Ministry of Agriculture in Egypt in 1921. During the voyage across the Mediterranean, he witnessed a migration of painted lady butterflies. The most remarkable examples of migration were, however, seen in Tanganyika, where Williams spent the years 1928–9 on locust research. One such flight lasted 16 weeks, and was still in progress as he left the country (Williams C. B. 1958a, Wigglesworth 1982).

In his book on *The Migration of Butterflies*, published in 1930, Williams drew on an extensive search of the world's literature, as well as his own observations and those of numerous friends. He demonstrated that, in addition to the large number of butterflies scattered willy-nilly by wind currents, there was also 'a very large, irregularly regular, deliberate, self-contributory distribution of insects by long unidirectional flights' (Williams C. B. 1926, 1930). As Elton remarked, the flights were further evidence that animals were 'constantly moving about, choosing and rejecting the situations' in which they found themselves.

It seemed that the real 'life' of animals was the outcome of a number of fundamentally different processes. It was fruitless to attribute the structure, physiology, and behaviour of animals to any one or two methods of evolution. Predetermination, sex, materialism, free will, destiny, originality, and tradition all had a part to play. The ability of animals to migrate, and thereby choose between good and bad surroundings, greatly increased 'the possibilities of adaptive radiation'. As well as there being a large chance element in all that happened, it suggested that there was a role for tradition, or the handing down of ideas born out of contact with new situations (Elton 1930).

In his contribution on 'animal ecology' to the *Encyclopaedia Britannica*, Elton emphasized the need to study 'the *whole* animal community living in one habitat'. It was otherwise impossible to understand the way any particular animal was affected by other animals. There were at present insufficient data available to express in quantitative terms 'the balance between size, rate of increase, and density of numbers', but, in Elton's words, the day would presumably come when ecologists would be in 'a position to predict the phenomena of animal inter-relations in almost as accurate a form as those in chemistry' (Elton 1929b).

2.5.3. *Journal of Animal Ecology*

By the early 1930s, there was tangible evidence of the burgeoning interest in animal ecology. Elton wrote an elementary introduction to the subject in the Methuen series of 3/6d monographs on biological subjects (Elton 1933b). Called *The Ecology of Animals*, it was later translated into Russian for use in the universities of that country. In a review for the *Journal of Ecology*, Tansley (1934) described it as 'welcome evidence of the increasing popularity and importance of Animal Ecology'. Not only was this long overdue, particularly in the light of the American experience, but it represented a dramatic widening of the vistas for British ecology.

In the same way as the establishment of plant ecology owed much to the support of academic figures whose reputations had already been made in other fields, so animal ecology drew strength from such patronage. The contributions of two distinguished scientists may be cited—those of Arthur Boycott and Patrick Buxton.

Arthur E. Boycott (*1877–1938*) was the first animal ecologist to be elected president of the British Ecological Society in 1932. He led 'a sort of scientific double life'. As well as being an eminent pathologist, serving as editor of the *Journal of Pathology* for over a quarter of a century, he was also a distinguished naturalist. Over half of his 180 published works related to British land and freshwater mollusca. Whatever the particular field of enquiry, 'he brought the same informed and disciplined mind to bear upon problems' (Martin 1939).

It was following his appointment to the Graham Professorship of Pathology in London University in 1915 that Boycott took up residence in Hertfordshire. He had always been interested in natural history, and in snails in particular. He soon joined the active Hertfordshire Natural History Society, giving the presidential address of 1918 on the freshwater mollusca of the parish of Aldenham (Salisbury 1938). Working conchologists had long used the known relationship of different species to different kinds of places as a method of predicting where the individual species might be found. In his presidential address to the British Ecological Society of 1933, Boycott sought to put this knowledge on a more systematic basis. In his view, it was possible to relate the geographical range of land mollusca to two types of environmental factor, namely moisture and lime (Boycott 1934). Some years later, he published a complementary paper on the distribution of freshwater mollusca. The Council of the British Ecological

Society made a grant of £60 towards the cost of printing such an 'important and exceptionally long paper' (Boycott 1936)[23].

The contribution made by Patrick A. Buxton (*1892–1955*) lay in a very different direction. Having obtained a second MB, and a first class in the natural sciences tripos, Buxton was elected to a fellowship at Trinity. In 1915, he joined the Royal Armoured Medical Corps, and was posted to Mesopotamia. After a short period back in Cambridge, he became entomologist in the Medical Department in Palestine, where he spent over two years studying malaria. A temporary post in the London School of Tropical Medicine secured him the chance to lead a research expedition to Samoa. On his return in 1926, he was appointed head of the entomology department in the London School of Hygiene and Tropical Medicine. Within a short space of time, he had persuaded the board of management to provide both the staff and resources necessary for establishing 'a science of insect physiology' (Wigglesworth 1956).

Buxton's misgivings as to the state of applied entomology were set out in a letter to *Nature* in 1926. Although entomologists had met with some success in safeguarding mankind, domestic animals, and livestock from insect attack, the results had been generally disappointing. In Buxton's view, too much importance had been attached to producing results of 'some obvious practical importance'. For real progress, there had to be much more time and labour given over to the study of the fundamentals of insect physiology. This meant investigating 'not only the functions of isolated parts of an organism, but also the extremely difficult task of understanding the relations between the living insect and the whole of its environment, inorganic and organic' (Buxton 1926).

Buxton had already drawn attention to the importance of these relationships in his book, *The Animal Life of Deserts*, based on his observations in wartime Lower Mesopotamia and north-west Persia. Deserts provided outstanding opportunities to observe 'the interaction of plant and animal upon each other, and the dependence of the living creatures upon climate and other physical conditions' (Buxton 1923). In a paper in the *Journal of Ecology* of 1924, Buxton emphasized the need to draw a distinction between the climate as measured by meteorologists under standard conditions and that which prevailed in the actual places where plants and animals lived. A visit to Palestine in the summer of 1931 afforded the chance to compare the temperature and humidity inside caves and small cavities with those on the desert surface, and to assess the significance of those differences for animal life (Buxton 1924, 1932).

The rising number of papers submitted on topics in animal ecology brought to a head discussions within the British Ecological Society as to the future role of the *Journal of Ecology* and of the Society itself. In doing so, the society was in effect responding to the rising stature of ecology. The Society was not only the obvious but, in practice, the only body that could respond to this upwelling of interest. It had an all-embracing title, and its membership was open to anyone with a genuine interest in ecology. The main question was how far the Society was willing and able to capitalize on the situation.

Membership of the society had risen from 138 in 1919 to 244 by 1929. As Salisbury pointed out, in his secretary's report for 1923, membership would probably have been much higher if it had been easier for 'those interested in Ecology' to attend the Society's meetings. About a third of members lived in London and the south-east, and another third in the north of England and Scotland. About a fifth were domiciled abroad.

One obvious response was to hold many more meetings in different parts of the country. The first Annual Meeting to take place outside London was arranged in 1922. An attempt was made to hold at least two field meetings each year. The excursion to Esthwaite Water under the leadership of Pearsall in 1920 was regarded as a great success, but a week's visit to Somerset in 1921 had to be abandoned through lack of support. As Salisbury remonstrated, 'the success of the Society's excursions depends even more upon the keenness and energetic support of the members themselves than upon the arrangements made by the Society's officials'. There was much better support for the day excursions, arranged for local members in 1922, to Delamere Forest in Cheshire (led by Adamson), Wicken Fen (Hamshaw Thomas), and a two-day visit to Chesil Beach in Dorset (Oliver).

Another way of raising membership and revenue was to increase the appeal of the *Journal of Ecology*. The journal doubled in size over the period 1919–29. Whereas Tansley complained of a dearth of suitable material in October 1921, there were so many papers waiting to be published in 1926 that Council decided to issue an enlarged number, using an anonymous donation given (by Tansley) for that purpose. There remained 'a great deal of material in hand', and, in 1929, Tansley again 'appealed to members to see that their papers were as brief as was consistent with proper exposition before sending them'. At the Council meeting of December 1930, he reported that several 'good' papers had been rejected because of a lack of space.

By 1930–1, the society's accumulated financial surplus was so large that much of Council's attention was focused on how it might be used. The position had mainly come about through the large sales of the journal to non-members (230 copies were sold in 1926), as well as the sale of back numbers. In reviewing future liabilities, Council noted how all the sub-editorial work was done by Tansley's personal secretary. The minutes of Council recorded how

> whilst the Council appreciated the Honorary Editor's generosity (in
> continuing to meet his out-of-pocket expenses), the present system
> tended to give a sense of false financial security.

Although Council approved plans to establish a library, taking advantage of an offer from John Ramsbottom (*1880–1966*) to accommodate it in the botany department of the British Museum (Natural History), nothing came of the venture. It was agreed to reprint photostatically the back numbers of the journal which were in short supply, and to issue an index to the first 20 volumes. Of the 750 copies of the index printed, only 150 were sold by 1937. It was not until 1950 that the initial outlay was recouped.

It was Charles Elton who ensured that the society's first priority was to repair the most obvious deficiency in British ecology, namely the neglect of animal ecology. He did so by using the growing pressure on space in the *Journal of Ecology* as the pretext for demanding that the Society should give tangible evidence that it did indeed purport to represent all sectors of ecology. That evidence came in the launching of a journal explicitly for animal ecology.

Elton was elected to the Council at the Annual Meeting of January 1931, and at the March meeting he drew attention 'to the increasing number of papers on animal ecology, especially in relation to animal numbers and economic aspects'. There was no obvious outlet for these papers, and the tendency was to publish them in a wide variety of journals. Because of the obvious editorial and financial difficulties in seeking a further expansion of the *Journal of Ecology*, Elton put forward two alternatives. One was to supplement the journal with a new journal, intended for 'the more specialized papers'. The other was for the Society to publish one journal for animal ecology, and another, the *Journal of Ecology*, for plant ecology and papers of 'a general bearing'.

In considering these proposals, the Council concluded that it would be wrong to embark on such a venture 'without the guarantee of some financial support'. Elton and Tansley were accordingly asked to make enquiries as to what support would be forthcoming for either (i) a 'Journal of General Animal Ecology', or (ii) a 'Journal of Animal Numbers and Migration'. At its meeting of May 1931, the Council approved their joint memorandum, which recommended a new journal under a separate editor. It 'would contain all papers dealing exclusively with animals or with the theory or methods of research having reference primarily to animals'. Elton was nominated editor, with Middleton as his assistant. As Elton recalled later, it was Salisbury who wisely persuaded him to drop plans to call the journal the 'Journal of Animal Numbers', or the 'Journal of Animal Population', and to adopt instead the much wider title of *Journal of Animal Ecology*.

The new journal was announced to the Society in July 1931. The rules were altered so as to enable members to take one journal, or both at a slightly reduced rate of 45 shillings. Elton raised the £100 needed to guarantee the Society against any deficit over the first five years. By New Year's Day 1932, there were 115 promises of support, and the next day the Annual Meeting of the Society empowered Elton to proceed to publication. The first issue appeared in May of that year. As well as including papers and book reviews, the new journal incorporated the abstracts of British papers on animal ecology, which Elton had previously prepared for the *Journal of Ecology*. From thenceforth, details of the Society's activities appeared in both journals.

The membership of the Society rose from 242 in 1932 to 305 a year later. Of the 69 new members, 15 elected to take both journals. Forty-nine took only the *Journal of Animal Ecology*, and five subscribed only to the *Journal of Ecology*. At the end of 1932, 113 copies of the *Journal of Animal Ecology* were distributed to members of the Society, compared with 243 copies of the *Journal of Ecology*.

Fifty-one members took both journals. An additional 70 copies of the new journal were sold to non-members. A motion was adopted at the Annual Meeting of January 1934, expressing 'the appreciation felt by the Society to Mr Elton for the high quality of the *Journal of Animal Ecology* and for its satisfactory financial position'.

2.6
Common Causes in Plant and Animal Ecology

The British Ecological Society was at pains to stress that the publication of the *Journal of Animal Ecology* did not mean the parting of the ways between plant and animal ecologists. On the contrary, the fact that animal ecology would now receive adequate attention was to the benefit of all ecologists, making it easier to assess the general relevance of concepts and techniques developed in any one particular field of ecology. No other body had so large an interest in extolling this kind of collaboration. Through its membership and meetings, the society was uniquely placed to help identify common goals, and to bring about the rapport necessary for realizing them.

An obvious question for all ecologists was how far they should acquire a specialist knowledge of the concepts and techniques being developed elsewhere in the sciences. For the British Ecological Society, the question took on a tangible form in the context of how far the Society's journals should accept papers that presupposed that readers had acquired this specialist knowledge. The need to strike some kind of balance was an obvious topic for a succession of presidents, as they came to give their biennial addresses to the Society.

In his address of 1923, Praeger recalled 'the glorious days of primary survey, when we ranged free over moor and mountain'. Now, the 6-inch map, binoculars, and pencil were being replaced, or at least reinforced, by instruments for measuring the amount and variation of light, heat, and moisture, and by 'the whole battery of the chemical laboratory'. As Praeger knew from his own experience, not everyone could take part in this new 'campaign'. He himself lacked a laboratory or a working knowledge of the methods of physiology. As an amateur, he could only follow with interest the advances made by others—advances that embodied concepts which were often difficult to grasp, particularly when obscured by formidable terminology (Praeger 1923).

Four years later, in his address to the Society, another founder member of the British Vegetation Committee, Woodhead, complained of the tendency to drift into 'unintelligible channels', and the way this would alienate many who might otherwise be attracted to 'an ecological outlook'. Woodhead did not, however, share the fears of those who believed 'ecology was in danger of losing its broadening and correlating value'. On the contrary, he believed intensive studies provided important links in the chain of evidence required for understanding the cyclic changes in habitats and their repercussions on the vegetation. Only through a closer involvement, and therefore familiarity, with the different specialisms could ecologists fulfil their role in identifying and creating the bonds that held together the different departments of their science (Wood-head 1929b).

2.6.1 *Statistical techniques and theory*

An obvious point of contact between plant and animal ecologists was their growing sense of frustration at the lack of precision in comparing and characterizing communities. Better ways had to be found for taking censuses of the different populations. As well as drawing on the considerable experience of agronomists, and most notably Stapledon and Jenkin in their censuses of grass-land populations (White J. 1985), a great interest was taken in the techniques being developed overseas.

The frustration was not confined to British ecologists. Paul Jaccard (*1868–1944*) of the Federal Polytechnic of Zürich wrote of the difficulties of comparing tracts of Alpine pasture, with their discontinuous distribution of even common species. In the first quantitative assessment of its type, he compared the flora of two meadows by calculating the ratio of the number of species common to both areas to the total number of species encountered in each. Tansley was so impressed that he translated and republished (in the *New Phytologist*) a paper setting out the technique (Jaccard 1912).

Although Jaccard's method provided a simple and convenient way of assessing homogeneity, it could not in itself form the basis of a quantitative analysis of vegetation. A species might be rare in one population, but common in another. To overcome this difficulty, Christen Raunkiaer devised the percent-age frequency method, whereby a quadrat was thrown repeatedly over the plant community, and the presence or absence of species noted. The percentage of throws in which a species occurred was called its percentage frequency. Raunkiaer used a quadrat of $0.1 \, m^2$, and usually distinguished five 'valency' classes. Almost invariably a reversed J-shaped distribution was found. The method was taken up by many ecologists, and considerably elaborated by the Uppsala school of ecology.

Because of the obvious difficulties of studying the plant and animal associations of the marine environment, the marine ecologist was particularly conscious of the need for quantitative sampling techniques. In 1920, the *Journal of Ecology* reproduced a paper by Ellis L. Michael (of the Scripps Institution for Biological Research at the University of California), arguing that mathematical problems would have to be solved before many of the ecological ones could be tackled. It was invariably found that the mathematician knew too little biology to visualize the problem, unless it was 'translated into mathematical terminology, and the biologist, as a rule, finds himself unable to make this translation'. The only way out of the dilemma was for biologists to develop the same proficiency in mathematics as they were expected to achieve in English. For Tansley, this was going too far. In an editorial footnote, he wrote of how most ecological research would continue to make little demand on mathematical skills. Although there was a pressing need to provide adequate training in such branches as calculus and statistical theory for those biologists needing it, mathematics should be treated no differently from physical chemistry and the other specialist fields which supported biological work (Michael 1920).

It would be misleading to portray the adoption of statistical techniques in terms of a battle between those ecologists who had acquired some competence in that field and the remainder who fought a rearguard action. Some of the most telling criticisms came from the comparatively few who had some expertise. In praising the efforts of the Uppsala school, Pearsall pointed out that the associations which had been described between communities reflected not so much homogeneity in the vegetation but the subjective manner in which the quadrat data had been collected (Pearsall 1924).

Raunkiaer's percentage frequency method might have provided a rapid and approximate way of describing vegetation, but Eric Ashby (of the botany department at Imperial College, London), found it useless for classifying communities. Not only did the method lead to the splitting of communities into many ill-defined subgroups, but the results were very misleading. The J-shaped distribution of species was an indication of neither any exact numerical relationship between the species comprising a community, nor any degree of homogeneity. The curve merely reflected the extent to which the lowest and highest frequency classes embraced wider classes of mean area than the intermediate frequencies (Ashby 1935).

Statistical methods seemed far more useful in studying the distribution of individual species within a community. One of the pioneers of this field was Geoffrey E. Blackman (*1903–80*), who was appointed to the Jealotts Hill Agricultural Research Station in 1927, where he worked in the general sphere of crop nutrition and management. Over the next 10 years, he acquired evidence from 12 000 quadrats to support the hypothesis that if a species was distributed at random the logarithm of the percentage absence was directly proportional to the density. For the more abundant species, the percentage frequency of a randomly distributed species, as determined for a single-size quadrat, could be used as the basis for calculating that of any other sized quadrat (Blackman G. E. 1935).

Meanwhile, the scope for applying statistical techniques to animal populations was being explored by C. B. Williams who, following a three-year period at Edinburgh University (during which he wrote his pioneer study of butterfly migration), became head of the entomology department at the Rothamsted Experimental Station in 1932. In order to assess the part played by the physical environment in causing mass outbreaks of pests, he began to investigate the effects of weather conditions on insect activity and abundance.

As Williams later recalled, 'had I gone to Rothamsted direct from the University about twenty years before, I should have been almost forced into the technique of simplifying my experiments as much as possible'. Nearly all the work would have been done in the laboratory under 'controlled conditions'.

Although this might have led to greater accuracy, it would have been much more difficult to apply the results to actual field conditions. It would also have eliminated any chance of analysing the interactions between different factors in the environment. Williams believed the dilemma of having to choose between the two approaches could be resolved to a large degree by the adoption of statistical techniques, and particularly multiple regressions. This should make it easier to extend the number of variables being studied simultaneously, and to consider their interactions (Williams C. B. 1964a).

Instead of disappearing 'into a constant temperature room in the cellars', Williams embarked on a four-year trapping programme in 1933, sampling the night-time insect population under completely natural conditions. He described the results, together with those from a further four-year period of nightly trapping between 1946 and 1950, in a paper published in the *Proceedings of the Royal Society* in 1951. Using the standard methods of multiple regression and analysis of variance, he derived formulae by which population changes could be estimated from a knowledge of previous weather conditions. Rainfall was the most important factor in summer, and minimum temperature in winter (Williams C. B. 1951).

The means were emerging whereby hypotheses could be explored and tested mathematically. Ashby, Blackman, and Williams each acknowledged the very considerable influence exerted on their work by Ronald A. Fisher (*1890–1962*). Fisher was appointed to Rothamsted in 1922 in order to provide statistical guidance in the interpretation of field trials (Kendall 1963). The outcome was 'a revolution in the design of trials which spread throughout the world and involved every branch of experimental science'. It was also during the 1920s that mathematicians began to contribute to the theoretical content of ecology. Not only did it become easier to generalize from the empirical evidence acquired for individual species and localities, but concepts began to emerge that helped to explain why populations of organisms behaved in the way they did. There was at last evidence to confirm what had long been suspected by those engaged in pest control work, namely that 'good and bad years' for species populations were subject to definite, but highly complex, laws.

The years 1923–40 have been called the golden age, when mathematical ecology reached a level of conceptual and analytical sophistication that had no parallel in any other branch of mathematical biology (Scudo & Ziegler 1978). At much the same time, but independently of one another, Vito Volterra (*1860–1940*) in Italy and Alfred J. Lotka (*1880–1949*) in the United States began to focus their mathematical skills on the problems of variations and fluctuations in the numbers of individuals and species. Volterra investigated analytical models to identify the effects of external perturbations on pre-existing natural equilibria (Whittaker 1941). Much of Lotka's earlier work had been directed towards the oscillatory behaviour of chemical reactions and human population dynamics (Dublin 1950). In 1925, he published a book on the *Elements of Physical Biology*. It was the first major attempt to apply modern mathematics to biological problems, and soon became one of the foundation-stones of contemporary ecology (Lotka 1925, Hutchinson 1978).

Although there were important differences between the theories put forward by Volterra, Lotka, and soon other mathematicians, all were able to prove mathematically that two species would fluctuate in numbers, or fail to coexist, even though all physical conditions remained constant. These changes might come about in five ways: through intraspecific competition, interspecific competition, herbivore–plant relations, predator–prey relations, and parasite–host relations. Not only did the models appear to be precise and elegant, but they were clearly consistent with the empirical evidence that certain types of population change were related to biotic processes, rather than to changes in the physical environment (Thorpe 1974). As Elton (1942) commented, this confirmation of what had been discovered through field-work was of fundamental significance. From thenceforth, every population ecologist needed to have within his technical equipment an armchair and some knowledge of mathematics.

2.6.2 Centres for research

An obvious topic for debate was the scope for setting up new centres for research on plant and animal life. Both formally, and through individual members of Council and officers, the British Ecological Society played an active part in promoting the Bureau of Animal Population and the Freshwater Biological Association—bodies which became important not only for the ecological research carried out under their names, but also for the encouragement which they gave the Society in advocating something even more ambitious in later years.

The omens for such ventures in the early 1930s were far from promising. Financial worries continued to plague the studies being carried out on animal populations, under the direction of Elton at Oxford. Although a grant from the Empire Marketing Board enabled Elton and Middleton to set up what they called the Oxford Research Investigation, both this grant and another from the Hudson's Bay Company were terminated during the world economic crisis of 1931[24]. The position was only saved by moneys received from the New York Zoological Society. Without them, it would have been impossible to establish what became the Bureau of Animal Population.

Ever since the death of his brother in 1927, Elton had wanted to set up a memorial in the form of a research institute, which would put collaborative research on a permanent footing. The University agreed to the Bureau of Animal Population being established on a trial basis in 1932, located within the department of zoology and comparative anatomy (it was formally absorbed into the University in 1947). Elton's personal position was not, however, assured until 1936, when he was awarded a readership in animal ecology and a senior research fellowship at Corpus Christi College in Oxford[25].

At first, the Bureau consisted of only Elton, Middleton, and David H. S.

Davis, who was replaced by Dennis Chitty in 1935. There were, however, many instances of collaboration on specific projects. Between 1932 and 1939, Summerhayes (who had been appointed to the Royal Botanic Gardens at Kew in 1924) measured the impact of voles on the vegetation of forest rides at sites in Merionethshire and Roxburghshire, using exclosures and other forms of control. Members of the Bureau estimated vole abundance at six-monthly intervals. The collaborative study indicated that the feeding and tunnelling activities of the animal were intense enough to keep the plant communities so open as to encourage a much richer flora of higher plants and bryophytes than would otherwise have been possible (Summerhayes 1941).

The contribution of the Bureau in respect of collecting, collating, and synthesizing data on animal populations, and in carrying out field surveys and experiments of its own, was summarized in a volume published by Elton on the Bureau's 10th anniversary. It was called *Voles, Mice and Lemmings: Problems in Population Dynamics* (Elton 1942). Most of the Bureau's attention was focused on the vole (*Microtus agrestis*). Through a diligent search for records of vole plagues in the past, and by monitoring the populations at nine stations, distributed between Oxford, north Wales, and the Scottish Highlands, it became possible to predict the trends in population cycles. A paper in the *Journal of Animal Ecology* described how the population of the Newcastleton area of southern Scotland rose to a periodic maximum in 1933 and, as expected, was accompanied by a severe epidemic. Laboratory studies revealed the presence of a protozoan infection of the brain, caused by *Toxoplasma*, which was capable of being transmitted to stocks of normal, healthy voles (Elton *et al.* 1935).

Something of the frustration experienced by the Bureau in conducting these studies is conveyed by the applications made by Elton to the Royal Society for funds. In 1935, a joint meeting of the Royal Society's sectional committees covering botany and zoology was convened to consider an application for £400 per annum for three years to continue research on the vole population in Britain. The meeting decided to make the grant for one year only on the grounds that the society's Darwin Fund was 'becoming progressively exhausted' and that there was reason to be sceptical of the role of *Toxoplasma* infection in causing a widespread epidemic among voles. There had to be greater collaboration with pathologists.

In March 1939, another joint meeting was convened to consider an application for £350 per annum for three years. Whilst acknowledging the importance and interest of the Bureau's research, the meeting concluded that

> periodic applications to various grant-giving sources are not likely to give either the security, or the sufficiency of means, which is required; a more permanent source of income is necessary.

It was recommended that no grant should be made and that, in view of the economic aspects of the research, 'any further assistance might more properly be forthcoming from the Forestry Commission'. An amendment to the effect of making a grant was defeated by nine votes to seven, after being discussed by the Council of the Royal Society[26].

Historians of ecology have emphasized the lack of communication between terrestrial and marine ecologists—a separateness reflected in the development of the British Ecological Society. Alister Hardy was one of the very few marine ecologists to take an active part in the society—holding the office of vice-president between 1937 and 1939. It would, however, be wrong to exaggerate this separateness, both in a conceptual or institutional sense. It was through their lecture courses in marine biology that many zoology students first recognized the importance of studying the 'whole' organism in the field. The history of the marine laboratories served as a model for what the freshwater and terrestrial ecologists desired for themselves during the inter-war period.

Despite the obvious need for research on marine fisheries, and the example of the marine stations on the continent and in America, it was not until the 1880s that moves were made to establish a Marine Biological Association. As a result of vigorous lobbying, starting with an address by Ray Lankester to the International Fisheries Exhibition of 1883, the Government made the largest single contribution towards the cost of establishing a laboratory at Plymouth, which was opened in 1888 (Currie 1984, Southward & Roberts 1984).

Both the Marine Biological Association and its Scottish counterpart went through particularly difficult times until the 1920s, when the Development Commission provided the means to employ more staff, and to embark on improvements to accommodation at the Plymouth and Millport laboratories. Later recalled as the 'golden age', the inter-war years of the Plymouth laboratory were marked by major advances in research on plankton production, the role of light and nutrient limitation, and water movements: on the general biology of fish and marine invertebrates, and on the physiology of marine organisms.

Here was an obvious model for collaborative research in freshwater biology. A small laboratory was maintained by the Ministry of Agriculture at Alresford in Hampshire during the 1920s, from where studies were made of a variety of unpolluted water-courses as a preliminary to assessing more precisely how plant and animal life responded to different forms of pollution. A brief paper, in the *Journal of Ecology*, on the relationships of vegetation to the character of the stream bed of the River Itchen was followed by another, looking more generally at the distribution of macrophytic vegetation (Butcher 1927, 1933).

The venture did little more than highlight the extent to which Britain lagged behind such countries as Sweden and Germany—a point which Fritsch emphasized in his presidential address to the botanical section of the British Association meeting of 1927. In his address, which was in itself the first to focus on the simple Algae, Fritsch (1927) spoke of the need to study Algae under natural conditions. The main reason why so many botanists resorted to cultures was that they had so few opportunities for direct observation. His plea for better research facilities was endorsed by a joint session of the botanical and zoological sections, under his chairmanship, at the Association's next meeting in 1928.

The Council of the British Association appointed an *ad hoc* committee, with Fritsch as chairman, 'to enquire into the steps to be taken to establish a freshwater biological station'. The committee recommended that a station should be established under the aegis of a freshwater biological association, modelled on the lines of the Marine Biological Association. The response to the proposal was sufficiently encouraging for a meeting to be convened in June 1929, where proposals for a Freshwater Biological Association of the British Empire were approved.

There was, however, no assurance of the funds needed to establish the laboratory. News that the Fishmongers' Company had offered £100, with further grants of £100 per annum for three years, encouraged others to follow suit. The Royal Society gave £100, the British Association £40, and the British Ecological Society £10. The Manchester Water Board and the Metropolitan Water Board were the first of the water undertakers to contribute. After a long period of discussions, conducted by Fritsch, the Treasury agreed to the Development Commission making a grant of £450, with the possibility of an increase to £600 in the second year. By mid-1931, the Freshwater Biological Association had attracted 115 individuals and 20 organizations as members (Fritsch 1937).

A booklet was published, stressing how the Association intended to tackle fundamental scientific questions, rather than pursue *ad hoc* fishery or water-supply problems. It was drafted by two members of the Council, Pearsall and John T. Saunders (*1888–1965*). Pearsall was the principal advocate of the laboratory being located in the Lake District. Not only was there a variety of lakes and tarns, but conditions were likely to be very different from those being studied on the Continent. Saunders was a lecturer in the zoology department at Cambridge, and responsible for introducing the first course in hydrobiology given in any British university. Most of the early staff in the laboratory came from his department.

Although the sum raised was insufficient to enable a purpose-built laboratory to be erected, it was possible to rent and equip rooms in the National Trust property of Wray Castle, standing on the shore of Lake Windermere. The laboratory opened as Britain came off the Gold Standard in September 1931. With Pearsall as honorary director, it comprised two assistant naturalists and a research student. The locally recruited 'attendant' was warned by Saunders that there was no guarantee of 'long employment'. The first of what became the famous Easter classes was held in the following year (Le Cren 1979)[27].

By the time of the Annual Report of 1933, the laboratory was intensively used by visiting research workers and students in the university vacations. Pearsall later recalled how the research of the resident staff quickly established a tradition for experimental ecology. The work on flatworms was almost a model of what ecological investigations should be, defining the problem in the field and then taking it into the laboratory to solve under controlled conditions. At a time when the only other important centre for research was the Bureau of Animal Population at Oxford, the impact of the laboratory on animal ecology was particularly marked, especially in respect of what Pearsall (1959) called 'habitat ecology'.

As at Oxford, times were far from easy. Despite the active support of key figures in the water industry, money remained desperately short. Such was the state of the nation's finances that the Development Commission reduced its promised grant of £600 to £540. It was not until 1936 that the Commission was persuaded to send a Visiting Group to inspect the laboratory. It was so impressed that funds were provided for a salaried director and a total of five full-time staff, coupled with a guarantee that the grant would be maintained for five years. The Department of Scientific and Industrial Research (DSIR) agreed to commission research on bacteriology. With funds raised jointly by the National Federation of Anglers and the National Association of Fishery Boards, work also began on the biology of coarse fish.

In these extremely modest ways, the ecologist began to gain insights and experience in the management of collaborative research during the 1930s. Enough was achieved to highlight the value of a permanent and well-equipped base, from which field-work could be pursued and the results collated and retrieved, as required.

2.6.3 Overseas ecology

Although British ecologists devoted most of their energies to studying the plant and animal life of Britain, great importance was attached to encouraging overseas workers to join the British Ecological Society, and to contribute to its meetings and journals. Such a policy brought benefits of two kinds. First, it helped to ensure an adequate flow of papers and, through increased membership and sales, a reduction in the unit costs of publishing the journals. Secondly, the mixture of home and overseas contributions not only helped to disseminate the results of British research worldwide, but enabled the British membership to keep abreast of new concepts and techniques, from whatever quarter they came.

The meetings and publications of the society played a large part in maintaining contacts with British workers whose careers had taken them overseas. John W. Bews (*1884–1938*) spent four years in university posts in Britain, before being appointed to the chair of botany at Natal University College in 1910, where he embarked on a study of the region's vegetation. He was a frequent contributor to the *Journal of Ecology*. Francis Lewis had been appointed in 1912 as the first professor of botany in the University of Alberta, where he remained until his retirement in 1935 (Thomas 1954–5). The *Journal of Ecology* for 1926 published a paper by Lewis and a colleague in the department, describing the retrogressive changes taking place in the muskegs of central Alberta, and the extent to which they could be correlated with long-term fluctuations in rainfall (Lewis & Dowding 1926). Charles Moss attended the Annual Meeting of the Society in January 1929, during his last visit to Britain—taking part in the discussion on a paper given by Christo A. Smith (*1898–1956*) on 'The importance of ecology in the solution of the economic problems resulting from the degeneration of the Grass-veldt of South Africa'[28].

South Africa held out a particular attraction for younger botanists from Britain. A prime instigator of this link was Harold H. W. Pearson (*1870–1916*), who obtained a scholarship to Cambridge in 1896, where he achieved a first class in the natural sciences tripos. This was followed by appointments in the herbaria at Cambridge and Kew, before returning to South Africa in 1903 as the first occupant of the Harry Bolus Chair of Botany at Cape Town. There, he realized his ambition of many years to establish a national botanic garden. He was made honorary director of the State Garden at Kirstenbosch in 1913 (Seward 1917, Gunn & Codd 1981).

The outstanding opportunities offered by South Africa were publicized not only by Pearson himself, but by his students, Edith L. Stephens (*1884–1966*) and Margaret R. B. Michell (*1890–1975*), while they were studying at Cambridge. Among those who were impressed by what they heard was R. Harold Compton (*1886–1979*). He accepted the post of director of Kirstenbosch (following Pearson's death in 1916), and became the first occupant of the Harold Pearson Chair of Botany at the South African College (Rycroft 1979). By this time, Edith Stephens and Margaret Michell were lecturers in the College.

A particularly severe loss to British ecology was Adamson, who, after five years in Cambridge, had moved to the botany department at Manchester in 1912. Despite attempts by Tansley and others to persuade him to stay in Britain, Adamson accepted the Harry Bolus Chair of Botany at Cape Town in 1923—cutting short his second period of office as a Council member of the British Ecological Society (Anon. 1965)[29]. He published a preliminary account of the plant communities of Table Mountain in the *Journal of Ecology* in 1927. At the Annual Meeting of 1928, he showed 'a series of photographs of the vegetation of Eastern Rhodesia', illustrating the important part played by rainfall in determining the different types of vegetation encountered (Adamson 1927).

The traffic was not all one way—those remaining in the mother country stood to gain much in the way of ideas and experiences. As Fritsch remarked, there were many aspects of plant life which could only be fully understood by turning to striking examples of the phenomena in the tropics. In his own case, it was only when Fritsch visited Ceylon in the early 1900s that he began to appreciate fully the role of Algae, as both humus-producers and disintegrators (Fritsch 1907a,b,c).

It was often through the accounts of these overseas experiences given to meetings of the British Ecological Society that British workers learnt of phenomena and techniques of potential relevance to their own work. The competitive advantages of alien and native plant species were highlighted by an account of the vegetation around Adelaide, given to the Annual Meeting of 1920 by T. G. Bentley Osborn (*1887–1973*), who left his post as lecturer in economic botany at

Manchester in 1912 to become professor of botany at Adelaide. The vegetation around Adelaide was described as being mainly composed of aliens, largely as a result of the introduction of grazing animals, cultivation, and other disruptive changes brought about by the white settlers[30].

Turning to techniques, the Annual Meeting of 1921 was the first to see the potential of mapping vegetation 'by aeroplane photography'. The talk was given by the palaeobotanist, H. Hamshaw Thomas (*1885–1962*), who had directed the air survey of Palestine during the war (Harris 1963). Keen to apply the experience gained, he persuaded the Cambridge University Aeronautical Department and the RAF Special Experimental Flight at Duxford to make three surveys of Blakeney Point in 1921. From the vertical photographs taken, a mosaic was constructed. A further sortie in 1922 ended with a forced landing on the spit. The plane had to be dismantled, and taken back to Duxford by RAF lorry[31].

Using lantern slides of the Jordan valley and Blakeney Point, Hamshaw Thomas illustrated how the plough furrows of the Palestinian fields, and single bushes of *Suaeda fruticosa*, could be distinguished on photographs 'taken from a height sufficiently great to comprise a considerable tract of country in a single plate'[32]. In the *Journal of Ecology* for 1925, the geologist and geographer L. Dudley Stamp (*1898–1966*) used examples from a photomosaic of the Irrawaddy Delta Forest to highlight the value of vertical air photographs, taken for stereoscopic use. Commissioned by the Government of India, the sortie demonstrated the potential of such photographs for vegetation mapping (Stamp 1925).

The most outstanding pioneer of plant ecology in the dominions was Leonard Cockayne (*1855–1934*). Soon after graduating from Owen's College, Manchester, he left for Australia, and settled in New Zealand in 1880, where he taught for four years. He had always been interested in natural history, and soon set up a botanical garden on the sand dunes near Canterbury, where he studied how the form of plants changed in response to environmental factors. He was, as one writer put it, 'already an excellent ecologist waiting for the term to be adopted' (Hill A. W. 1935, Laing 1936). His survey of the vegetation of a saddle over the southern Alps, carried out in the summer of 1897–8, was not only the first of its kind in the Empire, but the records and photographs provided the basis for a comparative study made some 34 years later in what had become by then 'a carefully guarded national park' (Cockayne 1898, Cockayne & Calder 1932).

The self-contained and distinctive character of New Zealand, with its intense endemism, extreme isolation, extensive tracts of 'virgin' vegetation, and varied climate presented outstanding opportunities for the ecologist. In over 170 publications, Cockayne used these opportunities to explore and extend many of the concepts in plant taxonomy and vegetational development, and particularly those of natural hybridization (Cockayne 1911a, Cockayne & Allan 1927). He was unique among British and Empire botanists in being invited to contribute to the series of monographs, *Die Vegetation der Erde*. Completed in March 1914, proofs were exchanged with the German publisher throughout the war (via Eugene Warming in Copenhagen). The monograph on New Zealand eventually appeared in 1921 (Cockayne 1921). Elected to the Royal Society of London in 1912, Cockayne was awarded the Society's Darwin Medal in 1928. Together with Henry Cowles, he was elected an honorary life member of the British Ecological Society in January 1934.

Although Tansley, Fritsch, Yapp, and other early British ecologists derived a great deal of inspiration from what they saw and read of the tropics, there was little in the way of a co-ordinated or sustained effort in tropical ecology. There was so much to investigate closer at hand, and so few university departments in the tropics. Many workers were either deterred or overwhelmed by the magnitude of the task. It was almost 50 years before Schimper's general account of the tropical rain forest was joined by another, published in English. It was written by Paul W. Richards, and was based on a total of two years of intensive study in three continents (Richards P. W. 1952).

The compilation of Richards's book highlighted the continued importance of the university expeditions between the wars. With the active encouragement of its chairman, Charles Elton, the Oxford University Exploration club sent a biological expedition to British Guiana in 1929 (Hingston 1930). Two of the 11 members were Owain Richards and his brother, Paul, who was then an undergraduate in the Botany School, Cambridge. In a joint presentation to the Annual Meeting of the British Ecological Society in January 1931, Owain Richards described how a prime objective of the 15-week expedition had been to discover whether there was 'a real tree-top zone of animals'. This was soon confirmed in respect of birds, but insects proved a much more formidable problem. Through the use of light traps at different levels, and by felling selected trees, it was possible to draw some comparisons with the ground fauna. A number of particularly striking forms of invertebrate appeared to be restricted to the higher levels.

In his contribution to the Annual Meeting, and in a paper subsequently published in the *Journal of Ecology*, Paul Richards described how, through collaboration with officers of the local forestry department, it had been the principal aim of the botanists on the expedition to carry out the first intensive study of a tropical rain-forest. Salisbury suggested that the complicated structure of the forest might be depicted in the form of a profile diagram, based on the exact measurements taken of the trees felled along narrow transects (Figure 2.7). It was a device that came to be much used in vegetation studies. Far from being floristically homogeneous, the small area chosen for the study near Moraballi Creek on the Essequibo was found to contain five distinctive types, each related to a particular soil type (Davis & Richards 1933).

The scientific expedition to the Dulit area of central Sarawak in 1932 provided Paul Richards with the chance to study a rain forest in the eastern tropics (Harrisson 1933). He found a remarkably close resemblance between the struc-

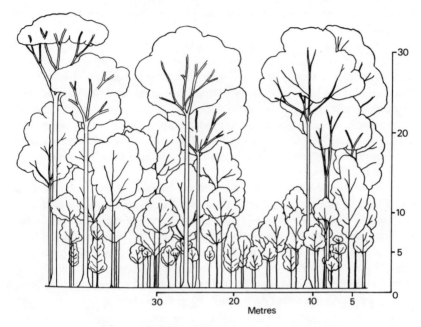

FIGURE 2.7 Diagrammatic profile of a strip, 7.6 metres deep, on a clear-felling plot (from Davis & Richards 1933).

ture and floristic composition of the Bornean 'heath forest' (one of three types occurring on Mount Dulit) and that of the Wallaba Forest-type of British Guiana (Richards P. W. 1936). A few years later, Richards was himself the leader of a Cambridge University Botanical Expedition to the Shasha Forest Reserve of southern Nigeria in 1935, where the first (top) and second storeys of the rain forest were found to be much more open, and the flora poorer, than those encountered in British Guiana and Sarawak (Richards P. W. 1939).

During the six-month expedition to the Shasha Forest Reserve, studies were made of various environmental factors and the course of secondary succession in what later turned out to be a very long-established forest, rather than a strictly 'primary' forest (Ross 1954). Measurements of temperature, humidity, light intensity, and carbon dioxide were made by G. Clifford Evans, as part of a larger autecological study of the undergrowth flora. Not only was very little known about the role of environmental factors in tropical forests, but, as Evans later recalled, no other higher plant environment lent itself so readily to laboratory simulation. The damping effect of the forest canopy on environmental change, combined with low average wind speeds, led 'to slow and majestic cycles of change singularly free from the short period fluctuations observed in the natural environments of temperate regions'. The general level of illumination on the forest floor could be simulated in the laboratory (Evans G. C. 1939, 1975).

At the Annual Meeting of the British Ecological Society in January 1938,

Joseph Burtt Davy (*1870–1940*) of the Imperial Forest Institute at Oxford introduced a discussion on 'the classification of tropical woody vegetation types'. He circulated a draft conspectus, which was based on his own first-hand experiences of forests in eastern and southern Africa (Davy 1938). In the discussion that followed, there was 'a strong feeling that standardization would be extremely useful', but that it would be wrong to make 'premature commitments ... where principles were not yet understood'. At its next meeting, the Council of the British Ecological Society supported the appointment of an *ad hoc* committee, with Tansley (the president), Paul Richards, and Alex Watt as members. The committee's report was later published in the *Journal of Ecology*, and as a paper of the Imperial Forest Institute.

In its report, the committee endorsed the need for stabilization in the nomenclature of tropical forest vegetation. Forest officers should be encouraged and assisted to collect and record data, which should be filed and correlated by a central institution. From such 'a sound foundation', it might be possible to develop a 'natural classification'. The committee put forward 42 points of primary importance which the forester should record systematically (54 fewer points than in Burtt-Davy's scheme). They covered structure and composition, and habitat and succession (Richards P. W. *et al*. 1940).

2.6.4 *The employment of ecologists*

Among the common causes for both plant and animal ecologists was the pressing need to find bread-winning jobs for the young ecologist. In America, there were plenty of openings for graduate botanists from schools specializing in ecology. One of Clement's students at Nebraska, A. W. Sampson, had been appointed to the U.S. Forest Service as early as 1907 (White J. 1985). The situation was very different in Britain. In their search for career openings, ecologists still laboured under a severe disadvantage, compared with mycologists and even plant physiologists or geneticists (Tansley 1939b).

Just as plant and animal ecology were far from being discrete entities in ecology, so also it was impossible to draw a dividing line between pure and applied ecology. Tansley alluded to the more practical considerations of running a botany department in the inaugural lecture he gave as the Sherardian Professor of Botany in Oxford in 1927. As he remarked, a university science department was much more expensive to run than one devoted to the humanities. The only way to overcome this 'most serious handicap' was for the science department to develop 'some effective practical orientation'. As Tansley (1927) recounted, the Oxford department had met the challenge by training young men

> in the foundations of scientific knowledge of plants and in the right
> attitude of mind towards the problems they will encounter in helping to
> conserve and to develop the resources of the Empire, depending as they
> do very largely on plant-life.

In Tansley's view, a knowledge of ecology was especially relevant to forestry, particularly in tropical climates where natural or seminatural forests were the rule. At the Annual Meeting of the British Ecological Society in 1932, he outlined the course he gave to forestry students, describing how he took them straightaway into the field, introducing them to examples of the well-known plant communities growing around Oxford[33].

In order to provide some ecological guidance to foresters, agriculturalists, entomologists, and others already employed in the different parts of the Empire, the Imperial Botanical Conference of 1924 appointed a committee for that purpose (Brooks 1925). Called the British Empire Vegetation Committee, it consisted of Tansley (chairman), Oliver, Salisbury, and Ramsbottom. The outcome was a volume on the *Aims and Methods used in Vegetation Surveys*, published in 1926. Having looked at a range of factors affecting vegetation generally, it dealt in turn with the natural regions of the Empire and the types of vegetation encountered.

A key role was played by the secretary of the Committee, Thomas F. Chipp (*1886–1931*), whose lifelong interest in forestry dated back to 1910, when he secured a post as assistant conservator in the Forest Department of the Gold Coast. Before taking up the appointment, he was sent first to Germany and then to the Malay States to gain experience in forest management and survey. After war service and a period as assistant director of the botanic gardens in Singapore, Chipp returned to the Gold Coast as deputy conservator of forests. Much of his subsequent research was brought together in a memoir on *The Gold Coast Forest: A Study in Synecology*, published in 1927. He became assistant director of the Royal Botanic Gardens at Kew in 1922 (Chipp 1927, Burkill 1931–2).

In a preface to the volume compiled by the British Empire Vegetation committee, the editors (Tansley and Chipp) wrote of how it was

> hardly possible to conceive a property owner or stores manager carrying on the management of a large estate or general store and yet being unaware of the stock at his disposal, the extent of his supplies or their nature.

And, yet, that was essentially the position in respect of the natural vegetation of the tropical colonies, protectorates, and even, to some extent, the dominions. In the case of living and active 'stock', such as vegetation, it was never enough to record its existence and extent. If it was to be managed intelligently and to great advantage, the behaviour and potential of the different organisms had to be studied before exploitation could begin (Tansley & Chipp 1926).

Having looked at a range of factors affecting vegetation generally, the volume dealt in turn with the natural regions of the Empire and types of vegetation encountered. Chipp and Dudley Stamp wrote on the tropics. A chapter on forest ecology in India was provided by Robert S. Troup (*1874–1939*), the professor of forestry at Oxford. Other contributors included John Bews (the vegetation of South Africa) and Theodore Osborn (the arid parts of Australia).

Leonard Cockayne and C. W. Howe wrote on New Zealand and Canada respectively. Since the laws which governed the formation and changes of vegetation were the same the world over, the editors hoped the book would be of value beyond the confines of the Empire.

Outstanding examples of how ecological research could be applied to practical issues of public concern had been provided by Oliver and Cockayne. Following the enactment of the Sand Drift Bill of 1907, Cockayne was invited by the Minister of Lands to advise on the character and management of dune areas which, in New Zealand, covered almost 300 000 acres, and threatened the agricultural productivity of much adjacent land. In a report of 1911, Cockayne not only described the affected areas in detail, but drew on experience in New Zealand and elsewhere to identify how the dunes might be stabilized, and their commercial and scientific value enhanced (Cockayne 1911b, Hill A. W. 1935).

Historians of ecology have drawn particular attention to the wider repercussions of the Great Drought, which began in 1933 and turned large parts of the Southern Plains of the American continent into a Dust Bowl (Worster 1979). The American experience seemed especially relevant to the African reserves of Kenya, the heavily populated parts of upland Tanganyika, and the intensively cropped areas of Uganda. In a continent already afflicted by the severe depression in world trade, this combination of real and imaginary crises, affecting both the economy and environment, led to an even closer scrutiny of the effects of African farming practices on the longer-term productivity of the colonial territories. As Anderson (1984) observed, the outcome was a more interventionist policy on the part of the colonial administration, with its far-reaching political implications.

The global scale of soil erosion was demonstrated in a monograph, published by the Imperial Bureau of Soil Science at Rothamsted in 1938, that described the situation in those countries most seriously affected by erosion, and the steps being taken to eliminate the causes and mitigate the effects. The authors, Graham V. Jacks (*1901–77*) (the deputy director of the Bureau) and Robert O. Whyte, followed up their monograph with a widely acclaimed book with the evocative title *The Rape of the Earth. A World Survey of Soil Erosion*. In it they emphasized the role of the plant ecologist, working with the animal ecologist and population expert, in devising ways of revegetating the prairie and forest lands (Jacks & Whyte 1938, 1939).

In America, that role was both recognized and exploited. Worster (1985) wrote of how

> as a result of this environmental crisis of the dirty thirties, the most
> telling in our history, the new profession of ecologists found themselves
> for the first time serving as land-use advisers to an entire nation.

That was a role which still eluded the ecologists working in Britain. As Jacks and Whyte had discovered from their global survey, Britain was one of the few west European countries where torrent and erosion control continued to be matters of local, rather than of national, concern.

The period of reconstruction after the First World War had seemed at first particularly suited to the adoption of a more ecological approach to farming and land management practices. In his presidential address to the British Ecological Society in 1918, William Smith spoke of how the young science of ecology was taking a forward position in economic applications, providing a better understanding of 'the utilisation of land for economic purposes' through the application of 'the methods of the laboratory to the requirements of outdoor observation and experiment', supported, and frequently preceded, by vegetation mapping on an extensive scale (Smith W. G. 1919).

The figure most closely associated with this approach was George Stapledon, who was described, on the occasion of his knighthood in 1937, as 'an ecologist in practice and agriculturalist by profession'. In his review of the progress made in ecology over the first quarter-century of the British Ecological Society, Tansley (1939b) spoke of the many ecological problems that were constantly arising in the pasture work of Stapledon's Welsh Plant Breeding Station at Aberystwyth, and how 'a young postgraduate with ecological training could hardly do better than attach himself to the school and take up some of these problems'.

In the event, the impact of ecological concepts and advice on land husbandry was minimal in inter-war Britain. This may have reflected in part the difficulties of persuading farmers at that time to apply the results of any kind of research to their everyday management of land. Not only were there few advisers to interpret the results of trials at Rothamsted and elsewhere in terms of the requirements of individual holdings, but the depressed state of farming would have deterred many farmers from seeking such guidance. The exhilaration of post-war reconstruction soon gave way to 'the great betrayal' of the industry of the early 1920s. As soon as there were signs of economic recovery, and the Ministry of Agriculture began to give explicit encouragement to such ventures as grassland improvement, Stapledon responded by collaborating with a leading agricultural adviser, J. A. Hanley, in writing a guidebook to the subject, *Grassland: its Management and Improvement* (Stapledon & Hanley 1927)—only to see farmers further disheartened by the even deeper economic recession of 1929–33.

The progress made in the management of animal populations was similarly slow and fitful. In the Bureau of Animal Population at Oxford, the search continued for 'a system of basic principles' that would make it easier to apply the discoveries of pure research to the needs of medical investigators, economic entomologists, and game conservators. In a paper published in the *Journal of Hygiene*, Elton (1931) discussed how a closer understanding of epidemic diseases in wild animals might make it easier to unravel the mechanisms of epidemic diseases generally.

A grant from the Imperial Chemical Industries (ICI) made it possible for the Bureau to investigate periodic fluctuations in game populations. Using data from game preserves in different parts of the country, Douglas Middleton found that rabbit, hare and partridge populations were subject to eight-year cycles, and grouse to cycles of six years. In a paper given to the Annual Meeting of the British Ecological Society in January 1935, Middleton itemized the large number of factors affecting the size of the partridge population in the course of a year (Middleton 1934)[34].

The support of ICI for this kind of research arose from the company's Eley cartridge interests. Although considerable sums were spent on research into the behaviour of cartridge cases and everything else that went into the breech end of the gun, it was realized that nothing systematic was being done to improve things at the muzzle end. Matters came to a head in 1931–2, when a serious outbreak of partridge disease combined with a general decline in keepering caused home sales of cartridges to fall by 15 millions in two years. On the premise that game research might make 'two birds fly where one flew before', a member of the main board of ICI, Major H. Gerard Eley (*1887–1970*), obtained approval for the establishment of an ICI Game Research Station. An experimental keepered-shoot and a commercial pheasant, partridge, and bantam farm were set out at Knebworth Park in Hertfordshire. Douglas Middleton was appointed director. The main thrust of the research programme was on reducing the fluctuations in partridge population. By flattening the peaks, it was hoped to bring about a more consistent demand for cartridges[35].

In any attempt to interpret fluctuations in animal population, Elton (1943) stressed the need to take account of what he called the 'historic and irreversible breaking down of the great realms of plant and animal life'. Whereas the great continents were once practically cut off from one another—there had been minimal contact between them—'alien' species could now pass from one to another with the greatest of ease. There was plenty of evidence of the 'complex ecological problems' that arose. During the early 1930s, there was increasing concern over the damage caused to water-courses by the Canadian muskrat (*Ondatra zibethicus*), a small number of which had been imported to fur farms in the British Isles during the 1920s. Warned by the British Museum (Natural History) of the consequences of allowing wild colonies to become established, particularly in the light of German experience, the Government introduced a bill which led to the banning of the import and keeping of muskrats in Britain (Sheail 1987)[36].

Never before had the Government taken so major a role in pest control. A grant was made to the Bureau of Animal Population for a two-year investigation of the ecology of the animal, focused on the two main areas of infestation, the Severn valley around Shrewsbury and the valleys of the Earn and Allan in Perthshire and Stirlingshire. In a review of the species' distribution, published in the *Journal of Animal Ecology* in 1934, the author of the study, Tom Warwick, warned of how an eradication programme would be a lengthy, if not impossible, task. The muskrat was very adaptable and could spread easily to new districts along the network of waterways. Most parts of Britain could support a population (Warwick 1934). In the course of trapping in Perthshire, over a thousand muskrats were taken (the progeny of five pairs that had escaped from a farm in 1927) (Munro 1935). In the event, the national campaign was brought to a successful conclusion by 1936. As Warwick (1940) recounted, trapping had been well organized, and it had started before the wild colonies had become large and more numerous.

It was not only recent arrivals that caused concern. By the late 1930s, there was increasing anxiety over the impact of rabbit damage on crops, and the cruelty inflicted on rabbits and other wild populations as a result of the continued use of the spring or gin trap (Sheail 1984a). In the forefront of the campaign to eliminate the gin trap was the University of London Animal Welfare Society (ULAWS), a pressure group formed by Major C. W. Hume in 1926. He argued that little would be achieved until 'a technical investigation' had been carried out 'by trained ecologists' on the longer-term effects of the various forms of control on the rabbit population (Hume 1939). To that end, the Society provided the resources for the Bureau of Animal Population to carry out what was in effect the first detailed analysis of a rabbit population, and the first to use visual markings as a means of distinguishing members of a wild mammal population. By the use of 'huge numbered ear-tags', H. N. (Mick) Southern (1909–86) was able to record the feeding habits, reproduction, and predator–prey relationships of a warren population in unprecedented detail (Southern 1940).

In an article on 'alien invaders', published in The Times (6 May, 1933), Elton emphasized the need for regular and thorough surveys of animal life. So far the initiative for such studies had come from the universities, scientific societies, and individuals. A 'planned scheme of research', which was comprehensive yet detailed, would add another valuable dimension to the popularity that already existed for 'all kinds of regional survey'. Under the direction of Dudley Stamp, a Land Utilisation Survey was already under way, covering the entire land surface of Britain on a field-by-field basis, carried out largely by schools and other educational establishments. Both Elton and Tansley were members of an ad hoc committee, convened by the town and country planner, L. Patrick Abercrombie (1879–1957), which called for the setting up of a National Survey Commission (Sheail 1981).

Perhaps the most daunting aspect of these initiatives was the lack of any material support from the Government. The frustration was particularly evident in respect of proposals put forward by the geographer, Eva G. R. Taylor (1879–1966), for the compilation of a national atlas of Britain, as a basis for regional planning (Taylor E. G. R. 1939). Largely through the instigation of Tansley, the Council of the British Ecological Society gave considerable support to the venture. Tansley acted as chairman of a subcommittee convened to decide what species and vegetation maps might be included[37]. A note was published in the Journal of Ecology, inviting members to submit suggestions for consideration. The British Association and other learned bodies gave their support, but the Government was so lukewarm that the military authorities had no difficulty in bringing work on the atlas to an end in 1940, on the grounds that the maps would be of the utmost value to the enemy (Taylor E. G. R. 1963).

Part 3
The Second Quarter-Century, 1939–1963

3.1
Continuity and Change

Anniversaries are necessarily arbitrary affairs. The celebrations associated with the first quarter-century of the British Ecological Society brought little in the way of discernible change of direction in the affairs of the Society. Tansley had ceased to be an office-holder, but his pre-eminence became even more striking. He had shown what might be achieved. His writings provided the essential context for the next generation of research workers. Of great significance too was the stress which Tansley placed on providing some kind of institutional structure and support for ecology. Through his own conspicuous contributions as editor, Council member, and president, he had inspired others to play their part in the running of the British Ecological Society.

For Council members and officers, the affairs of the Society were yet another demand on time, to be somehow squeezed between those of teaching and research. Godwin became editor of *New Phytologist* in 1931, and was soon afterwards 'pitchforked' by Tansley into becoming honorary secretary of the British Ecological Society (West 1986). He replaced Salisbury, who had held the post for 15 years. Godwin later recalled how the post was so onerous that he could never have coped without the assistance of his wife. It involved at that time

> maintenance of a membership list of about 300, collection of membership fees and subscriptions, the regular holding of society and management committee meetings with preparation of accounts of all proceedings and the setting up of regular scientific meetings in London and provincial universities, together with field meetings in the better weather for the demonstration of sites and areas where ecological work was active.

As Godwin (1985) commented, there were very full attendances at meetings, and he and his wife, Margaret, came to know personally most of the British membership. The size of the membership grew from 305 in 1933 to 361 in 1939. After falling to 335 in 1940, it rose to 398 in 1944, and to 485 in 1945.

The comparatively small number of officers had to sustain the more normal activities of the Society, even during the grimmest days of the war. The burden falling on the editors was particularly severe. Elton and Chitty 'nursed a thin continuation' of the *Journal of Animal Ecology* through the war. Pearsall, who took over as editor of the *Journal of Ecology* in 1937, found himself so short of papers by 1941 that he had to solicit material. Summerhayes and Watt became joint treasurers in the place of Boyd Watt in 1938. Summerhayes suggested writing to those members of the Society in the more outlying parts of the

Empire, who might be in 'a better position to continue work and publish papers'. He forwarded a list of persons who might respond to an invitation, 'pointing out the necessity of keeping the ecological-flag flying'[1].

The device met with some success. The volume of 1942 also included a memorial supplement on East African vegetation. The author, Bernard D. Burtt (*1902–38*), had been the survey botanist of the Department of Tsetse Research in Tanganyika. Burtt and the director of the Department, Charles F. M. Swynnerton (*1877–1938*), were killed while taking part in an air reconnaissance. The cost of the supplement was met by friends of the two men (Davy 1939, Burtt 1942). The dearth of material proved only temporary. As a result of wartime work, and an eagerness to resume research once hostilities were over, the chronic problems of trying to squeeze more and more copy into each volume soon returned—this time never to go away again.

The Annual Meetings of the Society continued. Godwin's presidential address of 1943 was published, but those of Cyril Diver and Owain Richards never appeared in print. The reasons are far from clear—the minutes of Council are silent on the subject. Diver devoted his address of January 1942 to the theme of 'The limits of tolerance'. It broke with custom by being followed, at Diver's request, with a discussion on the ideas put forward. In November 1943, Diver contributed to a symposium on the 'Interrelations of plants and insects', organized jointly with the Royal Entomological Society. In discussing the way in which insects were limited by 'the general conditions imposed by plant communities', he drew heavily on the observations he had made over many years on the South Haven Peninsula in Dorset (Riley 1943).

The minutes of the Annual Meeting of January 1945 described Richards's address on 'Ecology from the view-point of the entomologist' as being 'extremely stimulating'. Richards himself remarked that 'the editors didn't think there was much in it, so it wasn't published' (Southwood 1987). Richards had left Oxford for Imperial College, London, in 1927. Ever since his study of the fauna of fence-posts in Bagley Wood, he had become increasingly sceptical of the value of applying the concept of the community to animals, as opposed to plants. In the course of research on the factors controlling the abundance of the small white butterfly (*Pieris rapae*), started in 1932 and published in the *Journal of Animal Ecology* in 1940, he became convinced that the key to progress was in the reductionist approach, investigating the population dynamics of individual species (Richards O. W. 1940). However great the tensions engendered by the debate, Richards continued to be involved in the Society's affairs, serving as editor of the *Journal of Animal Ecology* during the four years leading up to his retirement as professor of zoology and applied entomology in 1967.

3.1.1. *Ecologists and the war effort*

Discussions as to the practical importance of ecology took on a new significance as war became likely and then broke out in September 1939. If ecologists had a role to play in securing a more competent and rational approach to the management of land and natural resources, that time had surely come. As the war continued, and particularly as thoughts turned to post-war reconstruction, the British Ecological Society took on a more vigorous role in promoting both the policies of the Society and the wider relevance of ecology.

The immediate response of the Council of the British Ecological Society to the outbreak of war was to abandon a proposed meeting on phytogeographic systems, and to hold instead a discussion 'on the application of ecology to war-time research' during the Annual Meeting already planned for Oxford in January 1940. The discussion, with Tansley in the chair, provided an opportunity to take stock of the preparations that had been made for war and of the changes that had come about since the beginning of hostilities.

Attitudes had begun to change in 1938. In October of that year, Stapledon was invited by the Ministry of Agriculture to tour the country, urging farmers to treat all grasslands as if they were a rotational crop. Under the slogan, 'Take the plough round the farm', he provided guidance as to how farmers could take advantage of the Government ploughing-up subsidy. In a new book, called *The Plough-up Policy and Ley Farming*, he dealt with such questions as soil fertility, seed mixtures, and the management of leys (Stapledon 1939).

It was also in the autumn of 1938 that the Ministry of Agriculture provided the resources for Stapledon and William Davies (*1899–1968*) to embark on a detailed survey of the grasslands of England and Wales, as a prerequisite to deciding which should be converted to arable as part of the wartime ploughing-up campaign. At the meeting of the British Ecological Society convened in January 1940, Davies described how floristic composition was used to assess productivity. Over 9 million acres (3.6 million hectares) were classified as being fit for the plough; much had been cultivated during the first world war and then abandoned during the depression years of the 1920s and '30s[2]. Drawing on the experience of his Cahn Hill Improvement Scheme, which had done so much to demonstrate the possibilities of reclaiming the waste hill-lands of Wales, Stapledon urged the Ministry of Agriculture to establish a farm where the scope for rehabilitating heavy clay soils could be demonstrated. The Ministry agreed to provide the funds for establishing a Grassland Improvement Station on the adjoining farms of Dodwell and Drayton, near Stratford on Avon.

Charles Elton described to the same meeting of the British Ecological Society how, following the outbreak of war, the Bureau of Animal Population turned all its attention to finding the most effective and economic methods of pest control, and to act as an information bureau generally. Most of the research focused on the rabbit, brown and black rat, and house mouse. Largely financed by the Agricultural Research Council, the Bureau quickly grew in size[3]. Douglas Middleton returned, and was responsible for a booklet, published in 1942, on 'The control and extermination of wild rabbits', describing how numbers could be reduced drastically by using a combination of existing methods of control, provided every occupier of land co-operated (Middleton 1942). The research on ways of controlling the rat and mouse populations was later collated in a three-volume study, published in 1954 (Chitty & Southern 1954).

Another contributor to the discussion meeting was E. Barton Worthington, who had been appointed the first director of the Freshwater Biological Association in 1937. With the outbreak of war, the scientists left at Wray Castle embarked on a search for substitutes for the fast-dwindling supplies of marine fish. While there was neither time nor resources to develop fish-farming on the scale practised in some parts of Europe, it was hoped to obtain worthwhile crops, using the natural productivity of the large lakes, reservoirs, and rivers. The value of 'manuring' the waters with sewage was investigated. Worthington made direct use of his African experience in seeking to resuscitate the local commercial fisheries (Worthington 1983). Whilst the yields from such lakes as Windermere soon fell to disappointingly low levels[4], the applied studies were to prove of considerable long-term significance in monitoring changes in the balance of fish populations and their effects on the ecology of lakes (Le Cren 1979).

In the discussion that followed the presentation of papers, Bentley Osborn (who had succeeded Tansley as Sherardian Professor of Botany at Oxford in 1937) wanted to know much more about how 'members of the Society could utilize their special knowledge' in contributing to the war effort. Harry Godwin deplored the gap that still existed between the working scientist and the Government's Scientific Advisory Committee. Much more pressure had to be put on Government departments to make greater use of biologists. To that end, the meeting passed unanimously a resolution:

> That this Society, either alone or in co-operation with other scientific bodies, should take steps to urge upon the Government a more active utilization of the services of biological research workers in the present emergency.

At their next meeting, members of the Society's Council agreed that any worthwhile initiatives could only come from concrete proposals put forward by biologists themselves. There had to be some kind of biological 'administrative department', with executive powers similar to the Department of Scientific and Industrial Research (DSIR). A committee was appointed to explore the possibilities, made up of the president (Diver), Godwin, Owain Richards, A. R. Clapham, and E. W. Jones[5].

On the basis of 'lengthy and valuable replies' received to a questionnaire, the committee drew up a memorandum, highlighting the growing sense of frustration felt among ecologists. Despite their being invited, in common with other scientists, to supply details of their qualifications to a central register, there was little evidence of their skills being called upon. This was in striking contrast to the immediate mobilization of engineers, physicists and chemists, who, by 1940, were working in large teams in such specialized fields as radio location, chemical warfare, and aeroplane design. In despair, some biologists had turned their knowledge of physics to good account, and had taken employment in radio location and other technical fields of the armed services.

Copies of the committee's memorandum were distributed widely. Discussions were held with the secretary of the Agricultural Research Council. At a meeting between representatives of the British Ecological Society (Godwin, Richards, and Clapham) and the Association of Applied Biologists (C. T. Gimingham, J. C. F. Fryer, and H. F. Barnes), it was agreed that some kind of organization was urgently needed, capable of sponsoring and developing both short- and longer-term research of an applied character. A 'whole-time energetic scientist of high standing' should be appointed to take charge of it.

The scientist soon emerged. At a meeting of the Association of Applied Biologists in December 1940, Geoffrey Blackman voiced the frustrations felt by many that, even after 15 months of war, there was still little evidence of biologists being recruited to help Government departments resolve many of the problems they so obviously faced. Blackman attributed this obvious neglect to 'the lack of personnel with the proper biological outlook and knowledge to appreciate that the problems were there and were capable of solution'. Just as the Air Ministry had a department to deal with suggestions and inventions, so there had to be a 'Technical Biological Committee', which would initiate research and ensure that there was proper co-ordination and co-operation between workers and research institutions (Anon. 1941a).

With his wide-ranging interests in biology and agriculture, Blackman soon emerged as the key figure in discussions between the Association of Applied Biologists, the British Ecological Society, and the Society for Experimental Biology. In the course of an 'active discussion' at the Annual Meeting of the British Ecological Society in January 1941, great importance was attached to developing contacts with the Government's new Scientific Advisory Committee (appointed towards the end of 1940). At the Council meeting of May 1941, Blackman reported that a joint committee had been formed between the three societies, with himself as secretary. Council members agreed that the Society should meet a third of the costs incurred, subject to their not exceeding £10⁶.

The first evidence of a positive response from the Government came in May 1941, when the Agricultural Research Council, DSIR, and Medical Research Council formed a small joint committee to consider how greater use could be made of biologists and biological knowledge in the war effort. The three societies responded by expanding their own joint committee into a Biology War Committee, so as to act as a clearing house in relaying problems and ideas to and from Government. The Biology War Committee was made up initially of 23 members, with Blackman as secretary. Godwin served as one of the five members of the executive committee (Blackman G. F. 1942).

Because of the confidentiality of much of their war work, it was far from easy to assess just how far ecologists were successful in becoming absorbed into the war effort. At the Annual Meeting of the British Ecological Society in January 1944, William Turrill gave a paper on 'Vegetation from a military stand-point', in which he described the kind of information which the military authorities were likely to seek from the ecologist. It included the density and height of the different communities, as well as

> the tallness and extent of the tree canopy, penetrability to men and vehicles of various kinds, the obstacles presented by climbers and spinous plants, by bogs and dunes, together with data on the way in which seasonal changes might affect the operation of these factors.

In order to provide such guidelines, the ecologist needed 'clear descriptive accounts', with measurements, maps, and fully annotated photographs of the different territories, studied at first hand. As Turrill pointed out, such material had rarely been acquired.

An obvious example of this kind of descriptive study was Turrill's own book, *The Plant Life of the Balkan Peninsula*, published in 1929. His interest in the peninsula dated back to 1917, when he was a member of the British Salonika Forces. He returned to Macedonia a further three times, and made full use of the unrivalled facilities available in the herbaria and libraries at Kew, and in London and Oxford. His book was the first to appear in a series, *Oxford Memoirs on Plant Geography*, edited by Tansley. In Tansley's words, it was an example of how studies of the distribution of species and vegetation could be 'most happily combined' (Turrill 1929).

With the outbreak of war in 1939, much of the Kew herbarium and library was transferred to the Bodleian in Oxford, where Turrill was well placed to contribute chapters on 'Vegetation and fauna' to the Geographical Handbooks being prepared under the aegis of the Naval Intelligence Division by geographers at Oxford and Cambridge. Although intended primarily for the Navy, it was expected that the armed forces and Government departments generally would make use of them. Turrill was responsible for the relevant chapters in the volumes on Albania, Algeria, French Equatorial and West Africa, Italy, Jugoslavia, Persia, Tunisia, and Turkey. Their preparation caused him to read and abstract every relevant publication available in the London and Oxford libraries. His notes and translations covered many thousands of foolscap sheets (Hubbard 1971).

In welcoming the creation of the Biology War Committee, the leader writer in *Nature* deplored the fact that it had taken over two years to achieve. The delay reflected both the poor status of biology in the school curriculum and its 'sterile academic flavour'. Far from encouraging those aspects of biology with a direct bearing on human welfare, many university biologists continued to look askance at applied biology. The leader writer hoped that the newly appointed Biology War Committee would help

> to raise the science of biology from the slough of complacency in which it has wallowed for decades to the firm road on which it rightly belongs—that of a science not only of cultural value but also of inestimable practical value to the progress and evolution of civilization,

a value on a par with that at present attached to the sciences of
medicine, engineering, physics and chemistry (Anon. 1942).

For those whose memories extended back to the first world war, and the controversy arising from the 'encyclical' published in the *New Phytologist* of 1917, the
exhortations contained little that was new. The main difficulty for the British
Ecological Society was to find a way of contributing to the war effort that
would be sufficiently distinctive as to arouse considerable interest and support
for ecology.

Some of the opportunities and frustrations of working as an ecologist in
wartime were highlighted by the career of John F. Hope-Simpson, one of the
very few research students to have been formally supervised by Tansley. From
1935 onwards, he resurveyed 47 of the grasslands which Tansley and Adamson
had recorded in 1920–1 (Tansley & Adamson 1925, 1926). Tansley helped him
to relocate the sites during a motor traverse of the South Downs in the summer
of 1935. Hope-Simpson quickly discovered that such a comparative study was
far from easy. Account had to be taken not only of errors inherent in the
methods used, but of the effects of annual fluctuations and seasonal changes in
the vegetation (Hope-Simpson 1940a).

The survey nevertheless highlighted the comparatively stable nature of
chalk grasslands. The dominant species had changed in only 13% of the sites
revisited, largely because the level of grazing had scarcely altered. With the re-
moval of rabbit grazing, some of the grasslands were likely to become Festu-
cetum rubrae or Brachypodietum (Hope-Simpson 1941). These findings were
outlined at the meeting of the British Ecological Society in January 1940. In a
paper subsequently published in the *Journal of the Royal Agricultural Society*,
Hope-Simpson made use of them in emphasizing the need for maintaining the
grazing pressure on chalk grasslands, especially those liable to invasion by
scrub or the useless and persistent tor grass (Hope-Simpson 1940b).

Hope-Simpson was transferred from the headquarters of the Agricultural
Research Council to the Grassland Improvement Station. Davies encouraged
him to set up a field trial on chalk pastures, where the bare patches created by
rabbit grazing were seeded with those useful pasture species native to well-
managed 'natural' downs, such as cocksfoot and red clover. It could, however,
only be a token encouragement on a station where the overriding priority was
to plough up as much grassland as possible—a policy for which Hope-Simpson,
with his background of studying attactive chalk grasslands, could have little
enthusiasm. He joined the Royal Air Force in March 1943, and served as a
photographic interpreter at Medmenham, where his knowledge of topography
and vegetation was put to good use interpreting from air photographs some of
the terrain likely to be encountered by Allied forces during the liberation of
France.

At Medmenham, Hamshaw Thomas was in charge of much of the specialist
photographic interpretation until his retirement later in 1943 (Harris 1963). At
his instigation, the British Ecological Society passed a resolution at its Annual

Meeting of January 1945, emphasizing the value of air photographs for scientific research and the need to establish a photographic reconnaissance unit on a permanent basis after the war. Six hundred copies of a memorandum were distributed to members of the Society and interested organizations, setting out the substance of the resolution. Discussions took place between the Air Ministry, Admiralty, and Royal Society, and in early 1946 it was announced that the newly established Joint Photographic Reconnaissance Unit of the RAF would consider any project for air photography put forward by the Royal Society.

Ecologists looked in vain for a distinctive role in post-war farming or forestry. At a joint meeting of the British Ecological Society and forestry societies, Tansley spoke from the chair of the need for foresters to understand more fully the processes by which natural succession took place. In his words, 'we have to understand nature before we can effectively control her'. While everyone agreed that there should be closer collaboration between forester and ecologist, later speakers found it difficult to identify what form it should take, particularly in view of the emphasis which the Forestry Commission was likely to place on the planting of softwood species, as opposed to the retention and establishment of deciduous woodlands (Anon. 1944a).

Speakers pointed out that foresters had been taking account of ecological factors for many years. An article on the relevance of ecology to forestry appeared in the journal *Forestry* as long ago as 1930 (Steven 1930). Harry G. Champion (*1891–1979*) had recently been appointed professor of forestry at Oxford. He recalled how, in an official report in 1933, he had advocated an ecological approach to the management of the sal forests (*Shorea robusta*) of India. By leaving some plots in their original state, it would be possible to compare the effects of climatic and edaphic factors, and the impact of fire on natural regeneration (Champion 1933).

The representatives of the British Ecological Scoiety vied with one another in expressing their humble status before 'the enquiring forester'. Eustace W. Jones spoke of how foresters, as practical men, already knew 'much of what the ecologist is still groping after'. Alex Watt described how forestry practice was far ahead of ecological theory in its understanding of how one species related to another. Ecologists still had a long way to go before they knew 'enough about the relations between species in the plant community to make generalizations' which would be useful in forestry practice. There had to be a much more intensive study of 'the static and dynamic structure of the plant community' (Anon. 1944a).

Whilst individuals might find ways of using their ecological training and research experience, there seemed little prospect of their making a distinctive and collective contribution to post-war reconstruction in forestry or agriculture. It was in this bleak context that a new and exciting prospect began to emerge in the form of nature conservation. Not only was it a topic of great intrinsic interest to many ecologists, but the involvement of ecologists in such ventures as the selection and management of a *national* series of nature reserves opened up unprecedented opportunities for promoting ecology. It was through their involvement in every aspect of nature conservation that ecologists saw their main (and perhaps only) chance of making a distinctive contribution to both science and the wider field of public debate.

3.2

The Protection of Wildlife

In order to understand how the British Ecological Society came to be so intimately involved in the conservation movement, it is necessary to trace the history of the movement back to Victorian times. As in the development of ecology, the 1890s proved a significant turning point, when, for the first time in Britain, it became possible to speak of a self-conscious movement dedicated to the protection of wild plants and animals for their own intrinsic importance (Sheail 1982b).

References were made not only to the impact of human activity on plant and animal life, but to the wider implications of allowing species to become extinct. The journal of the South-eastern Naturalists' Union published an article on 'The diminution and disappearance of the south-eastern fauna and flora' (Webb S. 1903). A note in the first volume of the *Journal of Ecology* drew attention to the value for vegetation studies of preserving the remnants of woodland bog in east Leicestershire. These instances of a 'vestigial flora' afforded the only opportunity for reconstructing the original vegetation of the countryside and for studying the genetic relationships among the plant formations (Horwood 1913).

3.2.1. *Nature preservation*

Not everyone was prepared to stand to one side as a passive observer. Some wanted to intervene and prevent the losses of species and sites. The Leicestershire naturalist, Arthur A. R. Horwood (*1879–1937*), asked, in an article on 'The extinction of cryptogamic species',

> What will be the effect in another hundred years, when many towns now
> only springing up into existence, will probably have expanded into
> veritable New Yorks and Chicagos (Horwood 1910)?

The form of response had to depend on the nature of the threat. Some losses were the inevitable result of the 'progress of civilisation'. Most commentators reserved their greatest wrath for the collector, who 'is ubiquitous, and all but omnivorous' (Boulger 1902).

Concern over the future of wild plant and animal species represented both a response to, and in part a strengthening of, the burgeoning natural history movement. According to Horwood (1910), it was important to identify and understand the reasons for the decline of species, not only from 'the biological point of view', but as a means of promoting the establishment of a body to preserve all endangered species. In an address of 1892, the president of the Norfolk and Norwich Naturalists' Society spoke of how one of the most permanently useful employments for such societies was to collect and record details of species becoming rare and extinct (Wheeler 1892).

In practice, there were three forms of response—legislation as encapsulated in the Wild Birds Protection Acts, educational campaigns as waged in the form of posters, public lectures, and lessons at school, and site protection in the shape of nature reserves or sanctuaries. The formation of the National Trust for Places of Historic Interest or Natural Beauty in 1894 seemed to hold out the possibility of a co-ordinated campaign to acquire and safeguard sites which were important for their wildlife. A member of the famous banking family, and a leading entomologist, the Honorable N. Charles Rothschild (*1877–1923*), was one of those to acquire parts of Wicken Fen for conveyance to the National Trust. In 1912, Rothschild established the Society for the Promotion of Nature Reserves (SPNR), with the express intention of identifying areas 'worthy of protection' and stimulating the National Trust and other bodies to accord those areas the necessary degree of protection (Sheail 1976, Rothschild 1985).

The preservation movement was by no means confined to Britain. In 1907, members of the British Vegetation Committee agreed to support a campaign being promoted by Professor H. Conwentz, the head of the Prussian State Department for Nature Protection, which was intended to stimulate wider interest in nature preservation. The Cambridge University Press published a book by Conwentz, called *The Care of Natural Monuments, with Special Reference to Great Britain and Germany* (Conwentz 1909). Early issues of the *Journal of Ecology* included two papers by him, the first taking the form of a brief description (in German) of a reserve in the Bohemian Forest, and the second comprising the text of an address originally given to the Berne International Conference for the Protection of Nature in 1913 (Conwentz 1913, 1914).

Oliver represented the British Vegetation Committee on the council of the National Trust. As Smith (1909b) remarked, the National Trust seemed to be widening its obvious interests in the historic and picturesque so as to include the scientific. The observation was put to the test in 1912, when the owner of Blakeney Point died, and there was a distinct possibility of the property falling into 'unsuitable hands'. Oliver had taken a lease on the site two years previously. With others, he raised the necessary funds to buy the property, which was later conveyed to the National Trust.

At the meeting of the British Ecological Society in December 1913, Oliver called upon ecologists to consider the function and management of nature reserves. Although it was probably too early to attempt 'a precise definition of a nature reserve', the time had come 'to encourage intelligent action by landowners'. So far, most reserves had been acquired by chance, rather than as a result of a considered policy. Careful thought had to be given to the management of reserve grounds. The extensive closure of areas might be resented by many people as 'a kind of game preserving' (Oliver 1914).

Both Oliver and Tansley gave their support to a survey of areas 'worthy of protection', organized by the SPNR—Tansley making it clear that his primary interest would be 'in preserving typical areas of characteristic native vegetation, whether including rare species or not' (Sheail 1976). A field excursion of the British Ecological Society in May 1914 included a visit to the nature reserve of Box Hill in Surrey, which had just been acquired by the National Trust. As Tansley reported, there was discussion as to how far the Society might undertake a 'systematic description of the vegetation of the various nature reserves belonging to the National Trust, provided suitable arrangements could be made for publication'. These descriptions might not only create interest in native vegetation among visitors to the reserves, but they would be of considerable scientific use as a record against which subsequent changes in vegetation could be compared.

Whilst the outbreak of the first world war destroyed any hope of the British Ecological Society taking an initiative, a provisional list of 273 areas 'worthy of protection' was submitted by the SPNR to the Board of Agriculture. Druce was responsible for suggesting most of the sites chosen on botanical grounds (Sheail & Adams 1980). Although supporting the concept of nature reserves, the President of the Board of Agriculture rejected any notion of the Board becoming involved in the designation of reserves. With so much land waiting to be reclaimed, there seemed no possibility of all the 'natural beauty spots' being eliminated (Sheail 1981).

Whether measured in terms of popular support, or conscious efforts to protect wildlife, the achievements of the inter-war years were meagre. Oliver may have alluded to the principal reason in his paper to the British Ecological Society in 1913. Because the country districts were not 'obviously and seriously threatened', the preservation movement lacked any 'strong popular appeal'. Oliver himself remarked on how 'those of us who accompanied the International Phytogeographical Excursion in Britain in 1911 received the impression that England is an almost empty country' (Oliver 1914). Although the amount of land converted to residential building reached unprecedented levels in the 1920s and '30s, much of the suburban and ribbon development seemed to be well away from the areas of outstanding interest for wildlife.

Among ecologists, Oliver continued to be the most ardent supporter of the preservation movement, joining forces with the indefatigable amateur naturalist, Sydney H. Long (1870–1939), in launching a successful appeal for funds to purchase Scolt Head Island off the north Norfolk coast. It was handed over to the National Trust in 1923. When the Trust refused to accept custody of a third area, the Cley Marshes, in 1926, Long appealed to local patriotism. A Naturalists' Trust was established as a special, non profit-paying company to hold and manage the marshes and other important sites as nature reserves (Anon. 1939). In the course of a discussion on nature reserves at the meeting of the British Association in 1927, Oliver prophesied that if the 'enterprise (in Norfolk) succeeded, one might look forward to the time when every county would have its County Trust' (Oliver 1928).

The impotence of the wider preservation movement was illustrated in 1929. There was no obvious successor to Frank Oliver in the preservation movement when he retired from the Quain Chair of Botany at University College, London, and soon left for an academic post in Egypt. The SPNR would have become completely moribund if it had not been for the exertions of its honorary secretary, G. F. Herbert Smith (*1872–1953*), a leading authority on gemstones and, until his retirement in 1937, the secretary of the central office of the British Museum (Natural History).

It was largely through Smith, and Geoffrey Dent (*1892–1978*), representing the Royal Society for the Protection of Birds, that the nature preservation movement quickly followed the lead of those concerned with amenity and out-door recreation in exploiting the opportunities promised by post-war recon-struction. In the spring of 1941, Smith persuaded the SPNR to convene a conference which would draw up a memorandum for presentation to the Government, setting out the objectives and organization necessary for the pres-ervation of wildlife after the war. Representatives from over 30 organizations were invited to the Conference on Nature Preservation in Post-war Reconstruc-tion, held in June 1941.

Those attending the Conference endorsed the need to incorporate nature re-serves in any post-war national planning scheme. An official body, representa-tive of all scientific interests, should be appointed to draw up detailed proposals. About 500 copies of the Conference memorandum were distributed. Despite the grim war situation, they attracted considerable interest and pre-pared the way for representatives of the Conference to join those from the Standing Committee on National Parks in discussions on post-war planning. The Government minister with responsibilities for post-war reconstruction told a deputation from the Conference in May 1942 that the principle of including nature reserves in any national planning scheme was readily accepted, but the Government was unable to give any priority to their selection. He suggested that the Conference should itself carry out the necessary investigations. With some reluctance, the challenge was accepted, and a Nature Reserves Investi-gation Committee (NRIC) was formally constituted in June 1942. It was a task that appealed to Herbert Smith and, largely through his drive and co-ordinating abilities, six memoranda were submitted to the Government by December 1945[7].

From 1942 onwards, Government ministers made increasing reference to the need to preserve amenity and wildlife, and to expand the opportunities for outdoor recreation. Not only did such aims catch the public imagination and help to bolster, albeit modestly, wartime morale, but these goals could be pur-sued with little danger of encroaching on the long-standing demesnes of other Government departments. One of the leading advocates of national parks, John Dower (*1900–47*), was commissioned to 'report on the general issues raised by the concept of national parks' (Sheail 1975). Wider discussions on the extent and character of post-war planning took on a new dimension, following the creation of a Ministry of Town and Country Planning in March 1943. Its task was to secure consistency and continuity in the framing and execution of a national policy with respect to the use and development of land throughout England and Wales. Similar planning powers were conferred on the Secretary of State for Scotland (Cullingworth 1975).

3.2.2. *Nature conservation*

At first, members of the British Ecological Society were simply content to contribute pieces of information to the survey being carried out by the Nature Reserves Investigation Committee (NRIC). Salisbury drafted in November 1942 a list of existing and proposed reserves, classified by habitat type. Pearsall exerted considerable influence over discussions on the scope and character of nature reserves in the proposed national parks of the Lake District and Peak District[8].

It was not long before the Society sought a more prominent role in discussions on the concept of nature reserves. In May 1942, Council decided to appoint a Nature Reserves Committee, which would support the work of the NRIC by putting forward a list of 'places suitable for preservation, with relevant information about them'. The Committee consisted of Tansley (chairman), A. W. Boyd, C. S. Elton, E. W. Jones, and O. W. Richards, together with C. Diver, J. Ritchie, E. J. Salisbury, and L. A. Harvey, each of whom was on the NRIC or its subcommittees.

Council soon found that this was not enough. The Society probably included 'the great majority of those best qualified to advise on the right areas to be reserved', and was uniquely placed to investigate the scientific requirements of reserve management. The Society should not only support the NRIC, but put forward its own, separate, recommendations to Government. To that end, the Council approved in April 1943 a proposal from Tansley that the terms of reference of the Nature Reserves Committee should be widened so as 'to consider and report on the whole question of the conservation of nature in Britain', with a view to the Society publishing 'a definitive report'.

A prominent role was played by Cyril R. P. Diver (*1892–1969*), both in promoting the Society's viewpoint and in maintaining close liaison with the NRIC (after an initially cool response from Smith). After the first world war, Diver took a post as clerk to the committees of the House of Commons, knowing that the very long vacations would give him plenty of opportunity to pursue his wide-ranging interests in natural history. As well as continuing his studies of the behaviour of web-spinning spiders and the genetics of mollusca (in which he collaborated closely with Arthur Boycott), Diver embarked in 1930 on a long-term ecological survey of the South Haven Peninsula in Dorset. Much of its success was due to his ability to win the co-operation of many distinguished specialists (Merritt 1971). Later described as 'an amateur giant who survived into an age of professional midgets' (Duffey 1970), Diver became a member of the British Ecological Society in 1932, and its president only nine years later.

Elton and Tansley had long recognized the need for ecologists to be intimately involved in the selection and management of reserves. Elton described the problems of reserve management in a series of radio talks that later formed the basis of a book (Elton 1933c). It was a theme which Godwin and Tansley developed in the concluding part of their monograph, 'The vegetation of Wicken Fen', published in 1929. The great variety of species on Wicken Fen was attributable to the presence of different phases in what the ecologists called the primary, secondary, and deflected successions. Far from helping to preserve the site, any attempt to prevent the cutting of the vegetation or clearing of the ditches would be 'the quickest way to exterminate most of the species it is desired to conserve'. Management programmes had to be devised, which would sustain the dynamic situation encountered on every reserve (Godwin & Tansley 1929).

In his presidential address of 1939, Tansley recalled the frustrations of having to abandon a long series of observations because of a change in the use or ownership of the site (Tansley 1939b). It was bitter experiences of this kind that caused the British Ecological Society to respond enthusiastically to proposals made by the director-general of the Forestry Commission, Roy L. Robinson (1883–1952), that ecologists should assist in setting up 'ecological reserves' on the Commission's properties. As Robinson explained, in an address to the British Association in 1938, the sites would provide opportunities to monitor changes and carry out trials. A joint committee was formed and, by the time war broke out, some 20 university departments were actively involved in choosing the sites (Sheail 1976).

As members of the Society's Nature Reserves Committee, Diver, Elton and Tansley played a major part in drafting a report on 'Nature conservation and nature reserves', which was approved by Council in October 1943, and published in both of the Society's journals in the spring of 1944. A further 750 copies were sold separately to the public. Although written from an ecological perspective, the report went to considerable lengths to emphasize the unity of purpose that lay behind the preservation movement. The most important aim had always been 'the maintenance for enjoyment by the people at large of the beauty and interest of characteristic British scenery'. The 'amenity values' of the countryside touched 'the deepest source of mental and spiritual refreshment, both conscious and unconscious'. The other values of nature preservation, namely the scientific, educational, and (less directly) economic, were at once secondary and essential to the first (Anon. 1944b).

The main reason why the reports of the British Ecological Society and the NRIC made so great an impact was their success in putting across a more positive and constructive message. The backward-looking word, *preserve*, was used less and less. Another 'ordinary' English word, *conserve*, took its place. First in America, and then in Britain, it came to imply not only protection but also the enhancement of wildlife populations. The familiarity, yet novelty, of the word reflected the history of the concept which it was meant to convey. There was nothing really new in the idea of statutory land-use planning and scientific management of land and natural resources, and yet these were perceived to be the exciting dimensions of post-war reconstruction.

The drafting of the reports and compilation of lists of proposed nature reserves ensured that there was a considerable body of data available to Government Ministers and their officials by the time thoughts turned to the implementation of wartime aspirations. The Nature Reserves Committee of the British Ecological Society found it extremely hard to decide on the criteria for recommending reserves. Efforts were made to include an example of each of the 67 principal habitat types identified. Tansley eventually imposed an arbitrary limit of 50 reserves, supplemented by a number of more extensive scheduled areas, where all activities inimical to wildlife would be banned or severely restricted.

The eventual lists included 49 national habitat reserves, 33 scheduled areas, and a further eight sites where wildlife was already protected. Nineteen of the reserves were categorized as being of 'outstanding importance', and their names are given in Figure 3.1. About half the reserves occurred east of a line linking the mouths of the Rivers Tees and Exe (the line usually taken to mark the division between lowland and upland Britain). The bias reflected both a lack of information on wildlife in the north and west, and differences in the character of the two parts. The vegetation of upland Britain was considered to be more uniform and extensive, and less threatened by large-scale disturbance.

The NRIC, which had begun its deliberations just after the worst of the London blitz, produced its own gazetteer of proposed national nature reserves and geological monuments as the V1 and V2 rockets fell on the city. The 56 reserves, of which eight were already protected, covered an aggregate 0.2% of the land surface of England and Wales. The wording of many of the entries in the lists of the British Ecological Society and NRIC was almost identical. The NRIC also incorporated the concept of scheduled areas, which were renamed 'conservation areas'. Twenty-five were identified, covering 2.6% of England and Wales[9].

Meanwhile, Dower had come to exert an increasingly powerful influence in the Ministry of Town and Country Planning. In the last days of the wartime Coalition Government of May 1945, the Minister (W. S. Morrison) sought the permission of the Cabinet's Reconstruction Committee to publish Dower's report on national parks, and to appoint (as recommended by Dower) a preparatory national parks commission. In stressing the close association between national parks and national nature reserves, the Minister outlined how the commission would be expected to appoint 'a committee consisting of one or two of their members with a sufficient number of experts in the main branches of natural history, to advise on all questions of wild life conservation and, in particular, on the possibility of national nature reserves' (Sheail 1976)[10].

FIGURE 3.1 Distribution of national habitat reserves and scheduled areas, proposed by the British Ecological Society.

Although agreeing to the publication of the Dower report (Dower 1945), the Reconstruction Committee regarded the appointment of a preparatory national parks commission to be premature. As the Minister of Agriculture warned, there was a risk of the Government committing itself 'to a larger programme than the limited available resources, especially manpower, could support'. It was suggested that the Minister and the Secretary of State for Scotland should instead appoint committees which would consider the question of national parks in greater detail, and identify where the first parks should be located[11]. With the benefit of hindsight, it was a recommendation that had the most far-reaching repercussions, both for the future of nature conservation and for the entire field of ecology in Britain.

The Minister had no alternative but to accede to the suggestion that a National Parks Committee should be appointed. In his letter of direction, the Minister asked the chairman, Sir Arthur Hobhouse, to assume that a national parks commission and series of national parks would be established. A parallel committee was appointed for Scotland, under the chairmanship of Sir J. Douglas Ramsay, which reported in June 1947, a month earlier than the Hobhouse Committee. The earlier recommendations made by Dower were endorsed. Detailed proposals were made as to the extent, management, and cost of the parks. Considerable attention was paid to the relationship of the national parks commission to central and local government (Hobhouse 1947, Ramsay 1947).

The Hobhouse and Ramsay Committees provided both the timetable and administrative context for further deliberations on nature conservation. It was expected that the Committees would appoint a nature reserves subcommittee, 'both on its own merits and so as to keep on good terms with the "Nature Reserves" and wild life interests'. Dower, who was a member of the Hobhouse Committee, insisted that if 'a committee of first class scientists' was to be assembled, it was essential that the committee should have both a high status and wide terms of reference. His view prevailed, and a Wild Life Conservation Special Committee was appointed under the chairmanship of Julian Huxley who, at Dower's instigation, was already a member of the Hobhouse Committee (Sheail 1976).

It was through this Committee, the Huxley Committee, that ecologists obtained a remarkable opportunity to publicize and stress the essentially scientific context of nature conservation in Britain. The long-term outcome was to transfer the question of nature conservation from the planning to the science sector of Government.

3.2.3. *The quest for an ecological research council*

As Diver remarked, at the Annual Meeting of the British Ecological Society in January 1946, the Society found itself extremely well placed 'to exert considerable influence' on the Huxley Committee. In appointing a committee with 'a broad ecological approach', officials in the Ministry of Town and Country Planning chose four members straightaway: Tansley and Elton to cover plant and animal ecology respectively, E. Max Nicholson to cover the ornithological aspects, and Diver as the 'all-round field naturalist'. They were joined by Edmund Ford (an ecological geneticist in the department of zoology and comparative anatomy at Oxford) and John S. L. Gilmour (*1906–86*) (director of the Royal Horticultural Society's Garden at Wisley). A proposal to include a marine ecologist was dropped so as 'to avoid offence to the Ministry of Agriculture and Fisheries'.

In its report of July 1947, the Huxley Committee accepted most of the recommendations made by the NRIC for national nature reserves and conservation areas (which were renamed scientific areas). Further sites were investigated. After the summer meeting of the British Ecological Society in 1946, Tansley and John Hope-Simpson motored round the Lincolnshire and Yorkshire Wolds, where they found 'plenty of chalk grassland demanding exploration'. Because so little survey work had taken place in Scotland, the special committee appointed under the chairmanship of James Ritchie, now professor of natural history at Edinburgh, was unable to make its detailed recommendations until 1949 (Huxley 1947, Ritchie 1949).

Right from its first meeting, the Huxley Committee debated whether nature conservation should be the responsibility of a national parks commission, or whether it should be placed under the aegis of a research council. In making such a choice, much more was at stake than the competent management of nature reserves. If ecologists could establish themselves as the primary body responsible for devising and implementing a national programme of nature conservation, they would have gone a long way towards transforming themselves into a new and powerful force. They would have secured what had so long eluded them.

In its earlier report of 1944, the British Ecological Society had endorsed the view of the NRIC that 'the Government should take formal responsibility for the conservation of native wild life, both plant and animal'. It had, however, rejected any notion of the Nature Reserves Authority operating in parallel with the National Parks Authority under the Ministry of Town and Country Planning. Nature conservation could not be treated as a mere adjunct of statutory planning. The reserves would be a failure unless there was a permanent scientific staff and trained assistants. A large research programme would be needed to answer even the most elementary questions as to how the reserves should be managed. The only way to secure the manpower and resources would be to set up a wild life service under the Lord President of the Council, the Government minister with responsibility for science.

A key figure in developing this approach within the British Ecological Society was Charles Elton. He had been involved in the drafting of a memorandum circulated by the Universities Federation for Animal Welfare (the successor to ULAWS). This had advocated the appointment of a statutory wild life service which would stand above sectional interests in adjudicating how animals should be treated. It was suggested that the new body should be placed under the Lord President of the Council. In the course of research on rodent control, carried out by the Bureau of Animal Population, Elton had become impressed by the potential of a research council for co-ordinating and leading the research effort in this and other fields of biological investigation. Not only could a research council take up initiatives in the theoretical and practical aspects of science with the minimum of 'red tape', but it had, unlike most Government departments, responsibilities on both sides of the Scottish border.

With the publication of its report on 'Nature conservation and nature reserves' in 1944, the British Ecological Society turned its attention to the much wider issue of 'wild life conservation and ecological research'. A draft memorandum began to take shape, which proposed that a biological service and a new ecological research council should be formed. The biological service would administer the national nature reserves. It would also promote 'fact-finding research' and disseminate advice on the conservation, regulation, control, and management of plant and animal populations generally. The report envisaged at least one institute for terrestrial study, equipped with laboratories 'for the comprehensive study of ecological and population problems both pure and applied, including the techniques of regulating populations, and the development of humane methods of pest control'.

It was envisaged that the proposed ecological research council would have a similar status and constitution to the existing research councils. As well as being responsible for the biological service, it would promote ecological research generally. It would be free to spend grant-in-aid at its own discretion, undertake research either in its own research institutes or elsewhere, and act as a clearing house for ecological information.

It was already clear from discussions with the Royal Society and the Agricultural Research Council that there would be opposition to the formation of a new research council. The meeting of the Society's Council in January 1945 considered submissions from Pearsall and Robert C. Maclean (*1890–1981*), the professor of botany at Cardiff, putting forward less contentious courses of action. The Council decided to persist in calling for 'the full Ecological Research Council', but proposed that the alternative suggestions might be held 'in reserve as a second and possibly more palatable string'. In an addendum to the Society's memorandum, it was stressed how the aim of the proposed council would be to *extend* the field of biological research, and not to *usurp* the functions already allocated to the existing councils and Government departments. The revised memorandum was issued to all members of the British Ecological Society, separate from the journals, and sold to the public for sixpence a copy.

Meanwhile, Tansley had finished writing a book which gave considerable prominence to the views expressed by the Society. Published by the Cambridge University Press, this extremely attractive, slim volume was called *Our Heritage of Wild Nature: a Plea for Organised Nature Conservation.* The typescript, completed in October 1944, was read by Elton, Diver, and Godwin. So as to give 'concreteness' to the arguments developed for the conservation of wildlife, the book described the main kinds of vegetation and those animals over which there were 'conflicting human interests' (Tansley 1945).

Tansley became the chairman of the Huxley Committee following the appointment of Huxley to UNESCO in May 1946. In coming to the unanimous conclusion that nature conservation should be made the responsibility of the Lord President of the Council, the Huxley Committee added a further complication to what was already becoming an extraordinarily confusing interplay of personal and institutional initiatives. It was the Committee's responsibility to advise the Minister of Town and Country Planning—not the Lord President of the Council, who had his own advisers, namely the Cabinet's Scientific Advisory Committee. The latter had already begun to consider the question of nature conservation, prompted by the report of the British Ecological Society on 'Wild life conservation and ecological research', and more particularly by a submission from Salisbury, in his capacity as biological secretary of the Royal Society.

In that submission, Salisbury had emphasized the importance which the Royal Society attached to the provision of nature reserves, and how their preservation was 'dependent upon highly technical considerations and can only be assured if this is carried out under expert guidance'[12]. In putting forward these views, Salisbury was able to draw on both his own first-hand involvement in research and conservation issues (he had become director of the Royal Botanic Gardens at Kew in 1943), and on the wide-ranging discussions that had been taking place within the Royal Society on the post-war needs of biology.

The Council of the Royal Society had earlier appointed a committee to review the post-war needs of biology. The chairman was Fritsch, and the members included Pearsall and Salisbury. In its report of January 1945, the committee found research facilities, except in the traditional fields of morphology and physiology, completely inadequate. It had become increasingly evident that fundamental research had to be directed towards the study of living organisms and their relationship to natural surroundings. It was essential that facilities were established for field research in terrestrial ecology, similar to those already available for marine and freshwater ecology at Plymouth and Wray Castle respectively. Within the universities, there was need for field stations and courses in field survey and experimentation. All these research and teaching activities would depend for their success on access to nature reserves, where the flora and fauna could be studied under natural conditions.

The Cabinet's Scientific Advisory Committee responded to such pressures by inviting one of its members, John C. F. Fryer (*1886–1948*), to compile a note on 'Nature conservation and research on the British terrestrial flora and fauna'. Fryer had been appointed secretary of the Agricultural Research Council in July 1944. In his note of September 1945, he emphasized the close interdependence of nature conservation and 'the scientific study of biology'[13]. Here again, the personal interests of a key figure can be discerned. The son of a famous naturalist, Fryer had been one of the few active members of the SPNR. In his attempts to re-establish the large copper butterfly on the Woodwalton Fen nature reserve in Huntingdonshire from 1927 onwards, he had experienced at first hand the difficulties of managing a nature reserve (Edelsten 1950).

Fryer's note formed the basis of a report which the Cabinet's Scientific Advisory Committee submitted to the Lord President of the Council, Herbert Morrison, in July 1946. The report emphasized how 'a comprehensive national policy on biological conservation and control' would benefit the field of pure science and a wide range of practical problems in forestry, agriculture, land drainage, water supply, and freshwater fisheries. To be effective, such a programme required a series of national nature reserves, the fostering of research through an ecological research institute, and the creation of a service of fully qualified scientific officers to manage the reserves, act as a clearing house for information, and provide an advisory service generally.

These proposals made the task of the Huxley Committee much easier. Knowing that they were being put forward to the Lord President of the Council, the Huxley Committee recommended in its own report that a biological service should be established, responsible for selecting, acquiring, and managing nature reserves, and for carrying out the survey and research work required. It would serve as a central bureau of information, operate an advisory service, and encourage educational activities. The Committee endorsed the proposal to set up a series of ecological research institutes, recommending one in Scotland and four in England and Wales. One might be in the Isle of Purbeck, undertaking 'intensive pure research'. Another might be based on Cambridge where, through at least four out-stations, it would engage in extensive and comparative work. A third would be based on Oxford, and focus on fundamental research and act as a centre for postgraduate and training courses. The fourth might be in Snowdonia or the Lake District, where it would work closely with the Freshwater Biological Association (Huxley 1947).

The relationship between the biological service and these institutes was left rather vague. Although they were expected to work closely together, members of the Huxley Committee thought it would be wrong for the biological service to 'dictate the research to be undertaken'. Such a viewpoint reflected both the practical experience of such figures as Charles Elton, and a series of more fundamental objections. Drawing on the experience of the Bureau of Animal Population, Elton affirmed in 1947 that the most productive research groups were those where individuals were given 'the utmost freedom to try unpromising-looking ideas', to engage in 'fool experiments', and to break away from the 'hardening traditions' of their group.

From the late 1930s onwards, and with the example of Russia very much in mind, there was considerable debate as to how far some kind of central control should be created in Britain, so as to ensure that scientists contributed more fully to the life and needs of society. The publication of a book on *The Social Function of Science* by John D. Bernal (*1901–71*) in 1939 aroused great interest (Bernal 1939). One of the most forthright critics of the book, and of what was perceived to be its implied threat to independent research, was John Baker of the department of zoology and comparative anatomy at Oxford (Baker 1939). He was soon joined by Tansley and the Manchester scientist and philosopher, Michael Polanyi (*1891–1976*). Together the three men formed in 1940 the Society for Freedom in Science.

It was not the first time that Tansley had spoken out against what he perceived to be the trend towards totalitarianism in science. Although the natural sciences 'usually followed utilitarian *impulses*', Tansley insisted, in his inaugural lecture in Oxford in 1927, that there must be 'freedom from utilitarian *compulsion*'. In his Herbert Spencer Lecture to the University of Oxford in 1942, on 'The values of science to humanity', he again emphasized how 'the highest type of research, that which has been productive of the most fundamental discoveries, is essentially the work of individual minds, freely dealing with their chosen material' (Tansley 1927, 1942).

The Huxley Committee envisaged the biological service being 'attached' to the Government through a nature conservation board, made up of scientists and a few members with experience in the educational and recreational fields. Having decided that it was 'not practical politics' to press for the status of a full research council, the Committee concluded that the board should be made responsible to the Agricultural Research Council, with the expressed hope that it could soon become 'self-contained and independent'.

Throughout this period, a crucial role was played by Max Nicholson, who, shortly after becoming a member of the Huxley Committee, took charge of economic co-ordination within Herbert Morrison's office. Between 1942 and 1945, Nicholson had been head of the Allocation of Tonnage Division in the Ministry of War Transport. An Oxford history graduate, he had moved to London in 1931 for a career in journalism (Donoughue & Jones 1973, Pinder 1981). His interest in ornithology developed while at Oxford, where he played a leading part in founding the Oxford Ornithological Society and pioneering a series of bird censuses. He was the author of a number of books on birdlife (Nicholson E. M. 1926).

In April 1948, Morrison submitted the proposals of the Scientific Advisory Committee and the Huxley Committee to his own Lord President's (Cabinet) Committee. He wrote of how he was impressed by 'the broad unanimity reached by successive bodies of experts'. Their proposals could be justified by the fact that

> the Government is constantly taking action liable permanently to affect the fauna, flora and even the geography of the country without having at its disposal any channel of authorative scientific advice about the probable results of its action, such as is available in all other fields of natural science.

The Cabinet Committee approved the recommendations of the two committees[14].

The Government's acceptance of the need for an official nature conservation body was followed by what Nicholson described as 'interminable administrative and legislative difficulties'. It was not until February 1949 that Morrison announced to Parliament that the proposed biological service and nature conservation board would be subsumed into a single body, called by the 'more convenient title' of the Nature Conservancy. In a letter, Nicholson described how the dropping of the word 'board' was a further step to avoid giving any impression of its being 'an additional research council, or a new government authority in its own right'[15].

The most obvious way of giving the Nature Conservancy the necessary statutory powers was to incorporate them in legislation needed to implement the recommendations of the Hobhouse Committee. The powers to designate national nature reserves formed the most minor part of the National Parks and Access to the Countryside Bill of 1949. In a measure that 'bristled with contentious points', the modest wildlife clauses aroused very little comment in Parliament. Whereas the Bill was expected to lead to the designation of over 10% of England and Wales as national parks, the proposals for nature reserves affected only 70 000 acres (28 000 hectares), most of which had little economic value. The cost of acquiring them was estimated to be only £500 000.

By seeking to create a scientific body of apparently modest status and requirements, the nature conservation movement avoided becoming embroiled in the highly sensitive question of the balance of power between the National Parks Commission and local authorities, and between central and local government. Although determined to preserve and extend their responsibilities for planning, both the local authorities and Ministry of Town and Country Planning were relieved to learn that the new and ill-defined responsibilities for nature conservation had been given to others to discharge. Unlike the remainder of the National Parks and Access to the Countryside Act, the nature-conservation sections applied to Scotland, as well as to England and Wales (Sheail 1984b).

It was a legal point that finally settled the relationship of the Nature Conservancy to the Agricultural Research Council. In order to discharge its statutory powers under the Act, the Conservancy had to be 'an incorporate body in its own right'. By making it directly responsible to the Committee of the Privy Council, the Conservancy became in all but name a research council. Among the 15 charter members of the Nature Conservancy, seven had played a major role in ecology, namely Tansley, Elton, Ford, Godwin, Maclean, Matthews, and Pearsall. Tansley became chairman of the Conservancy, and received a knighthood in the New Year's Honours for 1950. Diver was appointed the first director-general. He was succeeded by Nicholson in 1952.

For the British Ecological Society, these were the first fruits of a campaign designed ultimately to give ecology a larger role in post-war Britain. Before recounting how far the great expectations invested in the infant Conservancy were borne out, it is necessary to look more explicitly at the affairs of the Society and the course of ecological research generally.

3.3
The Business of the Society

It becomes easier to discover how, why, and when the Council and officers of the Society took decisions and followed particular courses of action. As membership grew and activities increased, there was a need for more formal procedures. Much more had to be written down and kept. There was often 'no convention except tradition' for officers to follow. Drawing on his experience of over 10 years as Council secretary, David Le Cren wrote of how, 'provided things go fairly smoothly, the membership in general do not want to be bothered much about the administration of the Society'. No one had ever contested the Council's nominees for office[16].

The Society reached a significant milestone in its history when Tansley finally 'bade farewell' to the Council in 1947, having served for 33 years in a variety of capacities. He marked the occasion with a paper in the *Journal of Ecology* on 'The early history of modern plant ecology in Britain', in which he remarked on the striking contrast between the conditions of the late 1940s and the 'time of quiescence' that had followed the first world war (Tansley 1947a).

The membership of the Society continued to rise, passing 'the 500 barrier' in 1946, and reaching 768 in 1952. Both journals were taken by 124 members. The *Journal of Ecology* was taken by a further 412 members, and the *Journal of Animal Ecology* by 221 members. The Society's records indicated that there were 559 members resident in the British Isles, and 209 overseas. In a more detailed analysis of 1955, by which time membership had grown to 850, it was found that 617 lived in the British Isles (200 in London and the 'dormitory counties'), 36 in the rest of Europe, 75 in Africa, 60 in North America, 31 in Australia, and five in South America.

Membership continued to be 'open to all personally and genuinely interested in Ecology, and who wish to receive one or both of the Society's journals for their own use'. Their names, together with those of a proposer and seconder, were submitted for election at a meeting of Council. The annual subscription of 25 shillings entitled members to receive either journal (or both for 45 shillings), attend meetings, vote in elections, and be eligible for election to Council and offices. Those seeking associate status paid a subscription of 7/6d, which enabled them to attend meetings. It was the custom to grant honorary associate status to members of over 25 years' standing, who might otherwise retire from membership of the Society.

The policy of the Society was 'to view ecology as one science in its own right and to take a fairly catholic view as to what ecology includes'[17]. The presidency alternated between a botanist and zoologist. The duties of an ordinary member of Council were hardly onerous—the Society's rules simply required him to attend at least one of the three meetings held each year. In a conscious attempt to reflect the growth in membership, the maximum number of Council members was raised in 1949 from 20 to 25. In supporting the change, Watt spoke of 'the desirability of increasing the scope of membership of the Council, so that a wider range of related branches of science should be represented'. It was felt that 'foresters, soil and grassland workers, and the like', were particularly under-represented.

Not surprisingly, 'the longer-term planning of the society's activities' fell to those holding the posts of secretary, treasurer and editor, particularly where they held office for a long time—Summerhayes retired as treasurer in 1957, after 19 years in the post. The practice of appointing joint secretaries and treasurers helped to spread the workload, and increased the chances of some continuity in the event of resignations or temporary absences overseas. Some specialization evolved. The 'zoologist' secretary acted as secretary to Council, while the 'botanist' secretary served as membership secretary.

As the membership rose to over a thousand in 1957, it became even more anomalous that the day-to-day handling of membership records, subscriptions, and journals should be handled by honorary officers, 'who have plenty of more valuable ecological work to do'. Council members and officers complained of the growing number of agendas, minutes, and papers. Even more worrying was the growing number of instances where deadlines were missed at the printers. It was clearly time for the informal arrangements of the past to be replaced by administrative procedures of a more formal nature[18]. Serious thought was given to placing 'the background organization' of the Society in the hands of a permanent secretariat, such as the Institute of Biology, which was already providing a similar service for some other societies. There was no escaping the fact that this would involve additional costs, but, as Frederick H. Whitehead (one of the treasurers) commented in September 1960, the days were past when a learned society could expect its administration to be done on the cheap. The Society was 'sufficiently viable financially to afford such assistance'. Council remained unconvinced[19].

Together with the Ecological Society of America, the British Ecological Society represented 'a large section of the world's active ecologists'. The question was how far the Society was prepared to involve itself in international affairs. The founding of international organizations, and the holding of congresses, had never found much favour among members of the Society. As Watt commented, it was reasonable enough that some interchange of views and coordination of effort should take place. The holding of occasional congresses was one way of achieving this, but the published proceedings were often very disappointing. More solid gains would come 'from the publication of reviews in particular fields by active workers, or by the publication of a series of papers on a given theme by people working in the field'. Having canvassed the views of

Council, David Le Cren concluded that 'congresses seem almost but not quite universally unpopular'[20].

The occasion for further debate was a request that the Society should support a proposal to form an ecological section within the International Union of Biological Sciences (IUBS). At its meeting of January 1960, the Council of the British Ecological Society learned that there was a strong possibility of the forthcoming General Assembly of the IUBS agreeing to such an initiative. Given that possibility, the Society's practice of opposing the formation of such bodies had to be reappraised. As the Society's president, James B. Cragg, wrote, it was essential that 'academic ecology in the true sense of the word' should be adequately represented. The section was duly formed by the General Assembly in July 1961. As many in the Society forecast, it seemed to make little material difference to the fostering of international relations in ecology.

3.3.1. *The journals*

Sales of the Society's journals returned to their pre-war levels in 1947, with both editors reporting an abundance of good papers. The greater volume of wartime research in animal ecology meant that the *Journal of Animal Ecology* was far less dependent on overseas contributions than the *Journal of Ecology*. The problem was one of balance. In his editorial report, Elton called for 'more papers on invertebrates and on marine and freshwater ecology'. Pearsall's hopes of publishing papers within a year of their acceptance were thwarted in 1947 by the printing delays caused by the national fuel crisis.

Editors continued to appeal for shorter papers. An attempt was made to impose a limit of 15–20 printed pages on the *Journal of Ecology*. The new editor, Harry Godwin, drew attention to the practice of geologists in depositing the details of 'suites of rocks' in a central library, and of including 'only carefully selected instances of such descriptive material' in their published papers. Godwin suggested that the long descriptive lists of plant communities should be similarly consigned to a reference library. Not everyone agreed. When the length of papers was again discussed in 1956, Eustace Jones spoke of how 'it was difficult with descriptive ecological material to present the relevant scientific information in a short paper'. All sense of continuity was lost if a major piece of work had to be split into a number of shorter papers.

Spurred on by a warning of a further increase in printing costs, a standing committee was formed in 1955 to deal with 'the increasing business connected with the journals'. It was made up of the editors, Le Cren, Summerhayes, and Diver. Its main task was to discover whether the journals could be published more cheaply, but to equally high standards. The Council accepted the committee's unanimous recommendation that the journals should be transferred to the publishing house, Blackwell Scientific Publications of Oxford. The decision took effect in 1956, and the committee became a Standing Committee on Publications.

An increase in membership subscription could no longer be postponed. It had remained at 25 shillings per annum ever since 1932. For many years, it had been possible to absorb the higher printing costs by increasing the subscription rates paid by non-members (mainly libraries) for the journals. The year 1954 was, however, the last to show a surplus. Inflation and the need to increase the size of the journals led to a rapidly worsening situation. The annual deficit on the journals rose from £770 in 1954 to £1697 in 1956, and to £6911 in 1957.

The cost of printing 2200 copies of the *Journal of Ecology* in 1958 was expected to be £5300, compared with sales of about £3500 to non-members and an income of £500 from back numbers—leaving about £1300 to be found from membership subscriptions and other sources. The position of the *Journal of Animal Ecology* was not quite so bad. Against an outlay of £2200, there was likely to be an income from sales of £2000, plus £1300 from back numbers. Council tried to arrest the situation by increasing the annual cost of the *Journal of Ecology* to non-members from 50 shillings to 84 shillings, and the *Journal of Animal Ecology* from 45 shillings to 63 shillings. Membership subscriptions were also raised to a level 'more nearly approaching the costs of the journals received'. A rate of 40 shillings was levied on those taking one journal, and 70 shillings on those taking both. The fact that only 8% of members resigned was taken to be a sign that most members agreed the increases had been long overdue.

Without the sale of back numbers, the increase of subscription rates would have been even higher. One of the most urgent tasks of the Publications Committee was to decide how many extra copies of current issues should be printed for 'deferred sales', and how many old volumes should be reprinted. Librarians tended to buy back numbers for only as far back in time as complete runs were available. The Committee decided to have reprinted those volumes of the *Journal of Ecology* which had been printed in very low numbers as a result of wartime paper shortages. Complete sets of volumes would then be available for as far back as 1929. Together with the reprinting of three volumes of the *Journal of Animal Ecology*, the cost came to over £4000. This meant realizing some of the Society's investments. In the words of the secretaries' Annual Report, Council concluded that back numbers of the journals were a more valuable investment than gilt-edged stocks[21].

There was clearly no scope for replenishing the stocks of earlier volumes of the *Journal of Ecology*. In 1959, the Society sold the remaining stocks of volumes 1 to 30, together with the right to reprint them, to the Johnson Reprint Corporation of America. The income from this sale and a grant of £1000 for three years from the Royal Society, together with the fact that the *Journal of Ecology* made a loss of only £1017, enabled the Society to discharge all its 'printing and publishing debts' in 1959, leaving a deficit for the entire year of only £212.

The financial plight of the Society remained desperate as printing and publishing costs continued to rise. The production costs of *Journal of Ecology*, which had been £2983 in 1954, rose to £4861 in 1959, and to £7172 by 1961. The cost of the *Journal of Animal Ecology*, which was £1863 in 1954, had reached £2490 in 1959 and £4285 in 1961. The subscription rate for non-members was increased from 63 shillings to 126 shillings in 1962. The *Journal of Ecology* went up from 84 shillings to 110 shillings in 1960, and to 130 shillings in 1962. As a result of these moves, the Society managed 'to just break even in 1960'. A surplus of £700 was achieved in 1961, bringing the Society's reserves to a total of just over £4000. An overall surplus of £1690 was recorded in 1963, with the loss on the *Journal of Ecology* reduced to £900.

The editors' response to the financial crisis was to seek economies in the use of space in the journals. Changes in typeface and format helped. The list of members and their addresses was published every two or three years, rather than annually. A notice in the volume for 1959 announced that Council had decided 'to encourage the submission of concise papers, especially those dealing with experimental ecology'. Priority would be given to those of less than 10 pages of the *Journal of Ecology*. As the editor reported later, 'some good papers' had to be rejected, simply on the grounds of their length.

These savings did little to solve the more fundamental problem, namely that the only way to publish a wide spectrum of papers within a reasonable period of their submission was to increase the size of the journals. The need to tackle the problem of the size of the journals was highlighted in early 1961, when it was learned that a commercial publisher was considering whether to launch a new journal devoted entirely to insect ecology. Not only was such a venture likely to reduce the sales of the Society's journals, and possibly membership, but it could easily presage 'the breaking up of the subject into taxonomic units'. As David Le Cren emphasized in February 1961, the Society had always 'stood for ecology as one subject'. The tradition had 'been a very valuable restraining influence against the pressure of the splitters and specialists', but it could only be sustained if the Society's journals and meetings continued to fulfil the legitimate aspirations of the membership[22].

Despite the plight of the Society's finances, the decision was taken to enlarge the *Journal of Ecology*, and to publish each volume in three parts from 1957 onwards. The *Journal of Animal Ecology* was similarly expanded from 1962 onwards. Discussions were held as to how far the appointment of editorial boards might help to expedite the consideration of submitted papers, and ensure that each sector of ecology received adequate prominence. Such discussions soon led to misgivings of another sort. As Le Cren wrote, it would be wrong 'to have "sections" in the *Journal of Animal Ecology* or any agreement that would tie the editor to reserve say 30% of the space for entomological purposes'[23].

In the event, a compromise was effected, whereby editors were encouraged to draw on as much assistance as they felt necessary. As editor of the *Journal of Ecology*, Harry Godwin was joined by David E. Coombe in 1954. Paul Richards took over from Godwin in 1958, and both Coombe and Richards were replaced by Peter Greig-Smith in 1964, who drew on the 'assistance' of J. L. Harper, J. W. G. Lund, P. J. Newbould, P. W. Richards, and A. J. Rutter. Derek A. Ratcliffe was appointed 'editorial assistant'. The founder-editor of the *Journal of Animal Ecology*, Charles Elton, retired in 1952, and was succeeded by H. Cary Gilson, the director of the Freshwater Biological Association. On his retirement in 1963, Owain Richards and James Cragg became joint editors, drawing on the 'assistance' of D. J. Crisp, M. E. Solomon, E. D. Le Cren, and H. N. Southern (Taylor & Elliott 1981).

3.3.2. *Meetings and symposia*

For the Society, post-war reconstruction began in 1946 with two meetings. The summer meeting was the first to be held since 1939. It was convened at Sheffield, where 40 members spent three days visiting sites in the area, listening to evening lectures, and studying a 'considerable series of ecological exhibits'. The itinerary drawn up by the staff of the botany department included a visit, led by Verona M. Conway (*1910–86*), to the raised bog at Ringinglow. The first of two papers on the site appeared in the *Journal of Ecology* in 1947 (Conway 1947).

A meeting of a very different kind was held later in the year. Keen to restore the links with scientists of other countries, Harry Godwin secured sufficient funds from the British Council to invite '5 distinguished workers from West-European allied countries' to attend a meeting on the 'survival and extinction of flora and fauna in glacial and post-glacial times'. The opening paper, on 'The temperate Irish flora', was given by Knud Jessen, who published the final results of his inter-war investigations on the 'Irish post-arctic deposits' some three years later in the *Proceedings of the Royal Irish Academy* (Jessen 1949).

Although the honorary secretaries were closely involved in choosing the location and subjects for meetings, the more detailed arrangements were delegated to members of the host departments. In 1955, it fell to Charles H. Gimingham of the botany department at Aberdeen to organize the summer field meeting. In correspondence, David Le Cren suggested that preference should be given to visiting communities not likely to be found elsewhere, and those which had been described in recent publications. One or two research stations might be included in the itinerary. From past experience, members seemed 'to stand up pretty well to poor weather and rigorous walking provided they see interesting things'. It was useful to distribute beforehand a certain amount of literature on the various places to be visited[24].

A marked feature of the late 1940s was the number of meetings held jointly with other organizations. Not only did such gatherings bring together a much wider variety of speakers and discussants, but they fulfilled a wider role of attracting new members and drawing attention to the Society and its journals. The proceedings of the day meeting with the Royal Meteorological Society in May 1946 were reported not only in the journals of the two societies, but also in *Nature* (Day 1946). 'The organic matter of the soil' was the title chosen for a joint meeting with the British Society of Soil Science, organized by Alex Watt in March 1948. An equal number of papers was given by members from each society. In the words of the report on the meeting, the papers were of peculiar value in showing 'the extent of the gap which remains between the approaches to the subject by the soil scientists and the ecologists'.

The numbers attending Annual Meetings rose from about 80 in 1948 to 120 at the 'well-attended' meeting of 1952. It became a tradition to organize some of the papers around a theme. At the meeting of 1952, George Varley introduced a symposium on 'population problems'. The six speakers included Alex Watt on competition and community structure, David Lack on the population dynamics of birds, and Dennis Chitty on the population dynamics of mammals. The greater part of the Annual Meeting of 1956 was taken up with a series of papers on ecological–genetical problems, introduced by William Turrill.

Good attendances could not be taken for granted. Over 150 members came to the Annual Meeting of 1957, with a further 40 at the spring meeting at Cambridge, and 60 taking part in the summer excursions from Exeter. On the other hand, the autumn meeting to the Isle of Man had to be cancelled owing to the extremely small number of bookings. Only 15 members attended the summer meeting of 1958, and the autumn meeting again had to be cancelled. The disappointing numbers were interpreted as further evidence of the recent increase in specialist societies, and the competition arising from their activities. Council concluded that the best form of response was to continue promoting 'ecology as a whole science', and to arrange joint meetings with other organizations.

A questionnaire was sent to all members of the Society, both in 1954 and 1956, seeking suggestions for symposia topics. Only 57 members responded on the first occasion, and 114 on the second. One of the most obvious topics for a meeting in the mid-1950s was 'Production and the dynamics of biomass, food chains and energy relations'. A well-attended, two-day symposium was held on the subject in March 1956, organized by David Le Cren, whose own research at the Freshwater Biological Association (which he joined in 1947) was focused on the dynamics of perch populations in Windermere. Le Cren defined the main aim of the meeting as 'to review the state of knowledge of this aspect of ecology in the various rather discrete fields, to bring out some common ideas and principles, review the prospects for future work, and generally to stimulate interest'[25].

In order to make 'the whole symposium a meeting ground for those working on different habitats, both from fundamental and practical aspects', Le Cren sought to devise a programme made up of both botanical and zoological contributions on the aquatic and terrestrial environments. It was regarded as important to have some papers which gave factual descriptions of work done and quantitative results obtained, as well as a few which tended 'to be more theoretical and have a broader sweep'. The goal was largely achieved.

An initiative of a different sort came from a proposal made by David A. Webb that the British Ecological Society should mount an ecological investigation of the Burren region of County Clare, in southern Ireland. Elected to the chair of botany at Trinity College, Dublin, in 1950, Webb had published 'extensively on Irish floristics and ecology' (White J. 1982). He believed a Society-sponsored investigation of the Burren would not only benefit Irish research by encouraging many more ecologists to visit the country, but would provide an unparalleled opportunity for participants to study plants in an extremely oceanic climate. Council members were, however, far from keen, commenting that if the precedent of a sponsored study was to be established, there had to be a new and distinct approach. It had to set 'a new fashion in such studies'[26].

A subcommittee of Council made a four-day visit to the Burren in April 1957. It consisted of David Webb, Martin W. Holdgate, and C. Donald Pigott, who were clearly impressed by the way many of the ecological features occurred in 'a clearer and a more acute form' than anywhere else in the British Isles. A sufficient number of university departments responded to a circular to suggest that the mounting of an investigation would be worthwhile. The unanimous support of Council was finally obtained in January 1958. A grant of £25 was allocated to an enlarged committee, under the chairmanship of Roy Clapham. Further grants totalling £1000 were obtained from the Royal Society of London and Royal Irish Academy. An excursion was led by Webb in 1959, and a short symposium on 'The ecology of the Burren' was held during the Annual Meeting of the Society in December 1961.

In a progress report to Council in April 1964, Webb described the survey as having been 'half success, and half failure'. Not only did it prove difficult to find enough people to do the systematic ecological work, but the launching of the study had taken so long that, by the time work began, Webb himself could no longer act as organizer. Fourteen pieces of work had been carried out. There had been three expeditions from Nottingham University, and a student party led by Donald Pigott investigated the rooting systems of certain calcifuge species on the limestone. By the mid-1970s, the balance remaining in the Burren Fund was so small that the investigation was terminated, and the residue used to compile and publish a bibliography on the Burren. Nine of the publications listed had received financial help from the Fund (Malloch 1976).

The Council of the Society decided in October 1958 to appoint a small Programme Planning Committee, which soon decided that the optimum number of meetings in a year was three, namely a winter meeting devoted mostly to papers on all aspects of ecology, and usually held in London; a summer meeting for excursions concentrating on field ecology; and at least one symposium focusing on a subject selected for its topical interest and fundamental importance. By adopting such a policy, the Committee hoped to preserve the traditions of the Society's meetings and, at the same time, adapt to 'present day conditions when scientific meetings are frequent and interests tend to be specialized'.

The policy proved a considerable success. The Annual Meeting of January 1959 included the presidential address and a wide variety of papers, and was attended by about 100 persons. The spring meeting (with 70 participants) was devoted to a series of 10 papers on freshwater subjects, and included a tour of freshwater habitats in the vicinity of Reading. As Le Cren commented, comparisons made between this and an earlier meeting of 1952 illustrated how far there had been a shift of emphasis over the intervening period. There had been a move away

> from purely field observation on the distribution of different species towards a more detailed and realistically quantitative approach, dealing with one or a few species and involving laboratory and field experimentation as well as observation.

The summer meeting of 1959, which attracted 40 members, was based on Galway, where David Webb led a series of excursions, including a two-day visit to the Burren.

The initiative for what became the first of a much more ambitious series of symposia was taken by John L. Harper of the Department of Agriculture in Oxford. In a letter of April 1958, Harper wrote of his keenness to organize a symposium on 'The ecology of weeds'. In this, he was supported by the British Weed Control Council, whose members felt that public concern over the indiscriminate spraying of weeds on roadside verges and elsewhere had taken too little account of the biology and ecology of weed species. Inquiries revealed that between 18 and 25 people were working on 'the Ecology, Gene-ecology and Evolution of weed species'. By organizing the meeting under the aegis of the British Ecological Society, Harper believed it would be possible to reach 'the widest possible audience of interested persons'[27].

Apart from welcoming the general idea, Council was far from enthusiastic. As Le Cren reported, it was 'the Society's policy to discourage the division of ecologists into Botanists and Zoologists'. Although there would be no shortage of material for a symposium on weeds, it was essential that there should be a zoological content. An obvious course was to widen the scope of the meeting to include 'weeds and pest ecology'. The Society was already discussing the possibility of a joint symposium with the Association of Applied Biologists. Whatever the choice of topic, Council would be extremely reluctant to commit the editors of the *Journal of Ecology* to the publication of the symposium papers at a time when the journal was losing money and was so short of space. Harper persisted, and the three-day meeting was held in April 1959 at Oxford, where 160 participants heard 24 papers on 'The ecology of weeds'.

The obvious success of the symposium was attributed to 'the correct breadth of the subject that had to be covered', and to the presence of participants from a wide variety of backgrounds, both at home and overseas. Sections of the programme dealt with taxonomy and evolution, the dormancy and dispersal of seeds, population studies, interference and competition, autecological studies of weed species, and special weed problems. Because there had already been an 'abundance' of conferences on weed control, the question of chemical controls was deliberately excluded from the programme. At a meeting held during the symposium, Council authorized John Harper to proceed as quickly as possible in the editing and publication of the proceedings on the basis of a 'half-profit' agreement, already negotiated with Blackwell Scientific Publications, with the recommendation that 'the Society's name should be adequately displayed on the title page'. The volume of 256 pages was published in 1960, assisted by grants from five manufacturers and suppliers of chemical products. By the end of 1961, the costs of publication had been covered (Harper 1960).

In order to capitalize on the experience, the Programme Planning Committee was made a standing committee, and straightaway drew up a tentative programme of symposia for the next three years. There was no shortage of ideas. As Le Cren commented, the main need was to identify subjects that were on the point of becoming topical, and which cut across the usual divisions of ecology so as to bring together both people and ideas. From a practical standpoint, it was essential to identify a suitable organizer at the same time as the subject of a symposium was decided. By guaranteeing half the costs, and receiving half the profits of each symposium volume, the Committee hoped the series would soon become self-supporting.

The second symposium was organized by Le Cren himself, on the theme of 'The exploitation of natural animal populations'. In his invitation to potential contributors, Le Cren wrote, 'I am sure the time is ripe for the development of *comparative* population dynamics and for ecologists working with different types of animals to get together'[28]. The symposium, held in Durham during March 1960, was attended by 130 ecologists. The fact that 26 of them came from a total of nine countries made it one of the most international gatherings ever convened by the Society. The 21 papers covered subjects ranging from cockles to antelopes, and from mathematical models to Antarctic whaling. As an experiment, generous time was given to discussion, including two evenings and one afternoon session devoted to discussion alone. The papers were later edited by David Le Cren and the local secretary for the symposium, Martin Holdgate, and published as the second volume in the series (Le Cren & Holdgate 1962). Almost 1500 copies were sold in the first six months—a thousand copies to an American distributor.

In a report on the symposium, published in the Society's journals, Dennis Chitty congratulated the organizers, but had some harsh words for some of the speakers. In spite of instructions about keeping their diagrams simple, there were many instances of slides being poorly labelled and overcrowded with detail. Some speakers had obviously not rehearsed their talks, gave too many details at the start, and crammed their main points into the last few minutes, or left some of them out, or simply ran out of time. Far too few realized that it might be difficult for a continental visitor to understand English spoken to the blackboard or screen, mumbled, or delivered at high speed. Although these were faults common to most scientific meetings, it was, in Chitty's words, astonishing how 'much mutual suffering is inflicted through lack of attention to form'.

By the time the third and fourth symposia were held, the series and its associated volumes were regarded as thoroughly established. The third, organized by A. J. Rutter and F. H. Whitehead in April 1961, was on the subject of 'The water relations of plants'. It was particularly successful in securing a well-balanced group of contributors. Of the 26 papers read, half were given by overseas workers. A volume made up of the papers given to the symposium appeared two years later (Rutter & Whitehead 1963). The fourth symposium, on 'Grazing', took place in April 1962, and was attended by over a hundred members and visitors. As its organizer, D. J. Crisp, noted, both botanists and zoologists were well represented, with interests ranging from agronomy to freshwater and marine biology. Members of the British Grasslands Society took part (Crisp 1964).

It had long been the intention of the Council to celebrate the Society's golden jubilee in 1963 with a special symposium 'reviewing progress and prospects in the science of ecology'. The meeting finally took the form of six half-day sessions, opening with an account by Salisbury of the history of the Society, and a jubilee address by the chairman of the symposium, Pearsall, on the development of ecology in Britain. There followed sessions on five topics which, over the years, had become 'focal points of interest in ecology'. They were the relationship of ecology to conservation (chairman E. M. Nicholson), Quaternary ecology (H. Godwin), experimental and single-species studies (A. R. Clapham), production ecology (W. H. Pearsall), and the community concept (A. S. Watt). The papers were published in the form of a supplement to the journals, edited by the symposium organizers, A. Macfadyen and P. J. Newbould.

About 400 members and guests attended the symposium, which was held at Queen Elizabeth College, London. It was the largest number ever to attend a meeting of the Society. The history of the Society and examples of current research were displayed in the form of exhibits. A jubilee dinner was held at University College, London, where the Society had met so many times since the inaugural meeting in the botany department, 50 years previously. The symposium concluded with an excursion to Kingley Vale National Nature Reserve, where a memorial stone (a sarsen brought from the Fyfield Down National Nature Reserve in Wiltshire) had been erected to the memory of Sir Arthur Tansley in 1957.

3.4
Ecology at the Mid-Century

In the preface to his book, *The Tropical Rain Forest*, published in 1952, Paul Richards referred to the difficulties of defining the scope of ecology. It had even been said that the only definition was that 'it is the subject-matter of the *Journal of Ecology*' (Richards P. W. 1952). Appraisals of the scope and character of particular aspects of ecology were often included in the biennial addresses given by the presidents of the Society, drawn alternately from botany and zoology. It is the purpose of this section to illustrate how these addresses, and the papers submitted to the Society's meetings and journals, reflected advances being made in British ecology.

There was no mistaking the trend towards 'quantification' in ecology. In his editor's report of 1950, Elton commented on how 'more and more editorial labour was involved in checking the statistical methods employed by authors'. The title and content of some of the presidential addresses were a further indication of the increasing importance attached to measurement and accuracy of expression in ecology. C. B. Williams chose for his theme the 'Statistical outlook in relation to ecology'. He recalled how, despite the numerous attempts on the part of biologists to formulate general laws from their case studies, there were very few instances where these laws could be considered generally valid. Because uncertainty could never be eliminated, the only course left for the ecologist was to measure it. To that end, the rather oversimplified trials of earlier years had been abandoned for more complicated field experiments, which incorporated several variables (Williams C. B. 1954).

Presidential addresses were obvious opportunities to highlight the wider and longer term relevance of research done long ago, often in the most fortuitous circumstances. Williams provided such an instance. Recalling his years as the head of entomology at Rothamsted, he could speak with first-hand experience of the difficulties posed by biased sampling and the lack of replication in fieldwork. In 1940, he had received a letter from A. Steven Corbet, on the subject of the large number of butterflies which Corbet had collected over many years in Malaya. Like many of its kind, the collection was strongly biased in favour of the rarer species—it had not been randomized for abundance. Williams discussed the deficiency with Fisher, who had left Rothamsted in 1933 to become the Galton Professor at University College, London, but returned as a member of Williams's entomology department during the first three years of the war. In the course of discussions, Williams realized that his own light trap at Rothamsted had collected a similar number of individuals, but in an entirely randomized manner. The light trap did not stop catching individuals of a species because many had been collected already.

The two sets of data provided Fisher with the opportunity to devise and test a range of mathematical theories for sampling. In a joint paper, published in the *Journal of Animal Ecology* in 1943, the relation between the number of species (*S*) and the number of individuals (*N*) in random samples of a mixed population was expressed as:

$$S = \alpha \log_e \left(1 + \frac{N}{\alpha} \right)$$

The value α is a constant for the population and was referred to as the index of diversity (Fisher *et al.* 1943). The formula provided an apparently sound basis for calculating the relation between the number of species and the size of the sample, provided that the samples were taken from the same population in the same ecological association. Over the next few years, Williams sought to extend the scope of Fisher's formula by, for example, seeing what changes in the number of species were to be found when the size of the sample was increased far beyond the limits of single ecologically uniform populations (Williams C. B. 1944, 1947).

Williams was later described as 'the first real quantitative empirical ecologist'—a naturalist who became numerate (Wigglesworth 1982). He developed so strong an intrinsic interest in the statistical aspects of ecology that he set out to discover how far the methods found useful in studying the structure of animal populations might be applied to that of plant populations. Whereas detailed studies of animal populations had been largely based on numbers of individuals, the plant ecologist tended to place greater emphasis on 'area', as expressed in terms of quadrats.

First in a paper in *Nature*, and then in the *Journal of Ecology*, Williams set forth techniques which made it possible to compare areas of similar size but different numbers of species, as well as areas of different size but similar species populations. On the premise that the number of 'individuals' was, in a small sample from a single association, proportional to the area of the sample, Williams (1943, 1950) found that Fisher's theory, which had produced a close fit between the formula and random collections of insects, birds, fish, and mammals, could be extended to cover plant populations as studied by means of quadrats. The richness or paucity of a fauna and flora could be measured, irrespective of the total abundance of a population or the size of the sample available for study.

These and other methods in quantitative biology were discussed by Williams in his presidential address to the British Ecological Society in 1953. Some years later there followed a book, *Patterns in the Balance of Nature*, which Williams wrote after his retirement from Rothamsted in 1955. As a pioneer study in mathematical synecology, the book foreshadowed many of the arguments that were to be advanced in respect of studies in multispecies populations, island biogeography, and the effects of competition on evolutionary genetic relationships (Williams C. B. 1964b, Wigglesworth 1982).

For his presidential address to the British Ecological Society in 1957, George Varley, the Hope Professor of Entomology at Oxford, chose as his theme 'Ecology as an experimental science'. He began by complaining of how ecologists had confined themselves so rigidly to making observations and comparisons, and had devoted so little effort to experiment. Even where major ecological experiments or, rather, experiments with ecological effects were being carried out (as in those parts of the world where farm crops or livestock were being introduced for the first time), ecologists were conspicuous by their absence in planning or recording what happened.

Not everything was black, however. There were already many indications of how experimental methods could be used as 'the most powerful tools' in investigating the basic principles of ecology as they affected plants and animals everywhere. In his presidential address, Varley (1957) referred to 'the series of wonderful monographs' by Patrick Buxton on the louse and tsetse fly, and by R. C. Muirhead-Thomson on mosquito behaviour, which demonstrated how 'experimental methods could be used in a multiplicity of ways to aid the understanding of the ecology of these animals'. The *Natural History of Tsetse Flies* was Buxton's *magnum opus*, based on his visit to northern Nigeria in 1933 and his extensive tours of Africa after the war (Wigglesworth 1956). In the preface to the book, Buxton (1955) wrote of how it was 'a source of pleasure to find that physiological work in the laboratory, and field experiments and observations, generally support one another'. Their remarkable correspondence suggested 'a valuable and fertile interplay between field and laboratory, and between observation and experiment'.

An outstanding example of the more large-scale kind of experimental approach was provided by Geoffrey Blackman and A. J. (Jack) Rutter, in their joint study of the seasonal growth of the bluebell (*Scilla non-scripta*). Blackman had left Jealotts Hill in 1936, and succeeded Eric Ashby as lecturer in ecology in V. H. Blackman's department at Imperial College, London. An award from the Royal Society in 1938 enabled him to appoint Rutter (a graduate in the department) to assist in extending what had already become a detailed investigation of the physiological ecology of the bluebell. After the hiatus of the war, the first of a long series of papers appeared, under the running title of 'Physiological and ecological studies in the analysis of plant environments' (Harley 1981). As Roy Clapham remarked, in his presidential address to the British Ecological Society in 1955, the papers demonstrated very strikingly both the complexity of field situations and the extent to which field ecologists could be misled by their preliminary observations, if they were not followed up by experiments.

In their first paper, Blackman and Rutter (1946) remarked on how it was only in recent years that statistical methods had been able to assist ecologists in untangling the complex factors that had to be taken into account. In order to distinguish the environmental factors responsible for the distribution of the bluebell, Blackman and Rutter (1947) made full use of the method of multiple regression in assessing the impact of light and nutrient supply, as observed in autecological studies conducted in both the field and laboratory. Measurements were carried out on the plant density and degree of shading at a large number of points on transects through three woods, using photoelectric cells. The evidence, together with that derived from artificially shaded experimental plots, indicated that light was likely to be the main environmental factor controlling the distribution of the bluebells in most closed woodland communities.

For his presidential address to the British Ecological Society in 1955, Clapham took as his theme the role of experimentation in autecological studies, highlighting the dilemma encountered in all such studies. Only the field ecologist was fully aware of the precise nature of ecological problems, and only he could 'guide the elucidatory experiment'. Field experiments were, however, conducted in far from ideal conditions, and called for the closest co-operation between 'the field ecologist and the laboratory ecologist'. Having seen at first hand the facilities available at the Boyce-Thompson Institute in New York State, and at Berkeley and Pasadena in California, Clapham (1956) concluded that little would be achieved in understanding the multifactorial nature of environmental control until at least one phytotron or, better still, a megaphytotron was installed in Britain.

The high cost of such an 'ecological machine', capable of exerting a much greater degree of environmental control, meant that it would have to be used intensively. In Clapham's view, the best location for such a machine was an ecological institute, preferably located in or near a university city or a national botanical institution, and certainly having an experimental garden nearby. Having taken the decision as to which species should be chosen and what kinds of control might be applied, on the basis of knowledge gained through primary field observations, ecologists would have to leave the rest to well-trained technicians.

3.4.1. *Vegetation surveys and phytosociology*

To outward appearances at least, there seemed little evidence of any kind of consensus emerging in the 1940s as to how vegetation should be described and classified. The methods of vegetation survey being applied in continental Europe, and increasingly in other parts of the world, were largely ignored by contributors to the *Journal of Ecology*. The eloquent 'silence' could in one sense be easily explained. Whereas the continental phytosociological system had grown out of a preponderant interest in classical taxonomy and phytogeography, British ecologists had drawn their inspiration much more consciously from plant physiology. If the overriding ambition of continental workers was to provide a complete description and classification of vegetation, British workers were much more concerned about 'the physiological relationship between plant and habitat' (Poore 1955a).

It would, however, be wrong to exaggerate these differences in perception. The early volumes of the *Journal of Ecology* displayed a considerable interest in the methods pioneered by pupils of Schröter at Zürich and Flahault at Montpellier. As Rübel demonstrated in a paper of 1920, there was close agreement between the schools (Rübel 1920). The most prominent 'pupil' to emerge was Josias Braun-Blanquet (*1884–1980*), whose search for an academic post caused him to leave Switzerland for Montpellier earlier in the century. The broad framework of Braun-Blanquet's concepts and methods of phytosociological analysis began to emerge during his doctoral study on the vegetation of the central Cevennes. A more formal statement appeared in 1921, and a textbook, called *Pflanzensoziologie*, in 1928 (de Bolòs 1982).

Tansley reviewed the most recent work of the two schools in the *Journal of Ecology* for 1922, commenting on how all ecologists had much to learn from one another. In the same way as 'our continental colleagues' might take greater account of the dynamic and more strictly ecological standpoints, British workers would benefit from the perception of plant associations as social units (Tansley 1922b). In his own contribution to a *Festchrift* dedicated to Schröter in 1925, Tansley used a case study of the vegetation of the southern English chalk to demonstrate the advantages of adopting both a 'developmental standpoint' and the numerical classification of the different communities. A further opportunity to combine these two approaches arose in a paper which Tansley published with Adamson in the *Journal of Ecology* in 1926, describing the preliminary findings of a survey of the chalk grasslands of the Sussex downs (Tansley 1925, Tansley & Adamson 1926).

Nothing further came of these initiatives. Adamson left for South Africa in 1923, and Tansley thereafter made little reference to European phytosociology. In the preface to *The British Islands and their Vegetation*, Tansley (1939c) wrote of how he had been 'unable to form an opinion as to the validity of or usefulness of the terminology of plant communities invented by Dr J. Braun-Blanquet'. He had accordingly made no attempt 'to consider its possible application to British vegetation' in the writing of his book.

Book reviews are often more revealing for what they say about the preconceptions of the reviewer than about the contents of the work under review. Salisbury was chosen to review for the *Journal of Ecology* the English translation of *Plant Sociology* by Braun-Blanquet in 1932. He was highly critical of the author's tendency to treat the character of vegetation units as if they were separate abstractions, with scarcely any emphasis on interrelationships. This sense of artificiality obscured the three-dimensional picture of the plant community. It was difficult to assess not only the relative importance of some characteristics, but also how far the significance of, say, successional status, physiognomy, or floristic composition might vary throughout a range of communities. As a result, any classificatory system based on so partial a view could only act as a temporary expedient. To be of permanent value, a system had to be built on a knowledge of the character and significance of all the interactions between plant and animal life and the edaphic, climatic, and historical factors in the environment (Braun-Blanquet 1932, Salisbury 1933).

Whatever the views of British ecologists, the Zürich–Montpellier approach was adopted more and more widely. At a Council meeting of the British Ecological Society in June 1939, 'a lengthy discussion' took place on the relationship of the Society to the newly founded International Association of Phytosociology. Although he had never used the approach himself, Alex Watt had taken 'a sympathetic interest' in it (Watt 1961). Together with Tansley and Clapham, he was chosen by Council to form a committee to decide which two 'visitors' might be invited to attend the next Annual Meeting, at the Society's expense. The greater part of that meeting would be given over to

a thorough discussion of the Braun-Blanquet system of vegetation analysis and nomenclature, at which visitors capable of expounding the system adequately should be asked to attend.

It was hoped to circulate written accounts beforehand, and for the discussion at the meeting to be continued in the *Journal of Ecology*. Tansley wrote explanatory letters to Braun-Blanquet and others; the outbreak of war prevented the meeting ever taking place[29].

In a preface to the volume of *Vegetatio*, commemorating the 70th birthday of Braun-Blanquet, Tansley (1954) wrote of how British ecologists had suffered from their relative isolation. In Tansley's words, 'we have admired Dr Braun-Blanquet's work from afar without attempting to take an active part in it or to harmonise it with our own'. In Tansley's own case, he had been invited to Montpellier as long ago as 1908, but had only been able to take up the invitation many years later. Although both he and Braun-Blanquet had attended the International Phytogeographical Excursion to Czechoslovakia and Poland in the late 1920s, 'we did not have a great deal of conversation, partly owing to the difficulties of language'.

As the experience of the International Phytogeographical Excursion to Ireland in 1949 made clear, personal contact did not always bring a meeting of the minds. Two of the 23 botanists to take part were Braun-Blanquet and Reinhold Tüxen (*1898–1980*). Tüxen had visited Montpellier in 1926 and soon became one of the most active exponents of phytosociology. In 1939, he became director of the Zentralstelle für Vegetationskartierung des Reichs in Hannover (Barkman 1981). Tüxen regarded Tansley as being directly responsible for persuading the British authorities to grant him repossession of the Institute after the war, when it was in constant danger of being closed down. He thanked Tansley when they met in Ireland in 1949—Tansley made light of it. Drawing on what they saw and heard during the 17-day visit to Ireland, Braun-Blanquet and Tüxen contributed a joint paper of almost 200 pages to the Excursion volume, which was not only the most comprehensive synthesis of Irish vegetation ever to appear, but which also provided the means of integrating the study of such vegetation into the well-established mainstream of European phytosociology (Ludi 1952, White J. 1982).

The Excursion report by David Webb provided ample evidence of the scepticism shown towards the continental approach. On the seventh day, Webb (1950) wrote,

> a spirited discussion took place on the classification of plant communities. It was introduced by Professor Braun-Blanquet who, with the enthusiastic support of Professor Tüxen, expounded and defended his system against some respectful but determined heckling, especially from the British and Irish members.

Webb expressed his own misgivings in a paper published in 1954, attacking those who compile 'an ever-growing list of names without precise specification', arrange them in a hierarchy according to some 'unspecified principle of relatedness', and then seek to impose such a classificatory system by international authority (Webb 1954).

Harry Godwin had also taken part in the Irish Excursion. It was one of his research students, M. E. Duncan Poore, who brought the methods of Braun-Blanquet to the forefront of consideration in the pages of the *Journal of Ecology*. One of the first recipients of a Nature Conservancy studentship award, Poore was encouraged to focus on the ecology of the Woodwalton Fen National Nature Reserve in Huntingdonshire, and to use it as a means of assessing the techniques of the continental schools of plant sociology. As it became clear that the vegetation of the site was largely the outcome of human activity, the main emphasis shifted to a study of the principles of vegetation classification. Stimulus came from two quarters, namely the work of Norwegian phytosociologists and the opportunity to spend two months at Montpellier (Poore 1954).

In 1942, Eilif Dahl had begun a survey of the Rondane, a group of mountains in southern Norway, on the premise that the simplicity of the monotonous vegetation cover would offer outstanding opportunities for developing techniques in phytosociology. It was clear, however, from the first summer season, that the problems were far greater than he had at first imagined. Because of the war, no further field-work was possible until 1946. In his monograph on the Rondane, Dahl (1956) wrote of how he had spent a year in Cambridge in 1951–2, on a British Council scholarship, and a further year in the United States (1953), bringing him into close contact with an 'approach to ecology very different from that usually employed in Scandinavia'.

Duncan Poore contributed a paper to a session on 'Plant-community classification', during the autumn meeting of the British Ecological Society in 1954. There then followed a series of papers in the *Journal of Ecology*, setting out the aims and potential of the phytosociological approach, and its relevance to British ecologists, 'from the point of view of one who believes in it'. The papers were regarded as being so important that the Council of the British Ecological Society investigated the possibilities of having them reissued as a reprint—until deterred by the printing costs.

As Poore (1955b) explained, the procedure was to take individual stands of vegetation, which should be as uniform as possible in character, and to list all the species present, together with some measure of their cover, and to record such details as area, height and cover of vegetation, soil profile, and the altitude, aspect, and slope of the ground. From the tabulated descriptions of similar stands, it became possible to distinguish what Poore called 'noda'. As he explained to the meeting of the British Ecological Society in 1954, choice was an essential part of the process: statistical methods were inapplicable. Whilst uniformity could only be regarded as a relative attribute of vegetation, particular grades of uniformity could be distinguished for particular purposes.

Because the system amounted to little more than a series of empirical treatments, Poore (1955c) argued that any constructive assessment had to be based on the results of practical trials, and to that end a study was made of the Breadalbane area of Perthshire in the summer of 1952. Much of the original and best continental plant sociology had been carried out on mountainous communities; the Breadalbane area was floristically rich by British standards, and could be expected to show species of high fidelity.

The field test revealed that vegetation units could be most satisfactorily distinguished by constancy and dominance—but not by fidelity, or by dominance alone. The relationships of the individual units were multidimensional, rather than hierarchical. Perhaps the most useful feature of the Braun-Blanquet approach was not so much the way it distinguished and named vegetation units for the purposes of classification, but rather the more precise survey methods which it brought to the definition and partial solution of ecological problems. It systematized the kinds of observation and lines of argument pursued by most ecologists in seeking to comprehend the inherent complexity of ecological situations. Based on inferences derived from exact observation in the field, the approach was just as applicable to an investigation of a few closely related communities in a small area as to another encompassing the entire range of a variety of vegetation types.

The requirements of the Nature Conservancy provided not only the pretext but often the resources to appraise and develop the concepts and techniques evolving in ecology. The appointment of Duncan Poore to the Conservancy held out the prospect of applying the experience gained in the Breadalbane district to the whole of the Highlands. A paper published in the *Journal of Ecology* in 1957, written jointly with Donald N. McVean, set out more fully the new approach to mapping the Scottish mountain vegetation, based on phytosociologically determined community types (Poore & McVean 1957).

McVean had also been a postgraduate at Cambridge in the early 1950s; Dahl acknowledged his assistance in preparing the monograph on the Rondane for publication. In 1954, Poore and McVean wrote to Rolf Nordhagen (*1894–1979*), asking whether they might come to Norway to see in the field the plant communities described in his several books, and to make themselves more familiar with his 'quadrat-method'. Nordhagen had published a new flora of Norway in 1940, and had become professor of plant systematics and phytogeography in the University of Oslo in 1945 (Nordhagen 1940, Faegri 1981). Although Poore left the Conservancy in 1956 (to become consulting ecologist to Hunting Aero Survey), the vegetation survey continued until McVean and his new collaborator, Derek Ratcliffe, were satisfied that they had covered a sufficient sample of areas to be representative of the whole Highland region.

Their findings were published in what became the first of a (very short) series of Conservancy monographs (McVean & Ratcliffe, 1962). In a review for the *New Phytologist*, Nordhagen (1964) described the volume as outstanding, not only because it provided so detailed a picture of the vegetation but because it was possible for the first time to make 'a rational comparison between Scottish plant communities and those of Scandinavia and central Europe'. For too long, British botanists had used their own methods to describe plant communities, and had regarded the phytosociological literature published in German and Scandinavian as something of a closed book. In Nordhagen's own words, 'my deceased friend Professor Tansley and his pupils were not interested in exact analyses of plant communities'. The authors of the new monograph had demonstrated convincingly what those in Scandinavia had long suspected, but had never been able to prove: there were strong links between the vegetation of the Scottish Highlands and that of the Norwegian mountains.

Meanwhile, further assessments of the Braun-Blanquet approach were being made in Ireland. The first came about as a result of a suggestion by Joseph Doyle, the head of the botany department at University College, Dublin, that it was time for a resurvey of the district south of Dublin, covered by Pethybridge and Praeger some 50 years previously (White J. 1982). The challenge was taken up by a final-year student, John J. Moore. Between 1952 and 1956, he devoted each summer holiday to compiling a new vegetation map, and assessing the usefulness of the Braun-Blanquet approach, as described in the second edition of *Pflanzensoziologie* in 1951. Moore (1960) found much to criticize, particularly in respect of the use made of faithful species in characterizing the different associations.

John Moore was shortly afterwards invited by Rheinold Tüxen to visit his cartographic institute at Solzenau-Weser in West Germany, where, in Moore's own words, the field exercises, weekly seminars, and 'almost overwhelming hospitality' of Tüxen quickly drew him into 'the world of phytosociology' (Moore J. J. 1981). His conversion was soon proclaimed in a paper published in the *Journal of Ecology*. Through a comparison of the earlier analysis of bog vegetation in the Dublin mountains with the more informed analysis carried out using Tüxen's methodology, Moore (1962) sought to illustrate how many of the serious criticisms of the Braun-Blanquet system were no longer valid. As developed by Tüxen, the practice of using faithful species to distinguish associations was perceived as simply a way of identifying associations already identified on floristic, ecological, and plant-historical grounds. The associations had already been recognized from analyses of a large number of exact descriptions of uniform vegetation.

With the publication of so eloquent a testimony of the merits of the continental system, the tide turned in favour of a calmer and more attentive
approach to phytosociology in Britain (White J. 1982). The most substantial
publication to arise from the Burren investigation, sponsored by the British
Ecological Society, was a paper by R. B. Ivimey-Cook and M. C. F. Proctor of
the botany department at Exeter, demonstrating how association analysis could
be a powerful tool for detecting faithful and differential species. Based on about
three months' field-work over the years 1959–63, the phytosociological survey
was the most intensive study to have been attempted of any small part of
Ireland (Ivimey-Cook & Proctor 1966).

3.4.2. *Pattern and process in the plant community*

One of the purposes of the biennial presidential address was to assess the progress achieved in ecology, and to identify the remaining gaps in research. In his
address of 1959, N. Alan Burges drew attention to the anomaly whereby a considerable amount of attention had been devoted to plant succession, but so few
papers had appeared on the most important parameter of any changing system,
namely the rate at which the changes took place. Among the very few examples
in Britain were the studies made by Salisbury on the dune systems of Blakeney
Point and Southport (Salisbury 1922b, 1925). For further examples of the role
of time as an ecological factor, it was necessary to look overseas (Burges 1960).

An Australian by birth, Burges's research career had taken him from
Sydney to Cambridge, returning to Sydney as professor of botany in 1947. He
became the Holbrook Gaskell Professor of Botany at Liverpool in 1952. It was
during his second period at Sydney that Burges investigated the large system of
parallel sand ridges in the Woy Woy district of New South Wales, where a transect running from the youngest to the oldest ridges represented a time-scale
from 0 to 4000 years. An abrupt transition from a sclerophyll forest dominated
by *Angophora intermedia* to one dominated by *A. lanceolata* could be correlated
with a change from iron to humus podsols in that part of the transect that corresponded with an age of 2500–2800 years (Burges & Drover 1953).

So far, most of the research on successional change had taken place in temperate plant communities, with their relatively small species populations and
well-marked single dominants. As Paul Richards pointed out in his presidential
address of 1963, the tropical rain forests provided some of the most striking
examples of rapid change. They offered unique opportunities for studying the
closely linked problems of what determined species richness, and under what
conditions a single species was able to become dominant in a community
(Richards P. W. 1964). The most frequently quoted example of plant colonization in the tropics was the restoration of vegetation cover to the island of Kraka-
tau, after the massive volcanic eruption of 1883. In a detailed account of the early
stages of secondary succession in the Shasha Forest Reserve of Nigeria,
Richards (1952) described how the new community was already approaching

the climax forest after only 20 years, although a long period of time would be required before that stage was actually reached.

The various studies made of the time-scale in successional change served to emphasize a number of aspects that had often been overlooked. As Burges (1960) pointed out, even under tropical conditions some successions required a very long time to reach the climax conditions, even where there was no evidence of any deflecting or retarding influences. Within the succession itself, one of the most important aspects was the number of generations involved in each stage of the succession. In its simplest form, a succession might be composed of single generations of each of the dominant species in each stage of the succession. On the other hand, each stage of the succession might be sufficiently prolonged to allow the development of what Alex Watt called a 'process and pattern phenomenon', namely a repetitive cycle of growth and breakdown within the community.

Both Burges and Richards were members of expeditions organized by the Cambridge Botany School, in which Watt played a prominent role. With the demise of undergraduate teaching in forestry at Cambridge in 1933, Watt had transferred to the Botany School, where he remained until his retirement in 1959 (Greig-Smith 1982, Gimingham 1986). Altogether, 22 members of the Botany School took part in expeditions to the Cairngorms in the summers of 1938 and 1939. Because of the war, there was a long delay in publishing the evidence collected on the altitudinal zonation of vegetation and the effects of exposure and snow cover on community composition (Watt & Jones 1948). Detailed studies of the mountain Callunetum and the Empetrum–Vaccinium zone were published by G. Metcalfe and Alan Burges respectively (Metcalfe 1950, Burges 1951). Each bore 'the imprint of Watt's distinctive approach to the investigation of plant communities' (Gimingham et al. 1983).

Alex Watt's attitude to vegetation studies was summarized in his contribution to the symposium on post-war forestry convened by the British Ecological Society in 1943. Tansley's book, *The British Islands and their Vegetation* had been a landmark in the history of ecology. It represented the climax to the phase in research on the relationship of one plant community with another. In Watt's view, there was now an urgent need for an 'intensive study and elucidation of the dynamic relations which hold between one species and another in the plant community itself' (Anon. 1944a).

Despite the adoption of a dynamic concept in ecology over 50 years previously, there was still little guidance available as to how individuals and species were put together, and what determined 'their relative proportions and their spatial and temporal relations to one another'. As a first step towards formulating laws and expressing them in mathematical terms, 'an acceptable qualitative statement of the nature of the relations between the components of the community' was required. It was with this purpose in mind that Watt gave his presidential address to the British Ecological Society in 1947. The title of the address was 'Pattern and process in the plant community'.

In presenting such a statement, Watt (1947a) illustrated how there was evidence of both orderliness and disruption at each phase in the development of vegetation. Despite an inherent tendency towards orderliness and persistence in the structural pattern of the community, a study of the distribution and form of both the plants themselves and their subfossil remains revealed many departures from the normal time-sequence, brought about by what Watt called 'fortuitous obstacles'. The outcome was a 'patchiness' in the vegetation, with each patch distinctive in its floristic composition, age of dominant species, and type of habitat. No matter how great the differences, the patchwork was part of the same space–time pattern. Each patch was an expression of the same cycle of change, reflecting how, at times, the balance between building up and breaking down was positive (the upgrade), and at other times negative (the downgrade).

In an analysis of the processes involved, Watt cited the evidence derived from two types of enquiry. The first consisted of the data collected by the various summer excursions. Those led by Harry Godwin to Tregaron Bog in 1936–7 had revealed, for example, the underlying uniformity of the processes involved in that type of community. From the sequence of plant remains in the peat profile, Godwin and Conway (1939) described a pattern of sequential change, whereby the open expanses of water were invaded by *Sphagnum cuspidatum*, which was in turn replaced first by *S. pulchrum* and then by *S. papillosum*. A hummock was formed, which was colonized by *Calluna vulgaris*, *Erica tetralix*, *Eriophorum vaginatum* and *Scirpus caespitosus*. As Watt pointed out in his presidential address, the speed and character of change on each patch in the space–time mosaic depended on events both on the patch itself, and also on those of its neighbours. Differences in the rate at which the level of one patch rose above that of another determined which hummocks were replaced by pools. Wherever a pool developed, the *Calluna* died, and its ground space was filled by another species, which had itself played no direct part in the demise of the heather.

For his next two examples of parallelism between spatial and temporal changes, Watt recalled the evidence found in the Cairngorms of small-scale dynamic processes, which resulted in various kinds of pattern or mosaic in the vegetation. In a foretaste of what was to be published by other members of the expeditions, Watt described how the prevailing wind on the higher slopes seemed to determine the direction of the erosion and spread of the dwarf Callunetum and Rhacomitrietum. Although there was steady erosion on the windward side of the 'complex units', compensating forward growth took place on the lee side. From growth ring counts, it seemed that the rate of movement was about 1 m in 50 years (Burges 1951).

It was during the course of the Cairngorms expeditions that Alex Watt became impressed by the striking similarities in the way morphological changes were brought about in such contrasting habitats as those encountered in the Cairngorms and the Breckland, near Cambridge. Already, by the late 1930s, Watt's name had come to be closely associated with the making of meticulously recorded observations, year after year, on the same sites. The area chosen for these detailed studies was the grass heathlands of the Breckland.

The merits of such an approach to the study of vegetation dynamics were already apparent in a paper of 1936, in which Watt suggested that the Festuco- Agrostidetum was a relatively permanent community, whereas others enjoyed only a temporary dominance as part of a longer-term cycle of change. It remained to be discovered whether their temporary occupation of the ground had a transient or permanent impress. It was, however, clear that Pickworth Farrow had exaggerated the extent to which rabbit grazing influenced the distribution of plant life. It was not enough to focus on the changes in the biotic or some other factor—the functional relationships of the entire community had to be taken into account (Watt 1936, 1937, 1938).

In the course of monitoring changes in what he described as 'grassland A type of vegetation' in Breckland, Watt (1940) discovered that each part of the surface of a typical community could be assigned to one of four phases, namely a hollow, building, mature, or degenerate phase. In order to understand the working mechanisms of such a community, it was important to elucidate how the phases related to one another. Careful monitoring over a long period of time revealed that the whole 'life' of the community centred round the life history of *Festuca ovina*, and its response to variations in microtopography and soil habitat. Seedlings became established among the stones covering the surface of the hollow phase. As the plant grew and spread, the level of the mineral soil rose inside the tussock, and a building phase commenced. As the height of the hummock rose, the plant entered the mature phase, when the number of inflorescences and length of the leaves decreased. During the degenerate phase, the hummock became increasingly exposed, and was removed by erosion. A hollow was once again formed, covered by flint and chalk stones.

The impact of habitat factors varied according to the phase reached by the individual patch within the community. During the winter of 1939–40, Watt noticed in Breckland how the rhizome apices of bracken were killed by frosts to a depth of 20 cm where the plant was in a pioneer phase, namely still invading the grass heath. Where the mature phase had been reached, and the bracken was dominant, the thick layer of litter protected the apices below a depth of only 6 cm (Watt 1947b).

It was the accompaniment of each phase by changes in habitat conditions that prevented any blending in the essential patchiness of communities. The floating differentia, as Watt called them, were first reactions to, and secondly causes of, the shifts from one phase to another. The differences found expression in terms of both space and time. Seeds and propagules, for example, could only develop in those phases when plants could survive, namely the gap phase, when there were restrictions as to not only the amount of space but also the time available for colonization. As the patch was invaded by plants, it became increasingly less receptive to more of even the same species. Time might be so short as to produce an even-aged population of the dominant species.

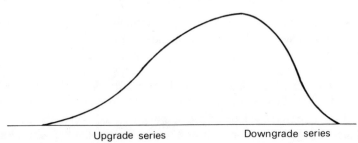

Upgrade series   Downgrade series

FIGURE 3.2 Changes in total productivity in the different phases of the cycle of change within plant communities (after Watt 1947b).

The effects of some event in time might persist long after its causes had ceased to exist. These were most obvious in the case of such violent events as frost or drought, and where the community affected was made up of long-lived dominants. In his studies of beechwoods, Watt (1925) found that regeneration was only significant in years of high seed production and unusually favourable weather conditions. Such years gave rise to even-aged stands of trees, which would persist like a tidal wave moving along the ages and could even influence the structure of the next generation of trees on the site. In such instances, the resultant pattern could only be explained in terms of that sporadic or exceptional event, sometime in the past.

Not only were these various observations of considerable relevance to the study of individual plant communities, but they had two wider implications. The fact that the total productivity of the sequence of phases in the upgrade and downgrade portions of the cycle of change could be plotted as a curve of the growth type (Figure 3.2) suggested that there was scope for mathematical inquiry. A phased equilibrium could be envisaged, where the community was in harmony with both itself and the environment. According to Watt (1947a), it should be possible to measure departures from that equilibrium, whether in space or in time, and correlate them with changes in environmental factors.

Watt's perception of change also had considerable implications for the long-continuing debate as to the validity of Henry Gleason's depiction of individualism in plant communities. Watt returned to this debate in his contribution to the jubilee symposium of the British Ecological Society in 1963. Most plant sociologists believed there was sufficient orderliness in the way plant associations succeeded one another for the term 'development' to be used in the Clementsian sense. They also supported Gleason's contention that each association was 'a fortuitous juxtaposition of plant individuals'. It was almost as if the processes of change from one community to another in the sere were different in character from those operating within the community itself.

According to Watt (1964), Gleason's main failing had been to overlook the role of the dominant species in forming a unifying pattern in a community. Drawing on his further studies of the competitive behaviour of bracken and heather, and the changes in the dominant species of his Breckland plots, Watt (1955, 1957) was able to highlight the extent to which these dominants determined the pattern of the community. They were the key to understanding the processes of change. Whereas environmental influences decided the relative areas occupied by the different phases in the vegetation (the spatial pattern), it was 'the heritable life-history of the individual' which determined the sequence of phases in the community (the temporal pattern), and thereby provided the basis for reconstructing what happened in the past and for predicting future changes.

By devoting greater attention to the social relationships in space and time within the plant community, the means could be found for accommodating Gleason's perception of the determinants of distribution with the dynamic approach of Clements. In calling for further experiments on the relationships that existed within the 'complex environment' of the plant community, Watt (1964) warned of how ecologists would have to reconsider the use of the terms autecology and synecology. As he pointed out to the jubilee symposium, the need was for assessments of individual species not as isolates, but in terms of their social status within the plant community.

3.4.3 A biological flora of the British Isles

One of the most outstanding examples of the British Ecological Society taking an initiative occurred during the darkest days of the war, when plans were again made for a British biological flora. The original suggestion made by Salisbury in the late 1920s had envisaged his being responsible for compiling and editing all the information sent to him by members of the Society and others. As Richards (1979) commented, it was hardly surprising that nothing came of the proposal. The new initiative anticipated that each account of a species or group would be made the responsibility of an individual person, working to guidelines laid down by the editors.

During the Annual Meeting of 1940, Owain Richards proposed that a further attempt should be made to launch a biological flora. He derived considerable support from Roy Clapham, whose first researches in plant physiology had been supervised at Cambridge by F. F. Blackman. There followed two years at Rothamsted where, under the influence of Fisher, Clapham developed both an interest and skill in the application of statistics to biological problems (Clapham 1936). In 1930, he was appointed university demonstrator in botany at Oxford, where his interests soon broadened to encompass morphology, taxonomy, and ecology. He succeeded William Pearsall as professor of botany at Sheffield in 1944 (Harley 1984).

Clapham and Richards were appointed editors of the new *Flora*. In an announcement published in the *Journal of Ecology* for 1941, they stressed how, at a time when field-work and longer-term investigations were so difficult to pursue, wartime conditions might 'not be so unfavourable to the sorting out of data which have, for the most part, already been collected'. Accounts of single species or small groups of species would be published in the order they were received, initially in the *Journal of Ecology* and later as reprints. As a stimulus to potential contributors, the same issue of the journal published accounts of four species of *Juncus*, compiled by Richards and Clapham (Anon. 1941b)[30]. A year later, a much longer list of intending contributors was published, together with accounts for *Cladium*, *Zostera*, and *Aster trifolium*.

Guidelines as to the use of standardized nomenclature and taxonomy were laid down in a memorandum published in early 1943, after considerable discussion by Council (Anon. 1943). In a further attempt to secure uniformity of nomenclature, a checklist of British vascular plants was compiled and published by Clapham in the *Journal of Ecology* in 1946. In addition to meeting its immediate object of allowing species names to appear without their authorities, the list marked a major advance in securing a consensus between taxonomic specialists (Anon. 1946).

Nineteen accounts of species and five of genera had been published by the end of the war. About 150 were promised[31]. Alarmed by the increasing length of the accounts, the editors issued the first of many appeals for briefer, interim accounts. As a general rule, they were not to exceed 4000 words. It was essential that each should be scrutinized by experts in mycology and entomology before being submitted. Specimens were then to be deposited either in the British Museum or at Oxford. A one-day meeting was convened by Clapham in October 1949, at which further guidance was given and speakers were invited to recount the kinds of problem they had encountered in preparing accounts of the different species.

Clapham, together with Thomas G. Tutin and Edmund F. Warburg (*1908–66*), provided in 1952 the first 'modern, trustworthy and comprehensive flora of the British Isles' in the form of a volume of almost 1600 pages—small enough to serve as a handbook in the field as well as a reference work in the library or laboratory (Godwin 1952). The volume included a foreword by Tansley, and was dedicated to Humphrey Gilbert-Carter (*1884–1969*), the director of the University Botanic Garden, in Cambridge. As the authors commented in a preface, there had been a marked change in the outlook and requirements of botanists since the standard floras of the 19th century. The rise of ecology, for example, had led to 'a demand not only for clear descriptions of species but for information of a kind not essential to identification, though of value to everyone interested in plants as living organisms'.

When the time came to prepare a second edition of the volume, in 1958, the authors were able to report further great advances, which had come about 'through a combination of intensive experimental, field and herbarium studies of variable species and "critical" groups'. Knowledge of the distribution of vascular plants had been materially improved, largely as a result of field-work carried out as part of a mapping scheme organized by the Botanical Society of the British Isles, with grants from the Nature Conservancy and the Nuffield Foundation (Clapham *et al.* 1962).

It was increasingly realized that a knowledge of the processes and circumstances that accompanied glaciation in the Quaternary Period was fundamental to an understanding of the biogeography, evolution, and ecological relationships of plants and animals at the present day. The alternation of glaciations and mild interglacials had led to extensive migrations, extinctions, and renewed opportunities for expansion among the natural populations of species. There were those botanists who envisaged a situation where species with a wide and almost continuous distribution in the late-glacial period of the last glaciation became confined to fragmentary, relict communities, following the spread of woodland. Others believed the present, discontinuous pattern reflected the varying degrees to which species had been able to spread since glaciation from areas where they had survived (Anon. 1935).

By the 1950s, Godwin and his colleagues had amassed so large a collection of late-glacial and full-glacial floral–plant remains that a factual basis was beginning to emerge for discussing the origins of the British flora. First at the Annual Meeting of the British Ecological Society in 1952, and then in a joint paper in the *Journal of Ecology*, Donald Pigott, of the Sheffield department of botany, and S. Max Walters of the Botany School at Cambridge demonstrated by reference to species characteristic of soils with a moderate to high base status in open habitats that widespread post-glacial reductions had occurred. They constituted the 'most striking distributional phenomena' to be found in the British flora. Two types of approach were required to find out more about what had happened, namely further investigations of the Quaternary Period and careful autecological studies of the relict species in question (Pigott & Walters 1954).

In 1938, Harry Godwin submitted a memorandum to the faculty boards at Cambridge covering biology, archaeology and anthropology, and geography, setting out proposals for promoting studies of the glacial and post-glacial periods. Although shelved because of financial stringencies and the worsening international situation, the proposals formed the basis of a renewed submission in 1943, following a request from the Government for a statement of the University's post-war needs. Godwin became a reader in the Botany School, and, in 1948, the director of a newly created sub-department of Quaternary research. He was elected to the chair of botany in 1960 (Godwin 1981).

Peat deposits yielded not only great numbers of pollen grains but all kinds of coarse plant material, such as seeds, fruits, leaves, hairs, and even whole flowers. As Clement Reid had recognized, and Knud Jessen had amply demonstrated in his studies of the Quaternary flora of Denmark and Ireland, such dated records could afford direct evidence of the former presence at specified places of many individual species within the British flora. While convalescing from an operation, Godwin completed and organized a card index of all the available and dateable records of Pleistocene plant material. Never before had the evidence been assembled on such a scale. What began as an intuitive exercise became the basis for a book, *The History of the British Flora* (Godwin 1956).

The book not only cited all the plant records available, but demonstrated how such evidence might be deployed in interpreting the changes that had taken place in the British flora. The fact that over half the flora was already present in the country by the end of the late-glacial demonstrated the variety of edaphic conditions, and consequently of ecological types, that must have been present already. There was ample evidence to suggest that the lowlands possessed a considerable flora that later became, in the main, characteristic of mountainous areas. Later chapters traced the spread of forest trees and the consequences of deforestation, the expansion of the bogs, and impact of human encroachment on the natural vegetation.

For many years, the vegetation historian was confronted with an insoluble problem. Whilst it was possible to relate the palynological and other forms of evidence to stages in the recurrent cycles of climatic change, it was difficult to date them absolutely, and therefore to be sure that the climatic changes signified by the evidence were worldwide and synchronous. Although the most recent evidence might be dated from its juxtaposition with artefacts that could be dated by archaeologists, the only absolute bases for dating the older material were the annual laminations (varves) deposited in Scandinavian glacier lakes, and the archaeological evidence derived from ancient Egyptian sources. As Godwin (1981) remarked, these were lengthy chains of evidence, losing much of their certainty and exactitude over the long distances involved.

It was in this context that the introduction of radiocarbon dating represented a major breakthrough. For the first time it became possible to determine the absolute date by applying a direct test to the samples of organic material being taken from the different pollen zones. The technique was devised by Willard F. Libby (*1908–80*) of the Fermi Institute of Nuclear Studies at Chicago University; it won him the Nobel Prize for Chemistry in 1960. In the course of development in 1949, Libby asked the Cambridge sub-department for some organic samples already dated by some other means. A close correspondence was found between the dates determined by the radiocarbon dating technique and the estimates based on the 'nexus of evidence' collected over two decades by Godwin in England and Wales, and by Jessen in Ireland (Godwin 1951). A detailed account of the technique and early successes was published by Libby (1952) a year later.

By 1960, Godwin could speak of 'a progressive systemization of knowledge of events of the last 70 000 years'. A firm pattern was emerging in which the basic element was widespread climatic change (Godwin 1960). A five-year grant from the Nuffield Foundation made it possible to install radiocarbon dating facilities in the Cambridge sub-department. Designed and built by Eric Willis, the laboratory was opened in 1955 (Godwin 1985, Willis E. 1986). The first series of samples to be dated came from a relict raised bog at Scaleby Moss in

Cumberland where all the known pollen zones were thought to be present. The site provided the first long British pollen diagram to be dated by radiocarbon assay, and went some way to affirming the synchroneity of the major zones on both sides of the North Sea (Godwin *et al.* 1957).

In a contribution to the jubilee symposium of the British Ecological Society in 1963, Richard G. West, of the sub-department at Cambridge, demonstrated how research in ecology and Quaternary palaeobotany were closely interlinked. Quaternary palaeobotany was essentially 'the extension of ecological studies backward in time'. It emphasized how the present-day plant communities were 'merely temporary aggregations', with no long history. For the palaeobotanist, a study of modern assemblages of plant microscopic remains in relation to the surrounding vegetation, and of the production and dispersal of pollen in modern plant communities, could lead to considerable advances in the inter-pretation of fossil assemblages (West 1964).

3.4.4   *The search for a* 'systema naturae'

The most long-awaited presidency in the history of the British Ecological Society was that of Charles Elton. It was not until 1948 that he agreed to serve. His address took the form of a plea for a more purposeful and cohesive approach to ecological research. Having attended many of the Society's meet-ings over the previous 25 years, he had become increasingly impressed by 'the extreme range and fragmentation of ecological knowledge'. Whilst it was essen-tial to focus 'deeply on limited fields', there had to be an ultimate goal. For those attending the Society's meetings, and for ecologists in general, there had to be a positive answer to the question, 'Why do we continue to meet at all and listen to each other's fragments of discovery?' (Elton 1949).

By piecing together the various fragments, it was possible to discern con-siderable progress in large fields of ecological enquiry. These might include the classification and understanding of plant communities, their successional rela-tions, ecological genetics, and vegetation history. At the same time, there had been 'an outburst of field research on the density of animal numbers, and on the age and sex structure, movements and fluctuations in animal populations, accompanied by some rather exciting experiments in the laboratory on self-limitation, competition, parasite–host and predator–prey relationships, backed by ambitious mathematical theories of various kinds'. There remained, how-ever, a tremendous region between these two fields of enquiry, where all the dis-coveries made 'should logically be synthesised to give us at least a working model of how whole ecosystems are arranged in nature, and how they work and interact'.

If an ecosystem was defined as 'a system that is strongly interlocked in its parts but shows some fairly well-marked boundary with neighbouring ones', every effort had to be made 'to discover the main dynamic relations between populations living on an area' which was small enough to make detailed investigations feasible, yet representative enough to enable comparisons to be made with the system as a whole. If the properties or, more specifically, the 'elasticity' of the biological network were to be discerned, it was important

> to map out, as it were, the species channels through which matter and energy move in the community complex, and seek at the same time to find how far oscillations in numbers of any one species are able or likely to extend their effects along these channels or even temporarily to exploit new ones.

For Charles Elton, the prospect of mapping such pathways of energy flow in terms of species networks, populations, and their associated biological phenomena implied a renewed commitment to ecological survey of *whole* communities (Elton & Miller 1954).

As Elton (1949) pointed out, there were very few models of how such surveys might be carried out. Over the first 16 years of the *Journal of Animal Ecology*, only 13% of the 312 papers published had been concerned with ecological surveys of whole animal communities, and only 8% with British communities. The two substantial models were those of Cyril Diver on the South Haven Peninsula of Studland Heath in Dorset, and Elton's own survey of Wytham Woods in Oxfordshire. The comprehensive, yet detailed, survey of the South Haven Peninsula began as an attempt to determine how far the fauna and flora of different habitats might be adapted to the peculiarities of their environment (Diver & Good 1934). During his period as director-general of the infant Conservancy, Diver drew heavily on his research experience in Dorset.

Elton's original intention had been to organize a 'continuing ecological survey of the Oxford region', investigating all kinds of habitat and all groups of animals, with each ecological event recorded on a punch card system. In 1943, however, the University acquired Wytham Woods and the adjacent series of limestone meadows, marshes, and streams. The rich variety of habitats contained about a quarter of the British non-marine fauna. It soon became clear that an ecological survey of the area by the Bureau of Animal Population would be of considerable intrinsic importance not only for studies in comparative ecology, but for identifying ways of improving the methods by which ecological events might be recorded and related to one another.

The Wytham Wood survey provided an outstanding opportunity for seeing how far a *systema naturae* might be devised, applicable to habitats of all sizes. As Elton pointed out, it was the kind of task peculiarly well suited to members of the British Ecological Society. As with taxonomy, the categories into which the habitats might be classified could be only approximations to reality, but, if properly designed in the light of general dynamic principles, they would provide a much better method of organizing field data than currently existed. Having carried out the preliminary survey and statistical work, the ultimate aim would be to focus on population processes and the balance between particular species. By measuring the productivity of the community or whole ecosystem in terms of energy paths and energy flow, a physiological and physicochemical interpretation could be added to the demographic picture (Elton 1949).

As Elton illustrated in his presidential address, and later in a joint paper with another member of the Bureau, Richard S. Miller (Elton & Miller 1954), it had become possible to perceive the ecosystem of any area as being

> composed of comparatively limited minor centres of action, each having certain distinctive characteristics as habitat that are reflected in their communities, and which have a considerable amount of interchange by lateral and vertical movements.

The extent to which the Bureau was beginning to unravel the workings of the different ecosystems in Wytham Wood and elsewhere was made clear in a volume called *The Pattern of Animal Communities*, published by Elton shortly before his retirement (Elton 1966). In a special number of the *Journal of Animal Ecology* to mark Elton's retirement, Alister Hardy described the new work as 'a classified natural history of all such niches available to animal exploitation'. It was concerned not only with the interlocking ecological links between species, but with those linking one community with another (Hardy A. C. 1968).

These were also years of a burgeoning interest in what came to be known universally as the trophic–dynamic approach, first set out in a paper by Raymond L. Lindeman (*1915–42*), published posthumously. Having completed a five-year study at the University of Minnesota on the biology of a senescent lake, Lindeman moved to a post under the leading limnologist, G. Evelyn Hutchinson, at Yale in 1941. Drawing heavily on Hutchinson's guidance and inspiration, he rewrote the last chapter of his thesis in the form of a paper, illustrating how the concept of the ecosystem could be used to integrate the seasonal trophic relations of organisms with the longer-term processes of community change. It was only after further revisions and the intervention of Hutchinson that the journal *Ecology* accepted the paper, despite continued doubts on the part of the referees as to whether there were sufficient data to support the theoretical model put forward by Lindeman (Cook 1977).

It was during his postgraduate research on aquatic food cycle relationships that Lindeman had first become sceptical of the validity of the clear-cut distinctions drawn between the living community and the non-living environment. The constant organic–inorganic cycle of nutritive substances was so completely integrated that to discriminate between the living organisms as parts of the biotic community, and dead organisms and inorganic nutritives as part of the environment, seemed arbitrary and unreal. It was in that context that Lindeman found Tansley's concept of the ecosystem so relevant—it embraced within one system both the biotic and abiotic environments. It was the purpose of Lindeman's paper to take the concept a stage further by developing what he called a trophic–dynamic viewpoint, whereby the functioning of physical, chemical, and biological processes within an ecosystem were perceived to depend on the transfer of energy (Lindeman 1942).

Earlier writers had identified the distinctive roles of 'producer' and 'consumer' organisms. As Lindeman recalled, Elton had demonstrated the role of animal size and population numbers in determining the general relationships of the higher food cycle level. Animals at the base of the food chain were relatively abundant compared with the progressively fewer, but larger, animals at the end of the chain. It was Lindeman's contribution to perceive these organisms within an ecosystem, where they were grouped into a series of more or less discrete trophic levels as producers, primary consumers, secondary consumers, and so on—each successively dependent upon the preceding level as a source of energy, with the producers directly dependent upon the role of incident solar radiation as a source of energy.

From the trophic–dynamic viewpoint, it was possible to express Elton's pyramid of sizes and numbers in terms of biomass. Since primary consumers were smaller than secondary consumers, they could increase faster in number. In all the ecosystems he analysed, Lindeman (1942) found that the progressive energy relationships of the food levels in the pyramid were epitomized by the productivity symbol λ, in the following manner:

$$\lambda_0 > \lambda_1 > \lambda_2 > \lambda_3 \ldots > \lambda_n$$

From that perspective, it was possible to envisage succession as a form of development within an ecosystem, brought about primarily by the effects of organisms on the environment, and each other, and directed towards a relatively stable condition of equilibrium. Lindeman included in his paper a hypothetical productivity growth curve of a hydrosere in a fertile, cold temperate region (Figure 3.3). The marked increase in productivity and efficiency during the early stages corresponded with the transition from oligotrophy to a prolonged eutrophic stage. Equilibrium and decline marked the onset of lake senescence, with a further rise during the terrestrial stages of hydrarch succession.

The controversy surrounding the decision to publish Lindeman's paper was reflected in fierce dissensions as to how far the concept of the ecosystem could be elaborated and applied. Among proposals for future meetings discussed by the Council of the British Ecological Society in January 1952 was one for a symposium 'On the classification of ecosystems'. A subcommittee, made up of Godwin, Elton and Poore, and the two honorary secretaries of the Society, was appointed to draw up a programme. Having reported that the botanists and zoologists held 'very divergent views' as to how the symposium might be organized, the subcommittee was instructed to devote a whole day to general questions and another to specific botanical problems. It all came to nothing. At a meeting of January 1953, the Council was informed of how 'no botanists were willing to discuss the subject so soon'[32].

A leading figure in the controversy was J. Derek Ovington, who had spent two years at the Macaulay Institute at Aberdeen, studying the ecological impact of afforestation on the Cublin and Tentsmuir Sands (Ovington 1950, 1951). This, together with his later responsibilities for research on woodland management in the Nature Conservancy, caused him to become increasingly dissatis-

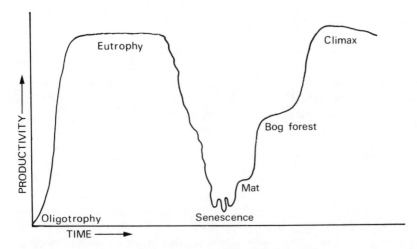

FIGURE 3.3 Hypothetical productivity growth curve of a hydrosere, developing from a deep lake to climax in a fertile, cold–temperate region (after Lindeman 1942).

fied with the way ecologists, confronted with the size and complexity of woodland communities, tended to confine their attention to the most important and easily recorded features. Such intensive and limited investigations failed to provide any kind of overall perspective. A unifying concept was required which, through its comprehensiveness, would bring a greater understanding of 'the functional processes of woodlands' and the extent to which these processes affected forest productivity. For Derek Ovington, the most obvious 'universal backcloth against which to show woodlands in their patterned complexity' was that of the ecosystem (Ovington 1953, 1962).

During 1957–8, Ovington obtained leave of absence from the Conservancy to become acting director of the University of Minnesota Terrestrial Ecosystem Project. This provided him with not only first-hand insights into current American thinking on the trophic–dynamic approach, but facilities for applying to plant material the kind of techniques previously used by animal ecologists in studying energy transfer and utilization. Using data collected from plantations at three experimental sites in Britain, Ovington was able to deduce from the measurements of dry weights for the trees, ground, flora, and litter that the level of efficiency in energy conversion was much higher than had previously been supposed, particularly for young dense plantations. At certain stages of development, woodland productivity might closely approach the maximum potential (Ovington & Heitkamp 1960).

Despite the impression of stability and permanence engendered by the longevity of trees, woodland ecosystems were essentially dynamic. They were characterized by a mass seasonal transfer of organic matter, energy, water, and chemical elements, the seasonal magnitude of which changed greatly as the trees matured and were replaced by trees of different species. There was a delicate and ever-changing relationship between woodland organisms and their physical environment of soil and microclimate. The ecosystem approach not only served as a basis for evaluating and bringing together the existing data derived from botanical, pedological, and zoological studies conducted at different scales, but acted as a guide to future research. By taking closer account of the size and mass transfer of material, it should become easier to construct a model of the interplay between climate, soil, and vegetation (Ovington 1962).

3.4.5 *'The struggle for existence'*

As an invited speaker, Nelson G. Hairston of the University of Michigan, reminded the jubilee symposium of the British Ecological Society in 1963, that there were two kinds of relationship which forced order onto natural communities. The first was energy transfer; the other much less obvious form was competition (Hairston 1964). No one denied that species competed with one another whenever populations were limited by the supply of a common resource. Controversy arose when it came to assessing how far such competition might be a primary force in deciding the fate of species populations.

As in the study of energy transfer, British ecologists drew heavily on the inspiration of overseas workers. It was during the second world war that the mathematical equations of Volterra and Lotka came to assume a further significance for animal ecologists in Britain, largely through the publications of the distinguished microbiologist at the Zoology Institute in Moscow, Georgyi F. Gause (*1910–86*). In a laboratory model of populations designed to measure the changes brought about in the numbers of two species of micro-organism which ate the same limited food supply, Gause found that one species was invariably eliminated by the other (Gause 1934, Vorontsov & Gall 1986). It was the first mathematical verification for the concept of 'ecological incompatibility between species', enunciated by Charles Darwin and put forward in terms of 'competitive exclusion' by Joseph Grinnell (*1877–1939*) in a series of studies on particular bird species (Grinnell J. 1904, Grinnell H. W. 1940).

Largely at the instigation of David L. Lack (*1910–73*), a symposium was organized under the aegis of the British Ecological Society in March 1944 to examine 'the ecology of closely allied species'. In a paper given to the symposium, he recalled how Gause had demonstrated the impossibility of two species with the same ecological requirements living in the same area. There seemed to be so many obvious exceptions to the rule among birds that Lack had originally been among the overwhelming majority of biologists to conclude that laboratory cultures of micro-organisms must behave very differently from higher forms of life. When, however, he began to look more closely at speciation among birds, the field evidence bore out the 'inescapable logic' of what Gause had written—ecological overlap was more apparent than real.

While still an undergraduate at Cambridge, Lack had spent a week of the long vacation in 1931 walking transects in every type of countryside in Breckland, recording breeding pairs of birds within a range of 20 yards. Many birds seemed to choose their habitat for reasons quite unconnected with survival or breeding limits. Psychological factors were by far the most obvious way of explaining how each species selected its 'ancestral' breeding territory, instinctively recognizing it by the conspicuous, but not necessarily the most essential, features (Lack 1933). When Lack put forward this interpretation at the Annual Meeting of the British Ecological Society in January 1932, there was, according to the minutes of the meeting, 'much difference of opinion' among those present[33].

After leaving Cambridge in 1933, Lack took a teaching post at Dartington Hall in Devon, where he made a study of the territorial behaviour of the robin. During 1938–9, he spent a year in the Galapagos Islands, where the endemic finches had been so well described by Darwin. On his return journey, he stopped over for five months at the California Academy of Sciences, studying the Academy's collections of finch skins (Lack 1973).

There was increasing evidence that most bird species had evolved via subspecies which had become differentiated as a result of geographical isolation, and remained distinct even after they came into contact with one another again. If such a mechanism was normal, Lack found it puzzling that closely related bird species should occur so often in the same region, but in different habitats. In his account of the Galapagos finches, hastily compiled in 1940, Lack attributed the differences in bill size to the preferences of groups of individuals within a species for different habitats, which ultimately resulted in the formation of species, each occupying one part of the habitat of the original species. Published as an occasional paper of the California Academy of Sciences, the account did not appear until after the war, by which time Lack had alighted on a very different hypothesis (Lack 1945a).

Having forwarded his name to the Central Register of Scientific Workers, Lack found himself drafted into military duties that involved touring the ship-watching coastal radar stations along the English Channel. Not only did this provide outstanding opportunities for detecting bird migration by radar (Lack & Varley 1945), but close association with working scientists, half of whom were biologists, caused Lack to reappraise the evolutionary aspects of his Galapagos data and, more particularly, how closely related species could resemble one another in their ecological requirements. Over the course of six months in 1943, Lack came to perceive the relevance of Gause's principle of competitive exclusion.

The pioneers of 'modern niche theory' were influenced not only by the mathematical population models of Volterra and Lotka, and the laboratory experiments of Gause, but also by the advances being made in population genetics (Jackson 1981). The publication of *Systematics and the Origin of Species* by Ernst Mayr (with whom Lack had stayed in 1939) affirmed that geographical location was the key factor in causing species to evolve (Mayr 1942). For Lack, however, there remained two difficulties. No matter how great the differences in habitat, there would always be opportunity for birds to come into contact with one another along the numerous border zones between habitats. No example could be found of incipient differentiation within bird species occupying adjacent habitats. At first, Lack had taken little notice of a hypothesis, set out by Julian Huxley, that size differences between related species in the same habitats might have evolved as a means of reducing competition (Huxley 1942). The hypothesis now took on a new significance.

Lack realized that where two forms, originally geographical forms of the same species, came into contact with one another, but remained distinct, they would compete at least to some extent ecologically. First in his contribution to the symposium of the British Ecological Society, and then in a paper in *Ibis*, Lack explained how a close study of species formation in passerine birds suggested that one of four situations could arise. One species might eliminate the other. Each might occupy separate geographical regions. Where they occurred in the same place, they might live in different habitats. If they occupied the same place and habitat, they would depend on different food (Lack 1944, 1945b). The reappraisal of the Galapagos data led to a new book being written, *Darwin's Finches*, which demonstrated how, over time, natural selection would have the effect of accentuating any ecological differences to the point where it was no longer necessary for two species to compete for essential resources (Lack 1947).

As the report of the symposium of March 1944 made clear, the support for explaining the striking differences in the ecology of species in terms of Gausian competition was 'fairly evenly balanced' (Harvey 1945). The vigorous attack made on Gause's concept was led by Diver, who claimed that the mathematical and experimental approaches had been dangerously oversimplified. Not only were there many examples of congeneric species living and feeding together, but there was little direct evidence to suggest that cohabitation or separation of related species was determined by competition for space and food. Other factors usually kept populations below the point at which serious pressures of this type could develop (Diver 1936).

Support for Lack's hypothesis as to the effects of competition on closely related species came from Charles Elton, both at the symposium and in a paper published later in the *Journal of Animal Ecology*. In an analysis of the published results of 55 ecological surveys of animal communities, and of 27 surveys of plant communities, Elton (1946) found that a very high proportion of genera in each habitat was represented by only one species. Where two species of a genus occurred together, they often proved to be of different ecotypes.

Considerable support for Gause's hypothesis at the symposium came from George C. Varley (*1910–83*) who, as a research student, had embarked on a study of the ecology of the knapweed gallfly (*Urophora jaceana*) in 1933. He had become a research fellow of Sydney Sussex College in 1935, and a university demonstrator and curator of insects in the University Museum in 1938. Throughout that time, he had been able to monitor the changes taking place in the nine herbivores, 15 parasitic Hymenoptera, and the predatory fly population making up the insect fauna of the knapweed flowers and seed heads of his census area near Madingley. The study provided a striking illustration of the value of acquiring life table information over as many generations as possible (Hassell 1983).

In the same year as Varley began his research, Alexander J. Nicholson (*1895–1969*) published a paper proposing a mathematical model for the inter-action of insect parasitoids with their hosts. The introductory paper stemmed from a larger study of the role of competition in animal populations, carried out in collaboration with V. A. Bailey, an assistant professor of physics at Sydney, where Nicholson had been before being appointed deputy chief of the Division of Economic Entomology in the Commonwealth Council for Scientific and Industrial Research (CSIR) in 1930 (Mackerras 1969). Nicholson (1933) demonstrated that wherever environmental factors enabled a species to exist the population would be held in a state of balance by a controlling factor that removed any surplus population. It achieved this effect by acting more severely against an average individual where population density was high, and less severely where it was low.

The only factor capable of exerting this kind of control was, according to Nicholson, the competition that occurred between animals seeking food or suit-able places to live, or between natural enemies that hunted for them. For each species, there was a particular density at which a balance existed. While such factors as climate and most kinds of animal behaviour whose action was not in-fluenced by the densities of animals could not in themselves determine popula-tion densities, they might have a significant influence on the values at which competition maintained these densities.

Although Nicholson's work received at first very little notice, Varley quickly recognized its value in interpreting the interaction between the various factors destroying the knapweed gallfly. By examining the quantitative effect of each factor separately, new insights could be gained into the insect community living in the flower-heads of the black knapweed. Particularly in its expanded form (Nicholson & Bailey 1935), the model supplied for the first time a key to analys-ing the mutual effect of parasitic and other factors of destruction on the popula-tion density of an insect, and was accordingly of considerable relevance to studies in insect ecology and the wider field of biological and insecticidal con-trol (Varley 1947).

The quantitative study carried out by Varley on the gallfly was not only among the first to highlight 'the inherent complexity of even a small fragment of the animal community', but it identified which of the food relationships was likely to be significant in controlling the numbers of a central species (Elton 1949). Varley believed such an approach could provide the common basis for resolving many of the problems arising in respect of the position of plants and insects in the ecosystem. In a contribution to an earlier joint meeting of the British Ecological Society and Royal Entomological Society in November 1943, he argued that the place of any species in an ecosystem would depend on its quantitative status, as expressed in terms of population density or weight per unit area. These measurements could only be deduced by closely investigating the reaction of each species to the physical environment, and its reaction with other species (Riley 1943).

It was Varley who had drawn David Lack's attention to the relevance of the work by Gause and Nicholson. They had first met in 1943 through Varley's wartime post in the Army Operational Research Group, studying centrimetric radar on the South Coast. In his contribution to the symposium of 1944, Varley explained how the outcome of competition between two closely allied species would be influenced by the relative severity of the two main types of mortality that were likely to affect them. Using the terminology first adopted by economic entomologists in America (Smith H. S. 1935), he drew a distinction between those forms of mortality which were density-dependent, and those that were non density-dependent. Where the latter failed to prevent an increase in population, numbers would rise until some density-dependent factor stabilized the population density at a higher level. As mortality factors were the agents of natural selection, any changes in the relative severity of their action was likely to change the direction of evolution.

By the late 1940s, Lack and Varley found themselves facing a new kind of criticism. No longer did critics argue so vigorously that closely related species could live together in the same habitat and have similar ecological requirements. Instead, they argued that there was no evidence to support the significance of competitive exclusion as a mechanism for ensuring that where related species occurred together, they differed in their ecology. The critics maintained that wherever any differences could be discerned, they had arisen quite incidentally as a result of the isolation of the species, one from another (Lack 1971).

Perhaps the best known critic of the importance attached to competitive exclusion and density dependence factors was an Australian, H. G. Andrewartha, whose interest in animal ecology had first been aroused by reading Elton's book *Animal Ecology*. Andrewartha believed that, despite the extraordinary breadth of Elton's writings, and the wealth of information provided by 'practising ecologists' in entomology and wildlife conservation, there had been a marked tendency for modern theoretical ecology to stress the interrelationships between animals at the expense of everything else. Andrewartha sought to redress the balance, first in collaboration with L. C. Birch, and then in a textbook written after he had been appointed to a readership in the University of Adelaide (Andrewartha 1961).

In their book, Andrewartha and Birch (1954) argued that it was scarcely possible or worthwhile to study whole communities of animals. They attempted instead to construct a general theory for determining the distribution and abundance of animal species by looking at each one at a time, with the other animals and plants being considered as part of the environment. It was an approach that succeeded in causing many ecologists to look afresh at aspects which had previously been obscure or neglected. It was, however, the authors' treatment of density dependence that caused the book to receive so much attention. They contended that the numbers of different animals were mainly determined by the shortness of the periods during which the rate of population increase was positive. Among the many agencies likely to affect the rate of population increase, the most prominent were weather and predators.

In a review of the book, written for the *Journal of Ecology*, Owain Richards conceded that the role of density-dependent factors might have been exaggerated, but it had always been obvious that any organism increasing geometrically could only be kept in ultimate check by a mortality which was dependent on the density of the species. The fact that a particular model happened to be wrong or 'unlifelike' in certain respects could not in any way undermine that conclusion. In their book, Andrewartha and Birch had emphasized the critical role often played by dispersal in determining how far available resources were used. No one denied the significance of the mosaic and changing nature of natural habitats, and the extent to which populations in large areas were made up of a number of small populations. What Richards (1955) did object to was the contention that these factors would somehow serve as a substitute for the periodic intervention of density-dependent factors. Again, no mathematical model was needed to prove what was intuitively obvious, namely that the interaction between a host and its predators would give rise to periodic fluctuations in their numbers, and that the bigger the complex of species involved the more likely it was that the amplitude of the fluctuations would be damped down.

In 1962, there appeared what David Lack called the third divergent theory of population dynamics, to take its place with those of Nicholson and Andrewartha and Birch. In that year Wynne-Edwards published a very full account of how the number and abundance of animals might be controlled through the social behaviour of the animals themselves. Wynne-Edwards had spent the greater part of his career at McGill University, Montreal, before becoming Regius Professor of Natural History at Aberdeen in 1946. While accepting the importance of density-dependent regulation, and the lack of food as the ultimate factor in limiting population numbers, Wynne-Edwards argued that it was through dispersive behaviour and restraints on reproduction that the animals regulated their own density well below the upper limit set by food (Wynne-Edwards 1962).

An outline of Wynne-Edwards's hypothesis had appeared three years previously in *Ibis*, when, in a series of examples, he demonstrated how conventional competition always took the form of a social phenomenon. The establishment of breeding territories and a social hierarchy (a pecking order) among bird species ensured that when some desirable resource such as food was scarce, the needs of an appropriate quota of the group were met, and the remaining individuals either went without or left the area. A very large class of social phenomena appeared to have evolved with the primary purpose of removing surplus populations and thereby preventing overexploitation (Wynne-Edwards 1959).

The wide diversity of theories reflected fundamental differences in the perception of evolution. As Lack remarked in his presidential address to the British Ecological Society in 1965, the principal differences over density dependence theory had arisen from the fact that Andrewartha and Birch refused to perceive it in terms of the survival of species. There was no place for evolutionary theory in their analyses. The dissension between Wynne-Edwards and himself arose from their interpretation as to how evolution was achieved. Wynne-Edwards argued that it came through group selection. Not only were there genetic arguments against group selection, but Lack believed such features as clutch-size and territorial behaviour could be explained much more easily in terms of natural selection (Lack 1965).

In Britain, research on the botanical aspects of population biology was largely confined to those working on agricultural problems. The principal exception was Salisbury. In his presidential address to the British Ecological Society in 1929, Salisbury described how the competitiveness, and therefore survival, of species increased as habitat conditions approached the optimum requirements of the individual species, in terms of light, water supply, soil reaction, supply of essential nutrients and other such factors (Salisbury 1929). During the 1930s, Salisbury came to regard the reproductive capacity of a species as being the characteristic of 'the greatest ecological importance'. It was not, however, until the outbreak of war, and 'an interlude of relative leisure', that he took up a suggestion of Ronald Fisher to write a book on the subject. The book was called *The Reproductive Capacity of Plants* (Salisbury 1942).

John Harper later recalled how, as a wartime student in the Botany School in Oxford, Salisbury's book[34], and visits made to the Bureau of Animal Population and Stapledon's Welsh Plant Breeding Station at Aberystwyth, provided the only intimation that plant populations might be counted, controlled, and manipulated, as a means of discovering how a plant community functioned (Harper 1977). For Harper, however, the implications of all this were for the distant future. It was not until the early 1950s that, as a member of Blackman's department of agriculture at Oxford, he secured a research grant from the Nature Conservancy, which enabled him to expand his earlier studies of the three commonest species of *Ranunculus* to include the five British species of *Papaver*, investigating such questions as the evolution of resistance to herbicides, ecological aspects of weed control, and the ecological significance of dormancy.

There evolved a small group made up of Harper, Ian MacNaughton (a geneticist), John Clatworthy (a Rhodes Scholar who focused on competition), and Geoffrey R. Sagar, who had just embarked on a study of the plant dynamics of three species of *Plantago*. They met regularly to discuss their own studies, and more usually 'to thrash out many of the more philosophical aspects of the evolution and ecology of closely related species'. A summary of these deliberations was published in the American journal, *Evolution*, having first been turned down by two 'prestigious' British journals, presumably because the treatment was so far removed from the current mainstream of ecological research in the plant sciences (Sagar 1985).

On the premise that populations of plant species were regulated or controlled by a reaction to density, the authors of the paper concluded that two or more species could persist together in a habitat only if their populations were subject to different controlling factors. These factors might include differences in site requirement for seedling establishment, susceptibility to parasites or predators, germination time, and requirements for breaking dormancy. Where isolation, differentiation, and divergence were followed by migration and their coming into contact with one another, the extent to which both species could persist would depend on how far their biological properties fulfilled complementary, rather than mutually exclusive, roles (Harper *et al.* 1961).

By the early 1960s, a burgeoning interest in population biology could be discerned in both plant and animal ecology. In many ways, plants made better material for study—they stood 'still to be counted and did not have to be trapped, shot, chased, or estimated' (Harper 1977). Whatever the species under close study, there were pitfalls to assuming that a statistically significant correlation always implied a causal relationship. There were so many different variables being measured. A biological mechanism had to be involved, as well as an arithmetical one (Varley 1957).

An obvious course was to mount long-term studies of individual populations. The role of birth and death rates became quickly apparent in a study of the tawny owls (*Strix aluco*) in Wytham Wood, conducted by Southern (*see* p. 117) and other members of the Bureau of Animal Population at Oxford over the period 1947–59. In terms of population control, everything hinged on what Southern (1959) called 'repression'. Although the owls tried to produce the greatest number of young they could rear, they were, except in crash years, in 'a perpetual state of over-optimism'. Density-dependent mortality began to redress the balance at an early stage, not only through the starvation of the young, but through the prevention of their being hatched or even conceived.

It was one thing to highlight the role of 'repression' or some other factor, but quite another to pinpoint how it was achieved. In his presidential address to the British Ecological Society in 1969, Southern recalled how 13 years of hard research had convinced him that knowledge about ecological processes came only gradually and painfully. In some way or other, each succeeding year proved unique. Only a long series of records was likely to include the extremes in situations that were so revealing to the ecologist (Southern 1970).

3.5
The Nature Conservancy

From the vantage point of the jubilee conference of the British Ecological Society in 1963, Pearsall described the creation of the Nature Conservancy as the most important post-war development in ecology. The combination of nature reserves and 'ecological research institutions' represented far more than 'the occasional nucleus of ecological research' that had characterized the 1930s. Out of the search for techniques for managing nature reserves there would arise innumerable opportunities for survey and experimentation. In appraising the merits of conserving a site, whether an oakwood or cotton grass moor, new concepts were required, that embraced the corresponding natural ecosystems, the processes that determined their existence, and the measures needed to regulate them (Pearsall 1964).

Under the terms of the Royal Charter, the Nature Conservancy was required

> to provide scientific advice on the conservation and control of the natural flora and fauna of Great Britain; to establish, maintain and manage nature reserves in Great Britain, including the maintenance of physical features of scientific interest, and to organise and develop the research and scientific services related thereto.

The Conservancy's capacity to respond to its Charter functions was seriously inhibited by the economies imposed on public expenditure. Annual reports warned of how, unless the budget was increased, the acceleration of reserve acquisitions would be 'throttled', and the management of existing reserves crippled. Fresh demands and commitments were arising all the time. The conservation and research programmes had to be adapted to take account of such unexpected events as the North Sea floods of early 1953 and the rapid spread of myxomatosis among the rabbit population in the following year.

Some insight into the difficulties faced by the Conservancy may be gained from contemporary correspondence. In a letter to William Pearsall in January 1950, the Conservancy's administrative secretary, Frederick Bath (who was on secondment from the Treasury), complained of a lack of 'agreed aims of a practical nature on policy, priorities and pace of actions'—the staff could not see the wood for the trees. As Bath remarked,

> the devil is the 'trees are not all trees'. There are bushes, creeping mosses, bogs, moors, vague mountains and mists, pits, shadows, baskets of acrobatic monkeys, &c, &c, all mixed up topsy turvy in the 'wood', all falling over each other all the time and producing a regular aladin lamp appearance of daily transfiguration[35].

Bath defined three important objectives for the Conservancy. The first priority was the 'declaration of nature reserves both for conservation as an end in itself and for permanent major field research laboratories'. The second aim should be research, both in-house and by supporting outside agencies. The third priority was advisory work and systematized information and other services. It was for that reason that so much importance was attached to supplying local planning authorities with county schedules of Sites of Special Scientific Interest, as required under the National Parks and Access to the Countryside Act of 1949. Once assembled, these schedules would serve as a 'foundation of sound information on places of scientific value'.

Many of the sites identified as important for wildlife in the 1940s were being threatened or destroyed as a result of changes in land use and management. In practice, scheduling conferred very little protection for sites. From the experience of the 1930s, the main threat was expected to come from building and industrial development. Under their new powers, it was hoped that planning authorities would pay considerable attention to the fact that a site was scheduled before deciding whether to allow development to take place. Not even that degree of uncertain protection was available when grassland was ploughed, or woodlands felled.

A very complacent view had been taken of the impact of farming on wildlife. In his book *Our Heritage of Wild Nature*, Tansley (1945) wrote of how

the great extension of agriculture during the war has not on the whole diminished the beauty of the countryside—rather the contrary is true ... In places no doubt heath has been destroyed and fenland drained and ploughed, and some of these changes have been deeply regretted by the naturalist and the lover of nature. But the total loss has not been very severe, and is offset by the gain in the agricultural area.

To Tansley, writing in 1944, it seemed 'scarcely probable that the extension of agriculture will go much further, for the limits of profitable agricultural land must have been reached in most places'.

It was therefore a great shock to discover that changes in the agricultural use and management of land were the most important single factor leading to the loss of sites in the 1950s, and that little or nothing could be done to regulate them. As the Conservancy's senior botanist, Verona Conway, told a meeting of the Linnean Society in early 1953, it was a major deficiency in the planning legislation that farming and silvicultural practices did not constitute forms of 'development'. The omission meant that even if a site was scheduled its scientific interest could be destroyed without the Conservancy being informed, let alone consulted (Conway 1953).

How well equipped were ecologists to respond to the range of challenges confronting the infant Nature Conservancy? At a Council meeting of the British Ecological Society in November 1948, Diver and Tansley appealed to experienced ecologists within the Society 'to give time to the survey of extensive areas of country ... even if it involved some personal sacrifice'[36]. Not only would the Conservancy experience 'extreme difficulty' in recruiting suitably qualified staff but, as Elton warned in his presidential address of 1949, ecologists had 'to do a great deal of hard thinking' in deciding how to survey the wild plant and animal life of both the reserves and the wider countryside (Elton 1949).

Many shared Pearsall's cautious approach to the Conservancy's responsibilities. Pearsall became chairman of the Scientific Policy Committee in early 1953, and was personally responsible for drawing up criteria for the acquisition and management of reserves, and for identifying ecological programmes of research[37]. Some insight into the kinds of experience which he brought to these tasks was conveyed in a lecture he gave to an audience in India in 1955, in which he extolled 'the conservative policy followed by the Freshwater Biological Association' in concentrating on one initiative at a time, namely the establishment of a laboratory, before attempting anything more ambitious. This approach had proved far more fruitful than that of the Council for the Promotion of Field Studies, which had become so overstretched that Pearsall, when he became chairman, was forced to carry out 'the unpleasant task of curtailing expansion and reducing administrative costs and aims to practical levels'[38].

It was not until the Conservancy's fifth and sixth Annual Reports that substantial progress could be reported. During 1955, the number of National Nature Reserves doubled to 35, covering a total of 32 000 hectares. Every part of England, except the Midlands, had a regional officer. A main research station was established at Merlewood in the Lake District, with a smaller station at Furzebrook in Dorset. The first significant increase in budget occurred in 1956, when it rose by just under a fifth to £280 000 per annum. After taking written and oral evidence (and visiting several reserves), the House of Commons Select Committee on Estimates issued a report in 1958, endorsing the importance of the Conservancy's work.

With hindsight, the years 1959–60 marked an important turning point when, as Max Nicholson remarked, 'recent events on this planet and also on the moon' created an atmosphere which was much more favourable for securing 'the necessary resources within the framework of an expanded scientific effort'. Where there had been only 115 staff in 1954, and 194 in 1959 (of whom 76 were scientifically qualified), the number rose to 278 by 1961 (of whom 111 carried out scientific duties). The Annual Report of 1961 identified 92 reserves, covering 72 000 hectares; the number of reserves rose to 100 in 1962, and to 111 by the time the Conservancy's annual report for 1964 was published.

Many of the acquisitions included sites that had been recommended on the basis of earlier research. In its account of the new National Nature Reserve of Rostherne Mere in Cheshire, the Annual Report for 1961 noted how it had been identified as worthy of preservation in the list drawn up by the SPNR in 1915. Studies on the biology of the mere by Pearsall in 1923 and by E. M. Lind in 1944 had shown an extremely high concentration of phytoplankton. The Nature Reserves Subcommittee of the British Ecological Society had recommended it as a national habitat reserve. The Huxley Committee described it as the last example of an unspoilt mere in Cheshire. The fact that detailed records of its birdlife had been kept for over 40 years greatly enhanced the value of the reserve as a wildfowl refuge.

An obvious source of expertise during the 1950s were those ecologists who had gained experience in the management of wildlife and natural resources in the Empire. Barton Worthington was appointed the Conservancy's first deputy director (research) in 1957, having spent the previous decade establishing regional research organizations in East Africa and latterly as the secretary-general of the Scientific Council for Africa South of the Sahara (Worthington 1983).

One of the earliest appointments to the Conservancy was that of Arthur S. Thomas (*1902–84*), whose papers in the *Journal of Ecology* had drawn on his detailed knowledge of Uganda, acquired as the senior botanist in the Agricultural Department of that country (Thomas A. S. 1941, 1944, 1945). Thomas had already begun to record the grassland vegetation of the national nature reserves of southern England when the first outbreaks of myxomatosis were confirmed. Over a period of eight years, the data derived from 100 000 point-quadrats taken from 10 sites in lowland England and south Wales emphasized the degrading effect which rabbit-grazing had exerted on the chalk flora, and the extent to which there had been a marked increase in floristic richness since the outbreak of the disease (Thomas A. S. 1960, 1963).

The continued difficulties in securing trained staff caused the Conservancy to investigate a variety of ways of extending the support given in the form of postgraduate awards (of which over 100 had been made by the end of the 1950s). In a memorandum on 'Higher education in ecology', Max Nicholson wrote in July 1959 of how the training of ecologists was 'exceedingly difficult and complex'. Besides having a genuine interest in ecology, trainees required 'a considerable depth and range of higher biological education' and some insight into the earth and social sciences. Above all, they needed adequate field experience. The most practical course was to form a partnership, whereby a university would initiate a one-year diploma or degree course in ecology, with the Conservancy providing salaries, maintenance grants, and field facilities.

Through the good offices of Pearsall (who had retired from the Quain Chair of Botany at University College, London, in 1957), discussions with the provost of the College led to the inception of a one-year postgraduate diploma course in conservation at the College in 1960, for which the Conservancy provided studentship awards and field facilities. The first year's students took up posts in the Conservancy (2), the Forestry Commission, a county planning authority, the Weed Research Organization, and, in the case of the sixth student, a Conservancy research studentship.

It was not long before a generation of ecologists who had come straight from postgraduate research began to exert an influence on the formulation of management programmes. Problems were particularly acute on the Woodwalton Fen National Nature Reserve, which had been leased to the Conservancy in 1953. By that time, much of the Reserve was covered by scrub. The distribution of plant communities in relation to differences in soil drainage and the history of the site was described by Duncan Poore in his thesis, and published in the *Journal of Ecology* in 1956 (Poore 1956). Poore was the obvious person to draft the management plan for the Reserve, outlining ways of coping with the rapid vegetational changes. By 1958, considerable progess had been made in re-excavating the dyke system and raising the water-table.

The regional officers of the Conservancy were expected to draw up detailed management plans for each reserve, in consultation with relevant experts and land agents, both within and outside the Conservancy. Having summarized the background data available, and defined the objectives of management, each plan set out a programme of operations and relevant research. Plans for 11 National Nature Reserves had been approved by 1959, and a further 19 were in preparation. In the words of one annual report, each represented 'a major exercise in conservation research', involving

> many branches of ecology as well as a knowledge of estate management
> and an appreciation of human as well as plant and animal relations.

It was confidently expected that the principles emerging from these plans could be applied not only to nature reserves, but to areas used for agriculture, forestry, recreation, amenity, and other purposes.

One of the most ambitious plans was for the Isle of Rhum in the Inner Hebrides, following its acquisition as a National Nature Reserve of 10 684 hectares. In a memorandum of October 1957, William Pearsall wrote of how the island would fulfil the same kind of role for upland management as the Rothamsted Research Station had provided for lowland Britain over many years. Except for the prolific writings of Edward Wyllie Fenton (*1889–1962*) (who succeeded William Smith as head of biology in the Edinburgh and East of Scotland College of Agriculture), very few detailed studies had been published on the management of Highland plant communities. An advocate of comprehensive biological surveys, Fenton had focused on the influence of man and domestic herbivores on hill grazings and upland pastures (Fenton 1937, Gregor 1964).

The need for a wide and integrated approach to the problems of the region had been given considerable publicity through the writings of Frank Fraser Darling (*1903–79*). His detailed field studies of the red deer, seals, and sea birds from the late 1930s onwards had demonstrated not only the close links between social behaviour and the population dynamics of individual species populations, but also the significance of human influence, both past and present. The impact of such bad management practices as the careless burning of Highland vegetation was strikingly illustrated in Fraser Darling's book, *The Natural History of the Highlands and Islands'*, published in 1947 (Darling 1937, 1947).

By acquiring the island of Rhum, it was the explicit intention of the Conservancy to set up large-scale trials to discover the reasons for soil erosion and deterioration of the vegetation cover, and methods by which the badly devastated habitats and fertility of the land could be restored. The ultimate goal was to embark on cropping systems and afforestation schemes, which would provide both a higher return and greater protection for the soils and vegetation[39].

All this was for the future. Such long-term aspirations did nothing to protect the Conservancy from charges of 'sterilizing' extensive tracts of farmland and deer forest, and of restricting public access. The fact that the grazing tenancy of the island had fallen vacant enabled the Conservancy to reduce the deer population straightaway as a prerequisite to introducing controlled experiments on the effects of grazing and burning on rates of vegetation recovery, the behavioural patterns of the red deer (*Cervus elaphus*), and the character of surface run-off of water. So as to ensure 'absolute freedom from disturbance', all rights of public access were withdrawn. The result was a storm of protest.

The Conservancy responded to criticisms by insisting that it would be a betrayal of its trusteeship if there was any trimming of scientific findings and management plans simply 'to satisfy sectional and opportunist views'. Conservation was 'for all time and should not be continually subordinated to issues of expediency which hold only for a decade or two, and often barely for that long'. The perceived hazards from public access were not imaginary. Some compromise was, nevertheless, called for. The Annual Report of 1958 included an access map, indicating where the general public could make day visits in summer. The map was taken from a leaflet distributed to visitors, setting out the Conservancy's aims and management policy.

Whilst there could be no denying the achievements of the first 10 years, there was also great concern over the management of all the reserves. That concern was encapsulated in an internal memorandum drawn up by two of the Conservancy's longest-serving regional officers, Eric A. G. Duffey and Norman W. Moore, in June 1958. They began by recounting the losses of species and habitats on the reserves in their respective regions, East Anglia and south-west England. Even where active management was carried out, there was only the most rudimentary understanding of the special needs of plants, let alone animals. With the acquisition of further reserves and the passage of time on others, the consequences of incompetence and inadequate resources could only grow worse[40].

With their increasing administrative responsibilities, the Conservancy's regional officers were finding it difficult to sustain even the few pieces of conservation research that had been initiated. Norman Moore had set up pilot schemes in 1957 to investigate the autecology of the Dartford warbler (*Sylvia undata*) and heathland reptiles, and the effects of heath-burning. The results of much of this survey work were brought together in a paper published in the *Journal of Ecology* in 1962, which chronicled the decline and fragmentation of the Dorset heathlands, and the implications of these changes for the abundance and distribution of 10 plant and animal 'indicator' species (Moore N. W. 1962). In their memorandum of 1958, Duffey and Moore recommended that an 'expert' should be appointed for each main type of habitat, who would not only carry out research on problems arising out of conservation management, but who 'would be responsible for giving general advice on conservation'.

As the Conservancy's 'woodland expert', Derek Ovington quickly discovered how little information was available on managing woodlands for conservation purposes. Although considerable importance was attached to setting aside areas to develop naturally, it was clearly impossible to reproduce exact replicas of ecosystems of the past. Even greater problems arose where efforts were made to sustain or revive woodland management practices. No one knew the precise effects of coppicing on invertebrate species in such localities as the Ham Street and Blean Woods National Nature Reserves in Kent. There was virtually no experience on which to base a programme for the natural regeneration of the native Scots pine forests in the Highlands (Ovington 1964).

In a paper to the jubilee symposium of the British Ecological Society in 1963, Ovington recalled how the first important step had been to recognize that all the woodlands on nature reserves were in effect seminatural communities. Once that important psychological step had been taken, it was easier to adopt 'an experimental approach to management', where solutions were likely to be found not only through empirical and experimental studies conducted on the reserves but over a wider woodland area. The Conservancy had embarked on a programme of fundamental research on woodland ecosystems, embracing not only living organisms and their remains, but the whole complex of the environment.

To a significant degree, the British Ecological Society had achieved its goal of establishing an ecological research council. Whatever the limitations of the Nature Conservancy in terms of its aims and resources, the foresight of those who played so large a part in its creation had been fully borne out. By 1959, nearly half the Conservancy's expenditure was on scientific work, of which about a fifth was used to promote research and training in universities and other independent bodies. The statutory responsibilities of the Conservancy had highlighted in the most tangible way possible the need for ecological expertise, in respect of managing both the nature reserves and the environment generally.

A working relationship was all the time evolving between the Conservancy and the British Ecological Society, whereby the officers, members of Council and ordinary members of the Society were strongly represented in all parts of the Conservancy, and reports of research done by the Conservancy's staff featured large in the Society's journals and meetings. Never in any sense formalized, it was a relationship which both bodies saw the greatest advantages in developing.

Part 4
The Third Quarter-Century, 1964–1988

4.1

Overhauling the Society's Affairs

A shift in emphasis may be called for in recounting the contributions and activities of the third quarter-century of the British Ecological Society. First, the considerable diversification of research interests, concomitant with ecology's expanding institutional base and the growing number of practitioners, makes even a concise historical appraisal doubly difficult (Duff & Lowe 1981). Too little time has elapsed for the advances in ecological thinking and practice, and the achievements of individual workers, to be seen in any kind of detailed, historical perspective.

Secondly, and more positively, there were subtle, yet significant, shifts in the weight given by the Society to its various responsibilities. The Society continued to be inward-looking (never before had members derived so much material support from the Society), but it also became a much more outward-looking body. As well as playing a crucial role in supporting scholarship through its journals and meetings, the Society was involved to a much greater extent than hitherto in promoting ecology more widely—trying to ensure that ecologists were heard wherever environmental issues were being discussed, whether in the corridors of power or beyond, at home or overseas.

The Society could take justifiable pride in the way it had retained its traditions and yet expanded those activities which befitted the modern trends in ecology. It was already one of the largest scientific societies in Britain. The number of members continued to grow. Within 11 years of attracting its first thousand members, the Society passed the 2000 mark in 1968. By the mid-1970s, the membership total had reached 3000, of whom 63% were resident in the United Kingdom. Eighteen per cent lived in North America, 7% in Australasia, and 6% on the continent of Europe. The total rose by a further thousand members in the next decade, reaching 4322 members by the end of 1985.

New categories of membership were devised. Persons of 65 or more years, no longer in full-time employment, and who had been members of the Society for at least 30 years (a figure later reduced to 25 years), became entitled to receive the journals at half the cost to ordinary members. The number of retired members reached 51 by 1985. Council finally agreed to a category for student membership in September 1973, whereby a bona fide, full-time student of any institute of higher education in the United Kingdom could become an associate or ordinary member for a period of three years (which might be extended to five), paying half the normal subscription rate. By the end of 1985, there were 649 student members, representing 15% of the total membership of the Society.

Nothing had come of the fears expressed at the Society's first meeting in

1913 that its scientific and professional status might be impaired by the fact that anyone who professed an interest in ecology could join the Society. Some 60 years later, there was, however, some cause for concern. Even the most complacent members of Council had to admit that the word 'ecology' was being used ever more loosely. There was a risk of 'non-ecologists' joining the Society with the explicit intention of using it as a vehicle for influencing 'political decision-making in the country'. A working group was appointed, made up of the president, honorary treasurers, and two members of Council. As a first move, it recommended that, in order to preserve the Society's traditional character 'as predominantly a body of working scientists', the constitution should be amended to make it explicitly clear that

> the objects of the British Ecological Society shall be the promotion and fostering of education, learning and research in Ecology considered as a branch of natural science.

The change was effected in 1976. Further alterations to the constitution were approved by the Annual General Meeting of January 1977. The category of associate membership was abolished. Council was to consist of the officers and 15 ordinary members, five of whom were to be elected each year to serve for three years (previously four members had been elected annually to serve for four years).

There was little point in increasing the turnover of Council members unless a conscious effort was also made to ensure that those elected represented a wide spectrum of viewpoints and interests. It was almost a tradition for the Annual General Meeting to accept without challenge or ballot the nominations made by Council. As the Society grew in size, and its affairs became more complicated, Council members were often unfamiliar with the scientific and other interests of potential candidates. It became increasingly common for Council to appoint 'search committees' as a way of involving more members in the choice of potential Council members and officers. In doing so, a conscious effort was made to secure a better balance between plant and animal interests, to reduce the heavy bias towards university academics, and to ensure that there was, for example, one member from Ireland. Progress was slow. Of the 30 members of Council in 1980, 21 were from universities, three from research institutes, three from industry/local authorities/Nature Conservancy Council, and one from secondary education.

The choice of honorary members also came under scrutiny. There were 14 such members in 1976. The honour was clearly intended to recognize service to ecology as well as academic excellence. Where the candidates were resident overseas, it was essential that they were esteemed in their own country as well as in Britain. Six of the nine honorary members elected in 1977 were from overseas. A proposal that the criteria should be widened to include those persons who had given signal service to the Society was rejected by Council in April 1982 on the grounds that such a change would produce 'a very large back log of worthy past servants of the Society, who ought in equity to be considered'.

There was little point in introducing a more open and systematic approach to choosing persons to run and represent the Society if there was no means of improving the flow of information between Council and honorary officers on the one hand and the wider membership on the other. By September 1969, the introduction of a house journal, to be issued to all members three or four times a year, had become 'the top priority consideration for Council'. The Society's editors gave their full support to the venture, recommending that the management and policy of the house journal should be kept entirely separate from that of the other journals. By the spring of 1970, the only outstanding question was the name of the house journal. On a vote, it was decided to adopt the title *The British Ecological Society Bulletin* rather than an alternative suggestion, 'Ecological News'. Because the *Bulletin* would not be a publication in the formal sense, Council decided it would not be appropriate for it to include comments or criticisms on papers published in the journals. The Society's constitution was amended at the following Annual General Meeting to enable members to receive the *Bulletin* at no extra charge.

From the earliest discussions, great stress was laid on the larger role of the *Bulletin*, namely that of bringing the Society into the late 20th century and ensuring that it survived as a unified society. As Palmer Newbould wrote in the first issue, there was much more that the Society should do, but the initiatives had to come from the membership rather than 'the establishment of the Society'. He recalled how the Society had started 'as a small club of all members knowing each other'. As it grew, its affairs passed into the hands of a self-perpetuating oligarchy. The Annual General Meeting became largely a formality. It was rare for more than 10% of members to attend any meeting, and it always seemed to be the same 10%. By producing the *Bulletin* four times a year, and requiring Council to report its decisions more fully within it, members would not only become much better informed, but they would have an opportunity, through the pages of *Bulletin*, to communicate with one another, fly kites, and solicit opinion.

The other important role for *Bulletin* was to help strike a balance between giving every possible encouragement to the specialist groups within the Society and maintaining a cohesiveness of interest and outlook among the growing membership. In an introduction to the first issue of *Bulletin* in June 1970, the president, Jack Harley, referred to the new concentrations of interest, as exemplified by the Tropical Ecology and Industrial Ecology Groups. 'The boundaries of the old areas of diversity' were being broken down, and ecologists with different skills and outlook focused on new problems. It would be the aim of the *Bulletin* to draw attention to the *full* range of activities going on within the Society. The editors hoped the more relaxed approach of the *Bulletin* would encourage a freedom and depth of discussion that were not generally possible in the 'official' publications of the Society—a sentiment borne out several years later, when Peter Hogarth of the biology department at York initiated a series of contributions on the ecology of dragons.

4.1.1 *A learned society run by a 'load of amateurs'*

Every president closed the last meeting over which he presided with an expression of gratitude for the support he had received. On this and other occasions, A. D. (Tony) Bradshaw spoke of his continued astonishment at how 'such a large and flourishing Society could be run successfully by a "load of amateurs"'. Not only did Council members and their officers lack formal training in the running of large organizations, but none could give his whole attention to the affairs of the Society. However great their commitment to the Society, their careers in research and teaching had to come first.

When Frederick Whitehead came to present his 10th and last Annual Report as one of the Society's honorary treasurers in 1965, he was able to record an increase in reserves to £18 000. The threat of bankruptcy had been removed and, in his words, the time had arrived when substantial sums could be set aside for reforming the administration of the Society and for embarking on initiatives which had previously been regarded as too expensive. A subcommittee was set up to consider suggestions as to how the increasing prosperity of the Society might be used to promote the subject of ecology and improve the quality of the Society's meetings generally. Its report of April 1966 confirmed that the greatest priority should be given to reducing the workload of the honorary officers. Although it would be wrong for the Society to commit itself to a permanent secretariat, and therefore the maintenance of an office, better ways had to be found for coping with routine administration, especially in respect of membership records.

After considerable hesitation over the charges imposed for providing such a service, Council agreed to the administrative office of the Institute of Biology taking over the membership subscriptions, changes of address, and general enquiries. In a further move to simplify administration, these duties, together with the dispatch of journals and all other membership business, were centralized in the more spacious office accommodation at Harvest House, which was leased from the Society of General Microbiology in 1972.

Whilst these changes represented an enormous improvement in the day-to-day running of the Society, a much more ambitious response was needed if the vacancies on Council and the posts of the honorary officers were to be filled by leading ecologists. In a further review of administrative arrangements in 1970, Clifford Evans (in his capacity as one of the vice-presidents) insisted that there should be no further delay in removing 'the worst burdens from the shoulders of the small number of the hardest-working members of the Society'. Despite the accummulated reserves of the Society having grown to £45 000, and the Council having far 'greater freedom of manoeuvre in managing the Society's affairs than ever before', there remained a marked reluctance to embark on any radical measures. Beyond agreeing to further secretarial and editorial assistance, the discussions were inconclusive.

Decisions could not be postponed indefinitely. Eric Duffey (who had succeeded David Le Cren as Council secretary in 1964) spoke of his increasing concern over the inadequacy of the Society's organizational structure to cope with existing demands, let alone any new activities contemplated. The growing public interest in the environment had brought the Society into much greater prominence, and had resulted in a considerable increase in correspondence and liaison with other organizations.

Council could no longer postpone making a choice between introducing a more centralized administration, involving the appointment of a full-time executive officer, or the alternative of strengthening the honorary officer system. The latter approach came to be known as the 'federal scheme of organization', and its merits were set out in a document circulated to Council. The federal scheme would be far more flexible, involve a greater number of members in the actual running of the Society, and would avoid tying the Society to a particular location or level of expenditure. Despite the dramatic increase in the Society's reserves, they were still less than the annual running costs—the failure of even one of the Society's publications would be enough to extinguish them. If the federal scheme were introduced, it would still be possible to introduce a greater degree of centralization, if this was found to be necessary in the light of experience.

At its meeting of March 1974, Council agreed to the adoption of the federal approach and the more detailed recommendations designed to ensure that individual items of business received adequate attention at the committee level, leaving Council free to examine the longer-term planning issues. Considerable importance was attached to a new committee, the Finance and General Purposes Committee, which would carry out the executive functions of Council between meetings and prepare a budget of expenditure. In deciding on the procedures to be followed by the different committees, the aim was to avoid too large a share of work falling on any individual officer and to ensure that the elected members of Council became much more closely involved in the work of the committees. In a review of the measures taken to streamline the Society's business, some three years later in 1977, it was generally agreed that meetings of Council had become 'shorter and executive action crisper'.

Although the primary purpose of the Society was to provide members with copies of the journals as cheaply as possible, good financial housekeeping demanded that membership subscriptions should cover the run-on costs. By 1969–70, there was a shortfall of over £1 on every member's subscription. A slight overall deficit in the Society's accounts for that year prompted not only a further increase in the rates to non-members, but an urgent review of membership subscriptions, which had not changed since 1958. Over that period, production costs had doubled, and administrative expenses had quadrupled, largely as a result of the doubling of the number of members. In view of the high rate of inflation and the need to delay any further increase for as long as possible, the existing rates were more than doubled from January 1973 onwards. For the first time since the war, there was a slight fall in membership (from 2754 in 1972 to 2726 in 1973).

Soon even these rates were inadequate. In the autumn of 1975, the membership treasurer warned of how, if the high rate of inflation continued and there was no increase in the membership subscription, every copy of a member's journal would have to be subsidized by about £4.50 from the sales to non-members. The total subsidy of £18 000 would exceed subscription income. The price of each journal to non-members had already been set at £30 per volume, twice the price of 1973. In order to reduce the increasing disparity between member and non-member rates, Council unanimously agreed to a membership rate of £9, which would include one free journal. A further journal could be purchased for an additional £5. The new rates took effect in January 1977. This time there was no fall in membership. Rather than resign, many members decided to economize by taking fewer journals.

The outcome of these prudent acts of housekeeping was quite unexpected, namely a large surplus of £58 829 (or the equivalent of £18 per member) in the annual accounts of 1977, and the appearance of a letter in the *Bulletin* demanding that subscriptions should be reduced or the level of support given to the Society's activities increased. In their response, printed in *Bulletin*, the president and honorary treasurers denied that it was Council's policy to amass surpluses of this magnitude. The Society had not been alone in making inaccurate forecasts. Under its rules, membership rates had to be set 18 months ahead of their being imposed. At the time the 1977 rates were set, inflation was approaching 20% and the dollar was strong. Not only was it better to err on the side of caution in adjusting rates, but a sense of perspective was required. Ever since the Society nearly went bankrupt in the late 1950s, it had been the policy of Council to build up reserves to a point where they would cover journal production costs for one year, on the premise that this would provide a 'breathing space' in the event of a crisis. The value of the Society's investments in mid-1977 was £122 154, compared with production costs for the year of £140 000. It would soon be necessary to divert funds into acquiring further assets.

In a report of September 1980, the Finance and General Purposes Committee set out four priorities in expenditure, namely:
1 Meeting the running costs of the Society.
2 Topping up the emergency fund.
3 Ensuring a high standard of provision for meetings and publications.
4 Sponsoring of research and education in ecology.
Having obtained the approval of the Annual General Meeting of 1979, the Society's constitution was further amended so as to confer both a charitable and company status on the Society. Forms were sent to members, together with a letter and self-addressed envelope, inviting them to resign from the British Ecological Society and to join the new company of the same name. Each mem-

ber would continue to pay an annual subscription, and would be entitled to attend meetings, vote in elections, and be eligible for election.

Meanwhile, discussions were taking place as to whether the Society should at long last acquire a headquarters. The advantages of having a central London address, and library and meeting facilities, were stressed. In April 1981, the Finance and General Purposes Committee recommended unanimously that the Society should proceed to merge its administration with that of the Linnean Society at Burlington House in Piccadilly—a move that took effect from January 1983.

Although it was now commonplace for Council minutes to refer to the Society being 'a very profitable organization', members were acutely conscious that this affluence depended on the sale of the Society's journals and the state of the stock market. The increase in membership in 1986 from 4322 to 4373 members was the smallest for some years. Journal sales made up 42% of the Society's income in 1985/6, compared with 27% from membership subscriptions. As a result of the policy adopted in the early 1970s of using a large part of the profits from the sales of journals to build up the Society's cash deposits and share portfolio, income from these sources made up 30% of the total income in 1985/6.

4.2

The *Journal of Applied Ecology*

In some ways, little had changed since the founding of the Society some 50 years ago. The most obvious way to discover what was going on in British ecology was to read the journals, and attend the meetings, of the British Ecological Society. One of the most effective ways of publishing research data was to give a progress or interim report to a meeting of the Society, and then, in the light of comments and discussion, to revise and expand the paper with a view to submitting it to one of the Society's journals.

The idea that the Society might publish a third journal, with the title 'Applied Ecology and Conservation', first appeared in the minutes of Council for March 1961, under 'Any other business'. In a paper to Council, the president, James Cragg, described how the purpose of such a journal would be 'to give form and direction to critical studies on the applied ecology of natural biological resources with special emphasis on terrestrial and freshwater habitats'. Previously professor of zoology in the University of Durham, Cragg took over from Verona Conway as Director of the Nature Conservancy's Merlewood Research Station in 1961.

Cragg identified eight subject areas that might be covered by the journal. These were:

1 Development of techniques of assessing animal numbers and plant cover in relation to conservation problems.

2 Management problems in maintaining and utilizing biological resources.

3 Production studies concerned with all types of habitat.

4 Experimental assessments of treatments such as burning, grazing, draining, and flooding.

5 Investigations on the establishment of animal and plant species as ecological tools for the modification and development of natural resources.

6 Studies on the selection and treatment of water catchments and problems of impounding water reserves.

7 Investigations on the grouping of biological resources within an ecological framework.

8 Theoretical and philosophical studies on the concept of conservation.

Although Council was generally in favour of a third journal primarily devoted to conservation, there was by no means unanimity. Joyce M. Lambert of the botany department at Southampton expressed her misgivings in a paper to Council in February 1962. While not questioning the ability of ecologists to play an important part in conservation studies, Dr Lambert believed it was of the utmost importance that the Society should maintain 'a complete independence of outlook on ecological matters'. It should not commit itself to any partial attitude or point of view in any of its permanent publications. This meant that if such a journal were launched the editor would have to be of a first-class calibre, independent of the conservation interests of any particular body. It was very unlikely that such an editor could be found.

Dr Lambert contended that it would be wrong to add to 'the increasing flood of periodicals' unless it could be shown that there was a large number of top-quality papers being written which were not being published because of the lack of a suitable journal. In point of fact, there was no difficulty in finding appropriate journals for papers falling into most of the categories identified by Cragg. Not only was there no obvious reason why conservation should be given primacy over the competing claims of other specialist interests within the Society, but there was no evidence to suggest that a conservation journal would be the best mechanism for relieving the pressure on space in the Society's existing journals. Dr Lambert proposed that if, after careful appraisal, Council believed a new journal was necessary 'the best general purpose solution' might be a 'Journal of Applied Ecology' (omitting all reference to conservation in the title), which would include such topics as certain aspects of forestry and agriculture, coastal defence, and ecological control of waterways, as well as wildlife conservation.

There followed a paper by Palmer J. Newbould, of the botany department at University College, London. On the premise that conservation proceeded mainly by the synthesis of ideas derived from a variety of disciplines, including those of ecology, geography, engineering, economics, and law, it merited a journal where these ideas could be brought together. Such a journal could act 'as a valuable two-way forum, placing ecological ideas before practitioners in other disciplines and vice versa'. There were only two bodies capable of launching such a journal—the Nature Conservancy and the British Ecological Society. Although it was likely that many of the Conservancy's staff would contribute papers and perhaps act as editors, Newbould thought it would probably be better if the journal remained independent of the Conservancy. As for the British Ecological Society, the Society was not only 'well placed to vet conservation papers scientifically', but it had an obvious role to play in helping ecologists to meet the challenge of producing data and ideas which had practical application in land use and management.

At the Council meeting of April 1962, it was decided that a journal with the title 'Journal of Applied Ecology and Conservation' should be published in two parts each year, with the first annual volume consisting of 500 pages—a decision reaffirmed by a large majority at the next meeting of Council in September of that year. Clapham and Lambert were invited to join an enlarged Publications Committee. A. Hugh Bunting, the professor of agricultural botany at Reading, and Vero Wynne-Edwards, the Regius Professor of Natural History at Aberdeen, were nominated as the joint editors.

Doubts as to the title and field of interest of the journal persisted. Later in

September 1962, the Publications Committee decided to shorten the title to the 'Journal of Applied Ecology', with the proviso that the publicity material should stress how the journal would include 'conservation in its widest sense'. Whilst recognizing that it was impossible to draw up any 'rigid definition', the Committee recommended that the *Journal of Ecology* and *Journal of Animal Ecology* should publish papers that 'added to the fundamentals of ecology'; the new journal would publish work that exemplified the application of ecology. The placing of borderline papers would have to be decided by informal consultation between editors. Provided that there were over 800 subscribers outside the Society, the journal was expected to be viable within two years. The first issue appeared in 1964.

By deciding to launch the *Journal of Applied Ecology*, the Society demonstrated in the most visible way possible its determination to remain at the forefront of ecological thinking and activity. It was no coincidence that the journal was launched at a time of mounting interest in the impact of pesticides not only on farming but on seminatural plant and animal communities. The assessment of the side-effects of these toxic chemicals on different organisms seemed an obvious focus of attention for the new journal, both for its undoubted scientific interest and as a topic of growing public concern. There are, however, penalties in being a pioneer. As the Society soon discovered, long periods of time can elapse between the identification of important topics for research and the publication of results.

For most ecologists, the realization that a chemical revolution of major proportions was taking place in the countryside occurred in a piecemeal fashion from 1950 onwards, beginning with complaints as to the damage caused to the flora of roadside verges by roadside spraying. It was, however, the adoption of an ever-wider range of pesticide compounds in agriculture that aroused the greatest anxieties. In 1957, a Notification Scheme was introduced whereby the pesticide manufacturers agreed to comply with regulations affecting the marketing and use of pesticides, as laid down by the Minister of Agriculture on the advice of an Advisory Committee. As a member of that Committee, the Nature Conservancy was expected to give advice on the wildlife aspects of the Scheme (Sheail 1985).

By the autumn of 1959, serious misgivings were expressed in Parliament. The Minister of Agriculture announced the appointment of a Research Study Group to assess whether sufficient effort was being made to discover the longer-term effects of the chemicals on consumers, farm crops, and stock—as well as on wildlife. The Research Study Group met under the chairmanship of Harold G. Sanders, the chief scientific adviser to the Ministry of Agriculture. The other members included Roy Clapham, who later recalled how the anxieties of the naturalist were dismissed as 'mere sentimentality'. The Group's attitude began to change, however, as reports of bird mortality increased, and the Press and Parliament called for remedial action. In its report of September 1961, the Research Study Group described how members had become impressed by the depth of feeling aroused, and by the difficulties encountered in seeking to re-solve some of the threats posed to wildlife by the use of pesticides (Research Study Group 1961).

The appointment of the Group marked the first occasion when the British Ecological Society submitted formal evidence to an official enquiry. Following an invitation to do so, Owain Richards drafted a submission, based on data provided by Geoffrey Blackman and Paul Richards. It identified eight topics where both short- and long-term research were urgently needed. These were: the autecology of weeds, the habitats from which weeds were thought to invade cul-tivated land, the effect of toxic chemicals on natural and seminatural vege-tation, the effects of current control practices on particular weeds, the improvement of the selectivity of toxic chemicals, the long-term effects of insecticides on crops, the impact of insecticides applied to crops on the fauna and flora of adjacent uncultivated land, and the frequency of gross overdoses of toxic chemicals and their lasting effects[1].

It had become increasingly clear that the only way the Conservancy could furnish the necessary advice to the Minister's Advisory Committee on Poison-ous Substances was to establish a research unit for that purpose. Discussions as to the character and location of that unit quickly became subsumed into wider proposals for

> a new and adequately staffed centre for the effective experimental study,
> testing, demonstration and dissemination of applied knowledge in
> animal and plant ecology, and the factors underlying successful
> management of the fauna and flora.

The outcome was the decision in 1959 to build an experimental station adjacent to the Monks Wood National Nature Reserve in Huntingdonshire, which would be made up initially of three sections, the largest of which would be a Toxic Chemicals and Wild Life Section under Norman Moore. By focusing on the longer-term implications of using some of the organochlorine compounds found in the new insecticides, the new Section was able to play a key role in identifying what proved to be the most dangerous side-effects of pesticide use (Sheail 1985).

All this happened before Rachel Carson's polemic, *Silent Spring*, caught the public attention in early 1963, and dramatized the implications for human health and the natural environment of using 'persistent' pesticides (Carson 1963). Recognizing the international dimensions of the pesticides problem, the International Union for the Conservation of Nature (IUCN) set up a commis-sion in 1961 to study the ecological effects of chemical controls. Norman Moore was secretary of the commission and, as a result of initiatives taken in that capa-city, he persuaded the North Atlantic Treaty Organization (NATO) to sponsor an Advance Study Institute on 'Pesticides in the environment and the effects on wildlife'. Attended by 71 scientists from 11 countries, both inside and outside NATO, the meeting was held at the Monks Wood Experimental Station in July 1965.

Following an approach from Norman Moore, as director of the Advanced Study Institute, the editors of the *Journal of Applied Ecology* agreed to publish the papers given at the meeting as a third part to the volume for 1966, subject to their being 'good reports of good scientific investigations'. It was agreed that each paper should be refereed and edited like any other paper submitted to the journal. As Bunting remarked, 'I should not be attracted by the idea of publishing sermons or philosophical disquisitions on the balance of nature and the sins of agriculturalists'[2]. Taken together, the papers that eventually appeared in the supplement highlighted the kind of material which was so badly needed by the *Journal of Applied Ecology* (Moore N. W. 1966).

It was not until the fifth year of publication that there was sufficient material for each volume to appear in three parts. After a promising start, hopes that the journal would soon become self-financing were dashed by a 'period of exceptional financial stringency for libraries', both in Britain and in the United States. Whereas the number of non-member subscribers rose from 374 in 1964 to 503 in 1965, and to 631 in 1966, there was comparatively little increase during the rest of the decade. This, together with the increasing size of each volume, was soon reflected in the accumulated deficit, which rose from £2836 in 1966 to £6566 in 1968. It was not until 1975 that the journal began to cover its costs.

Meanwhile, the journal was criticized for acting as little more than an overflow for papers from the other two journals. It had failed to develop a 'positive' image of its own, focused on land management and conservation. Far too little prominence was given to papers demonstrating the practical applications of ecology. Both the editors and Council were quick to point out that a journal could only be as good as the contributions it received. Whatever its limitations, the *Journal of Applied Ecology* was the journal most requested by people applying for membership of the Society.

The most radical ideas for producing larger and more frequent issues of the journals were put forward in a series of proposals drawn up by the editors themselves in 1970, designed to stimulate discussion among the membership. Instead of subdividing the field of ecology between three journals, they suggested following the multidisciplinary approach of the journal *Ecology* in America, *Oikos* in Scandinavia, *Oecologia* in Germany, or the Japanese *Journal of Ecology*. Whereas the Society's three journals had appeared in nine parts, made up of 2426 pages in 1968 and 2187 pages in 1969, the better approach might be to issue a monthly or bimonthly journal, with approximately 200 or 400 pages respectively in each part. This would reflect not only the Society's broad view of ecology, but also the current shift of teaching in the universities away from botany and zoology into more horizontal divisions.

The proposal was rejected on practical grounds. Not only would it be impossible for one or two honorary editors to produce a large, combined journal, but, more significantly, there was considerable scepticism as to whether the advantages of having a single, expanded volume would be sufficiently great as to warrant paying a much higher subscription. There were fears of considerable resentment among members at having to purchase and store a larger journal of 2400 pages, in which only a small proportion of papers might be of interest. In 1968, only 12% of members took all three journals (56% took one journal and 32% subscribed to two journals).

Although the concept of subsuming the three journals into an enlarged *Journal of Ecology* was rejected unanimously, there was far less accord in deciding upon an alternative course of action. Several editors argued that a full-time 'super-editor' should be appointed, who would be supported by the existing honorary editors and an editorial board. This would be much more cost-effective than to continue increasing the number of part-time sub-editors employed. It would also enable papers to be published much more quickly, thereby encouraging potential contributors to submit their manuscripts to the Society's journals[3].

Despite some fluctuations in the numbers of manuscripts received, the overall trend was upwards, so that by 1983 over 70% of the papers submitted to the *Journal of Animal Ecology* had to be rejected. Almost half the papers considered by the *Journal of Ecology* and over a third of the papers received by the *Journal of Applied Ecology* were similarly turned down. There was every indication that the position would get worse, if the journals were to remain within their 1000-page limits. The main casualties were likely to be those scientifically acceptable papers that did not fall squarely within the core discipline.

Beyond agreeing that the situation was wasteful and discouraging for all parties concerned, there was little unanimity as to how it might be resolved. Blackwell Scientific Publications reacted favourably to the idea of launching a new 'Journal of Theoretical Ecology', and it was generally agreed that such a journal would establish the Society as a major publisher of such papers. There were doubts as to how far it would relieve pressures on the existing journals. In a straw poll conducted at the December meeting of Council in 1983, 18 members favoured an increase in the number of pages in each volume, and 10 wanted a new journal. As an interim measure, each journal was allocated an additional 100 pages.

At its meeting of April 1986, Council approved the launching of a quite different journal, intended to take short articles of up to six pages on the many different aspects of physiological, theoretical, evolutionary, and experimental ecology. The new journal was to be called *Functional Ecology*, and its overriding aim would be to provide a medium for very rapid publication. By making use of a computerized system devised by Blackwell Scientific Publications, no more than five months should elapse between receipt of finished manuscript and publication. Although the journal would appear initially only four times a year, it was hoped this would rise to six later. By the end of the year, a pilot issue had been published, giving a flavour of what was to come.

4.3

The Annual Round of Meetings

The principal event of the year continued to be the combined winter and Annual General Meeting. After some discussion, Council affirmed in 1964 that the Meeting should continue to consist of a varied programme of papers. By 1967, there were so many contributions that parallel sessions had to be organized. Beginning with the winter meeting of 1970, those wishing to give a paper had to submit a summary of 50–100 words. A time limit of 20 minutes was imposed on each paper, with 10 minutes for discussion. The Programme Planning Committee obtained 'a suitable timing device to assist in a more rigorous adherence to the time allocated to speakers'.

During his period as president, Clifford Evans suggested that a biennial Tansley Lecture should be instituted, to be given in those years when there was no presidential address. The speaker, who would generally come from overseas or from outside the Society, would be invited to choose a broadly based ecological topic. Sir Harry Godwin was invited to give the first Tansley Lecture. He chose for his subject Tansley's association with the Society and his role in the development of ecology (Godwin 1977). The Tansley Lecture to arouse most debate and discussion was, perhaps, the fourth, given in 1981 by Paul Colinvaux, the professor of zoology at the Ohio State University. Its title was 'Towards a theory of history: fitness, niche and clutch of *Homo sapiens* L'. The lecture was reprinted as an appendix in the author's best-seller, *The Fate of Nations. A Biological Theory of History* (Colinvaux 1982, 1983).

The summer field meetings continued to provide opportunities to visit nature reserves and other sites of biological interest, as well as university departments and research institutes. In 1965, the Society held its first meeting outside the British Isles. The seven-day meeting in the Netherlands was organized by members of the Institute of Plant Ecology at Groningen, and consisted of visits to a large number of ecological institutes and research areas[4]. Two years later, a 14-day field meeting was held in Czechoslovakia at the invitation of the Academies of Science in that country. Led by Frederick Whitehead, the party of 29 ecologists covered 1500 miles by coach, in addition to those on foot. In the words of the published report,

> there were visits to medieval fishponds in southern Bohemia, steppe and woodland on the Bohemian–Moravian borders, alluvial forests in southern Moravia culminating in 4 days in the Tatra mountains of Slovakia. Here the tremendously rich flora coupled with the knowledge and enthusiasm of our guides kept even the weariest members going on mountain walks of up to 18 miles in a 15-hour day.

So great was the success of these foreign meetings that the Society went to great lengths to reciprocate the kindness and hospitality shown.

The annual symposium quickly fell into a routine. The meeting lasted three days and gave roughly equal prominence to papers in botany and zoology. It was usually held in the university of the organizer. The sixth symposium took place in Cambridge in 1965. The subject was 'Light as an ecological factor'. Twenty-two papers were delivered to an audience of 121 members, covering a diverse range of aspects (Bainbridge et al. 1966).

For the Society, the most valuable symposia were those that not only focused on a particular field of research, but looked more widely at the role of ecologists in research, education, and employment. A striking example of the growing stature of ecology was given in 1966, when the topic chosen for the annual symposium was 'The teaching of ecology'. It came almost 50 years after the publication of the famous 'encyclical' in the New Phytologist of 1917. The programme drawn up for the symposium, the number of participants, and the publication of its proceedings could not have served as a more graphic illustration of the great strides that had been made in 'the teaching of ecology' as the Society entered its third quarter-century.

The contributions to the symposium ranged from consideration of the basic problems, through the practical teaching of ecology in schools, colleges, and universities, to discussions of the opportunities available to those who wanted to become professional ecologists at the end of their formal education. Some papers were concerned with the underlying philosophical concepts, some provided first-hand details of courses in which the authors had taken part, and others were concerned with primary data or general facilities and opportunities for ecological work. As Joyce Lambert pointed out in the preface to the symposium volume, the development of ecological teaching was taking place at a time of far-reaching changes in general educational theory (Lambert 1967).

Although the difficulties of teaching ecology were freely acknowledged, far less diffidence was shown in respect of the students than had been the case half a century ago. Any concern that the students might be overtaxed and confused had given way to worries that there might not be enough challenge and stimulation. Whilst it was both common and sound practice to start teaching ecology with facts about plants and animals, there were felt to be penalties in reserving the synthesis of the facts for the latter part of the course and for the more advanced students. General theories, such as that of the ecosystem, were important, not only as a framework within which students could place a multiplicity of facts which would otherwise be incomprehensible, but as a stimulus to further research carried out by the students themselves.

Field classes had always featured prominently in the teaching of ecology. Students developed skills in taxonomy, encountered a variety of living organisms, and were able to correlate species distribution with features of the habitat, particularly where specialized or sophisticated sampling techniques were used. Field teaching had rarely been able to go further. Experiments were needed to answer questions as to why there were such correlations. Fortunately, the scope for carrying out such experiments had greatly increased during the 1950s and 1960s, following the striking development of the open-ended project in both school and university teaching, whereby students sought to answer questions to which no right or wrong answers could be given in advance.

The latter part of the symposium looked more explicitly at the institutional courses and outlets for employment available to ecologists. No British university possessed a department of ecology. The recent development of one-year postgraduate courses had come about entirely as a result of collaboration between existing departments in the respective universities. The pioneer one-year diploma course at University College, London, was soon restyled an MSc in Conservation. Three further one-year courses were established in 1965–6. The emphasis of those at Bangor and Durham was on specialist training, with students focusing much of their attention on selected options and research projects. The curricula at Aberdeen and University College, London, were much broader.

Turning to career opportunities for ecologists, Palmer Newbould outlined to the symposium the results of a recent survey. If an ecologist was arbitrarily defined as a member of the British Ecological Society, over half of those living in the United Kingdom were probably employed in education at university and other levels of education. Of those in universities, about three-quarters were lecturers—the remainder being research fellows or students. To date, there had been 161 recipients of a Nature Conservancy research studentship. Of the 144 students for whom records were available, 68 had taken zoology degrees and 30 of them were now university lecturers. Of the 64 who had obtained botany degrees, 35 were lecturers. Newbould estimated that 150 ecologists were produced in the United Kingdom each year and that there was a demand for at least 100. The overproduction could be fully absorbed by an overseas demand. Replies to a questionnaire sent to 47 organizations or research stations suggested that the demand for ecologists was increasing comparatively rapidly in the Nature Conservancy and other conservation agencies, and more generally within the fields of freshwater and marine ecology.

One way of attracting large audiences was for the symposium organizers to choose a very topical theme. The year 1970 had been designated European Conservation Year by the Council of Europe. It was also the 21st anniversary of the founding of the Nature Conservancy. Alex Watt had already suggested a meeting on 'the scientific management of animal and plant communities for conservation' when Eric Duffey put forward the idea of collaborating with the Nature Conservancy in organizing an international symposium on 'Current research on natural and semi-natural plant and animal communities', followed by a series of excursions with a special emphasis on conservation. The idea was approved by Council[5].

Nearly 400 people attended the symposium held at Norwich in July 1970. Duffey and Watt, as the joint organizers, drew up a programme of 41 papers, arranged under seven themes. Having examined some of the more fundamental aspects of wildlife communities, later sessions directed attention towards the more applied problems of land management, and the criteria upon which conservation policies might be based. Some of the liveliest discussions arose from the sessions concerned with the management of populations of large mammals in Africa, North America, and Scotland (Duffey & Watt 1971). The most disappointing feature was the cancellation of most of the excursions because of a lack of bookings. Among the few to go ahead was a six-day tour of lowland calcareous grasslands led by Terry C. E. Wells and Michael G. Morris of the Conservancy's Monks Wood Experimental Station. A 50-page tour booklet was published by the Conservancy, and later sold for the price of 7/6d[6].

Largely as a result of initiatives taken by Richard Southwood during his period as president, the Society began to play a significant part in organizing a series of European Ecological Symposia. Such gatherings were seen as the most effective way of developing contacts with overseas workers, and fostering the development of ecology in those European countries where it was not well represented or organized. A steering committee was already drawing up plans for what was to be the 19th symposium of the British Ecological Society, to be held in Norwich in 1977. By widening the scope of the symposium from salt-marsh ecosystems to include all coastal ecosystems, the topic and location seemed eminently suitable for the first of a series of symposia sponsored by the Society and other European societies. Details of the Norwich meeting were circulated in English and French to the various ecological societies on the Continent, and to prominent ecologists in those countries without a national body. The steering committee was enlarged, and a special effort was made to solicit the active support of members of the British Ecological Society resident on the Continent (Jefferies R. L. & Davy 1979).

The success of the symposium in attracting participants from 20 countries (with the largest overseas contingent coming from the Netherlands) encouraged the steering committee (now made up of representatives from seven participatory countries) to accept an offer by the Gesellschaft für Ökologie to stage a second symposium, on urban ecosystems, in Berlin in September 1980. A third, on plant–animal interactions, was held at the University of Lund in 1983.

4.3.1 *Ecology and the industrial landscape*

It was one thing to convene a meeting, and quite another to ensure that the interest and enthusiasm generated by that meeting were sustained and developed. The question increasingly arose as to how far the Society was content to remain a passive observer, and how far it sought a larger role in promoting research and debate. Some hint of an answer emerged in the aftermath of the fifth symposium, organized by Gordon T. Goodman of the botany department at Swansea in April 1965.

The theme of the symposium was 'Ecology and the industrial society'. In his opening address to the symposium, Roy Clapham described how the absence of precise data on the current size and distribution of different organisms placed ecologists at a serious disadvantage in trying to substantiate statements as to how far species had declined as a result of industrial and other changes made to the environment. There was an urgent need 'to obtain quantitative assessments' which were nation-wide and repeated at regular intervals. Clapham concluded by arguing that the detailed planning for such an ambitious programme of 'analytical ecology' was the kind of task which the British Ecological Society might be well suited to undertake, in collaboration with the Nature Conservancy (Goodman *et al.* 1965).

The reaction of the Council of the British Ecological Society to Clapham's proposals was almost entirely negative. It affirmed that, whilst the Society might offer its support and advice for 'a nation-wide and long continued set of quantitative observations to measure the nature and extent of changes in the British flora', any direct help would have to come from individual members rather than from the Society as a whole. Both the Burren investigation and the compilation of the biological flora had highlighted the difficulties which a society run by short-term honorary officers was likely to face. As the minutes of the Council meeting recorded,

> the Society was not constituted to initiate, supervise and finance long-
> term observation of this type, particularly because its officers serve for
> relatively short periods and it might be difficult to maintain the
> continuation of the work.

Such a venture seemed to fall much more obviously within the sphere of responsibility of a public body, such as the Nature Conservancy[7].

Meanwhile a different kind of proposal was being drafted by Derek S. Ranwell (a plant ecologist at the Conservancy's Furzebrook Research Station in Dorset), setting out the case for a landscape ecology research and advisory unit. As Ranwell commented, it was difficult to find a satisfactory name for such a proposed unit—the all-embracing words, 'land use' would run the risk of treading on too many toes. The draft was revised in the light of discussions with Eric Duffey, Gordon Goodman, and Joyce Lambert, and considered by Council at its meeting of January 1965. Having drawn attention to the increasing extent of derelict land and the demand for advice as to how it might be reclaimed, the paper set out the case for a new organization, with the twofold responsibility of collating whatever information was available and carrying out further research on landscape improvement. Established under the aegis of the Nature Conservancy, the best location for such a unit might be a university with a strong geography department and teaching facilities in landscape architecture[8].

The effect of this call for a new research unit was to shift the emphasis of discussion away from Clapham's earlier suggestion for a nation-wide survey, organized by the British Ecological Society, towards questions as to how far the Society might help in establishing a landscape advisory and research unit. At the suggestion of the president, David Lack, a subcommittee was formed to 're-view the need for landscape improvement advice and research, and to determine whether or not a case existed for a special organization to carry this out'. At its first meeting in March 1965, the subcommittee decided to set up 'a small work-ing group consisting of three people with special knowledge of the subject' to collect the facts and prepare a report on the scope for landscape ecology and the types of problem that required survey and research. Gordon Goodman, Derek Ranwell, and Edward Broadhead were appointed to the working group.

The benefits of an interdisciplinary approach had already been exemplified by the way in which botanists, civil engineers, economists, geographers, geolo-gists, and metallurgists had participated in a project, organized by the Univer-sity of Swansea, to study the severe problems of dereliction in the lower valley of the River Tawe, north of Swansea. In a paper to the British Ecological Society symposium in 1965, Gordon Goodman and other members of the botany department at the University, described how, within the three growing seasons available, they had sought to discover the most economical ways of revegetating the waste tips and infertile clays, both through short-term empiri-cal studies and by applying experience gained from reclamation schemes else-where. It quickly became apparent that little or nothing was known about the physicochemical factors controlling the availability and uptake by plants of heavy metals, and the way in which such metals were translocated and accumu-lated in litter.

Several of the speakers at the symposium emphasized how it was wrong to exaggerate the distinctiveness, in ecological terms, of the problems encountered on industrially derelict sites. Tony Bradshaw and other members of the depart-ment of agricultural botany at Bangor described the results of trials carried out on the spoil heaps around Bangor. Far from being something special, the con-ditions and processes at work in the industrial environment gave credence to Darwin's observation that 'evolution is silently and insensibly working wher-ever and whenever the opportunity offers'. Through a close study of the way in which such species as *Agrostis tenuis* genetically evolved a mechanism that en-abled them to tolerate individual metals in the soil, there was a possibility of dis-covering plant material which would be 'of value in coping with the more objectionable effects of industrial activity' in the environment.

Drawing on this kind of experience, and the replies to a questionnaire sent to 62 authorities involved in reclamation problems, the subcommittee drew up a report, recommending the appointment of an information and advisory centre, with some facilities for *ad hoc* research to be developed by an independent, but interested, body[9]. An amended version of the report was published in the Society's journals (Ranwell 1967): reprints were sent to national newspapers and members of both Houses of Parliament. A meeting of representatives from the Society and the Nature Conservancy in November 1966 agreed that the re-port presented a prima facie case for setting up an organization to collect tech-nical references, disseminate technical and scientific advice, and promote research on methods of landscape reclamation and improvement. As a pre-requisite to deciding on the structure of such an organization, the joint meeting invited the subcommittee to draw up a paper, setting out in more detail the kind of technical advice which landowners confronted with reclamation problems might seek.

In a further report, edited by Gordon Goodman, the subcommittee empha-sized how the first priority must be to decide on the ultimate goal of any recla-mation project. As well as the more obvious commercial reasons, account might be taken of the benefits to public health, recreation, and education. The next priority was to identify those environmental factors most likely to inhibit re-vegetation. For this, technical advice was necessary in the fields of meteorology, hydrology, physiography, pedology, microbiology, and ecology. Having identi-fied the type of treatment least likely to fail, further expert guidance was needed to ensure that the 'new land' was managed in the most cost-effective way pos-sible. Whatever course of action was decided, there was an obvious need for further basic research and closer liaison with contractors so as to ensure that the technical advice was translated into practice (Goodman 1967).

Gradually, as it became clear that the Conservancy was unable to set up a research and advisory unit, the Society was forced to make a decision as to how far it should go in creating such a body. Matters came to a head in September 1968, when a motion was put before Council, inviting it 'to nominate a Com-mittee to establish some means by which information on all aspects of industrial ecology be collected and disseminated'. In the words of the minutes taken at the meeting, members were very interested in the idea of setting up a landscape reclamation unit to give advice and generate the relevant research, but they were forced to conclude that 'the financial involvement would be very great and it would be more appropriate to foster research in this field in Research Stations and Universities'[10].

In January 1969, Gordon Goodman and Michael J. Chadwick followed up a suggestion that a meeting should be arranged with the chief research officer of the Ministry of Housing and Local Government, with a view to discussing whether the Ministry might be able to finance ecological research connected with derelict land reclamation. It was stressed how, for the expenditure of rela-tively small sums of money on relevant ecological research, large savings might be made in the cost of managing vegetation on reclaimed sites. The Ministry pointed out that it had no statutory authority to finance research investigations. The only course open to research workers was to write to their local authorities, indicating what expert advice was available, in the hope that the authorities might engage them as consultants[11].

The long-running initiative to create some kind of research and advisory unit had finally run out of steam. Whatever the impact on research in landscape ecology, the wider implications for the Society were clear. Although keen to help promote new ventures, particularly where the research was obviously rele-vant to issues of public concern, Council continued to draw an important distinction between instigating research and advisory work in ecology and the actual management of that work—a task which Council believed was much more appropriate to the Nature Conservancy, the universities, or some *ad hoc* body established for that purpose. Far from seeking to take the Society into a new dimension, Council was at pains to emphasize its traditional and support-ive role in ecology.

4.3.2 *The specialist groups*

Any follow-up to meetings and symposia had to come from the 'grass roots'. This was how a number of specialist groups came to be formed from the early 1960s onwards. Not only was Council lukewarm to the idea, but it was not until the third group was proposed that Council members began to consider the wider implications for the Society. Which fields of ecology led to the formation of a group? How far did these groups highlight the growth-points in ecology in the 1960s and '70s?

The idea of a tropical ecology group was taken up and presented to the Annual General Meeting of 1961 by Paul Richards. By its very nature, tropical ecology tended to be carried out in isolation, and the results published in scat-tered periodicals. The aim of the proposed group was to establish a forum for the exchange of information between older (often ex-Colonial Service) ecolo-gists and younger ecologists keen to work on the tropics, and between home-based and overseas workers. Following an encouraging response to a circular, sent to all members of the Society, suggesting that a working group on tropical ecology might be formed, an inaugural meeting was held at Imperial College, London, in April 1961, attended by 50 members of the Society. In the course of a lively discussion, it was decided to hold two symposia a year.

The wider benefits to the Society of forming a Tropical Ecology Group were soon appreciated. One of the first meetings was a one-day symposium held with the Royal Entomological Society of London, at which six papers were read on 'the effects of habitat on insect populations in the tropics'. In December 1961, a circular was sent to those known to be interested in tropical ecology, but not members of the British Ecological Society, inviting them to join the Society and thereby take part in the group's activities. The circular probably played a very large part in causing the intake of new members of the Society to rise from what had been about 80 per annum for several years to 126 in 1962.

The idea of forming a second specialist group developed during the early deliberations on the feasibility of setting up an advisory and research unit in landscape ecology. Gordon Goodman told the Council meeting of September 1967 that the aim of the new group would be 'to foster interest in the ecology of environments affected by urban and industrial conditions and other intensive forms of human use'. There was an encouraging response to a circular sent to all members of the Society. The inaugural meeting was attended by 180 people. The proposed name of the group, the Industrial Ecology Group, met with some criticisms on the grounds that it was misleading—the group would be concerned with land use generally. In the absence of a suitable alternative, the title was retained; Gordon Goodman was elected the Group organizer[12].

The idea of forming an energy and production ecology group first surfaced at the symposium held in 1969 on 'Animal populations in relation to their food resources'. Following the pattern of the two existing groups, it was suggested that up to two meetings a year might be organized, focusing on (i) methods for the study of energy flow and the calculation of production in ecosystems, (ii) the relationships between studies on energy flow and production and those on other aspects of ecology and population dynamics, and (iii) the application of ideas on energy flow and production to concepts of ecosystem function. Stuart McNeill of the Imperial College, London, Field Station and John H. Lawton, a member of the Animal Ecology Research Group at Oxford, agreed to act as joint secretaries in establishing the group.

Although no one disputed the importance of energy and production ecology, there were doubts among some Council members as to the wisdom of forming further specialist groups. In correspondence, Owain Richards argued that the establishment of smaller groups of enthusiasts was in general good for the Society. There was no evidence of their interfering with the success of ordinary meetings. Jack Rutter took a different line, commenting that the meetings on energy and production ecology were likely to generate so much interest that it seemed a pity to hold them at 'an odd time outside the Society's normal meeting times'. A better course would be to organize them explicitly for the whole Society, within the annual cycle of conferences and symposia[13].

How far was the formation of specialist groups compatible with what many perceived to be the Society's overriding function, namely to preserve the unity of ecology? David Le Cren warned of how they could lead to extensive fragmentation. Whilst it was 'a good idea to have an occasional specialist meeting and a group concentrating on production ecology', a mechanism had to be found for ensuring that contact between the different kinds of ecologists was maintained. A very important function of the new group should be 'to educate non-production ecologists in the importance of this approach to their subject'. Another member of Council, Francis H. Merton, also perceived advantages and dangers. Whilst it was easier for a small group, interested in a particular approach and methodology, to have more creative discussions, there was a risk of the rest of the Society suffering 'by the gradual withdrawal of contributions in an increasing number of fields'. In Merton's words, 'ecology is a synthetic subject; its strength lies in it being able to relate data from very diverse disciplines'. Council had to make it abundantly clear to the specialist groups that

>they should endeavour to return their ideas to the main body of the
>Society, refined and enlarged as we hope they are by intimate discussion.

An obvious way of achieving this was for the groups to provide, from time to time, sessions at the general meetings of the Society. At its meeting of September 1969, Council agreed that the secretaries of the groups should be co-opted onto the Programme Planning Committee and Council. Experience soon indicated that members only attended those meetings which were most relevant to their research interests—irrespective of who organized them. The Society's winter meetings continued to provide outstanding opportunities to disseminate information not only on individual pieces of research but on wider initiatives being taken within the biological sciences.

A morning session of the winter meeting of 1965 was devoted to papers introducing the International Biological Programme (IBP), the full title of which was 'The biological basis of productivity and human welfare'. Organized by a special committee of the International Council of Scientific Unions, its overall aim was to promote 'the world-wide study of (i) organic production on the land, in fresh water and in the seas, and the potentialities and use of new as well as existing resources, and (ii) human adaptability to changing conditions'. Within these very wide aims, it had been decided to focus on those basic biological studies related to productivity and human welfare which could only be properly made by international co-operation. The national committees of the participating countries would be responsible for the operational aspects. The Royal Society had agreed to be responsible for the British National Committee. It was expected that phase I (1965–67) would be focused on the development of methods and pilot projects. The main programme of co-operative projects would occur in phase II, 1967–72.

The considerable effort being invested in the IBP was soon reflected in the programmes of the Society's meetings. The Tropical Ecology Group devoted both its meetings in 1966 to themes associated with the IBP—the first meeting being on the 'Productivity of tropical ecosystems', and the autumn symposium on 'Human ecology in the tropics'. About 80 people attended the first meeting. The opening papers dealt with primary and secondary production of aquatic systems. The remaining papers considered a range of terrestrial situations: Donald Pigott spoke on the measurement of productivity of African savanna; Geoffrey Blackman on the limits of crop production; and Brian Hopkins on a comparison between productivity in forest and savanna in Africa[14].

There was no Easter meeting of the Energy and Production Ecology Group in 1973—it was assumed that most members would be at the symposium marking the formal ending of the IBP. As the symposium (jointly organized by the British Ecological Society and the British National Committee) demonstrated, the terrestrial IBP programme had been one of the most ambitious, expensive and concentrated programmes of ecological research ever to have been attempted in Britain. Enough data had been acquired 'to keep British ecology going for years'. The aim of the three-day meeting was to make a preliminary assessment of its achievements. As Palmer Newbould remarked, it was easy, with hindsight, to find faults. Objectives could have been more clearly defined. There was insufficient opportunity and expertise to study crucial aspects. The enormous problems of data handling had not been foreseen. More positively, the teamwork developed in the course of the IBP projects had helped to produce what Newbould called 'a new breed of ecologist' who had forgotten his (or her) own origins and background, and learned to work as an ecological team. The participating scientists had gained much from the multidisciplinary, and national and international, dimensions of the terrestrial Programme.

Many of the meetings of the specialist groups were held jointly with another society—it was a way of ensuring that there was a good attendance and a wide exchange of ideas. The second meeting of the Energy and Production Ecology Group was with the Fisheries Society. The spring meeting of the Tropical Ecology Group in 1970 was held in conjunction with the Challenger Society. Six of the nine papers focused on the marine ecology of the Indian Ocean. The experience of data-handling, modelling, and systems analysis, as part of the IBP projects, emphasized further the benefits of drawing on the professional expertise in other disciplines. It was out of this kind of dialogue that the Society's fourth specialist group came to be formed, the Mathematical Ecology Group.

The foundations for the considerable development of multivariate methods in plant ecology had been laid in the 1950s. David W. Goodall demonstrated the value of such techniques as factor analysis and the discriminant function in distinguishing and classifying vegetation units (Goodall 1952). In a series of papers in the *Journal of Ecology*, Brian Hopkins re-examined pattern and species–area relationships in plant communities, showing how species diversity varied continuously with area (Hopkins 1955). In 1959, William T. Williams and Joyce Lambert published the first of a series of papers on association analysis in plant communities, describing a method for sorting quadrats from a given area into a hierarchy. It became one of the most frequently cited papers in the Society's journals (Williams W. T. & Lambert 1959).

The practical limitations of hand computation imposed severe constraints on what could be achieved. Access to a computer enabled Williams and Lambert to carry out many more experiments on their heathland data sets, and it became possible to test the methods developed on more complex communities. Their account of how the Ferranti 'Pegasus' digital computer dealt with up to 76 species in large numbers of quadrats represented a further milestone in the quantification of ecology (Williams W. T. & Lambert 1960). Among the earliest examples of the application of computer numerical models to population dynamics in Britain was work carried out by George Varley on the winter moth and a paper published by Edward Broadhead and Anthony J. Wapshere (of the zoology department at Leeds) on two closely related psocid species and their common parasitoids. The University of Leeds installed its first computer in 1957 (Broadhead & Wapshere 1966).

As Pearsall remarked, in his address to the jubilee conference of 1963, the pages of the Society's journals had closely reflected the increasing use of quantitative techniques (Pearsall 1964). For many readers, however, quantitative ecology remained 'an esoteric discipline whose practice was limited to those fortunate enough to have access to its widely-scattered research literature'

(Williams W. T. 1965). In that sense, the publication in 1957 of Peter Greig-Smith's *Quantitative Plant Ecology* was an important turning point. In a review of the first edition, C. B. Williams wrote of how it provided an excellent introduction to the quantitative description of vegetation, with chapters on sampling, the correlation of vegetation with habitat conditions, and the classification and ordination of plant communities (Greig-Smith 1957, Williams W. T. 1958b).

It was also a time of growing interest in the theoretical content of ecology. As John Maynard Smith, the professor of biology in the University of Sussex, remarked, ecology could not come of age until it had developed a sound theoretical basis. Among an increasing number of publications illustrating the potential of a range of theoretical approaches was his own volume, *Models in Ecology*, which illustrated the significance of mathematical models in understanding the general properties of ecosystems. Models were developed to help explain such phenomena as predator–prey relationships, breeding seasons, competition and migration, stability and complexity at different trophic levels, and the evolution of specialists and generalists in the animal kingdom (Smith J. M. 1974).

The topic chosen for the annual symposium for 1971 was 'Mathematical models in ecology'. The idea for such a meeting came from John N. R. Jeffers, who, after 12 years as head of the statistical section of the Forestry Commission, had succeeded James Cragg as director of the Conservancy's Merlewood Research Station in 1968. What promised to be a relatively small gathering took on the dimensions of a large symposium—over 240 people came. Conscious of the failure of some previous conferences and symposia to generate much in the way of a dialogue between mathematicians and ecologists, a particular effort was made 'to provide an effective communication between mathematicians who devise mathematical models and ecologists who have problems which are capable of being solved by such models'. During the course of the symposium, proposals were made for a series of ecological workshops and the setting up of a study group, under the aegis of such bodies as the British Ecological Society (Jeffers 1972).

Although nothing came directly from these proposals, important initiatives were soon under way. In the course of correspondence with Michael B. Usher, a member of the biology department at York, the suggestion was made that there should be much closer liaison between the Biometrics Society and the British Ecological Society. The proposal came from Richard M. Cormack, the professor of statistics at St Andrews, and the secretary of the British region of the Biometrics Society. It was relayed to the Council of the British Ecological Society by Michael J. Chadwick, the meetings and membership secretary (and another member of the York department). Although Council welcomed the idea of establishing a joint steering committee of the two societies, there was little enthusiasm for the larger idea of forming a mathematical ecology group. At its meeting of April 1973, Council suggested that, before taking any further action, a half-day session should be held during one of the Society's winter meetings in order to assess the degree of mutual interest.

Meanwhile, 'a "workshop" group of mathematically inclined ecologists' had been set up by Mark H. Williamson, and met under the auspices of the University of York (where he was professor of biology) and Imperial College, London. Attendance was by invitation only; each of the 25 participants was actively engaged in the relevant field of research. Prior to a further workshop being organized in 1973, Williamson wrote to the president of the British Ecological Society, Amyan Macfadyen, asking whether there was any possibility of the group being recognized by the Society. Although the Society had always insisted that meetings should be open to all, and the group would want 'to be free to continue these closed meetings', there would be nothing to stop meetings of a different nature also being organized[15].

These overtures, together with the fact that a second successful workshop was held at York in 1973 (Usher & Williamson 1974), persuaded Council that there was sufficient interest in mathematical ecology to warrant a specialist group being formed jointly with the Biometrics Society. Michael Usher and Richard Cormack were appointed joint secretaries by their respective societies—Usher joined the secretaries of the other specialist groups as a co-opted member of Council[16].

The aim of the Mathematical Ecology Group was 'to cater for biologists with a quantitative interest and biometricians with an ecological interest'. Right from the start, meetings were 'a blend of formal papers and workshop-type sessions'. The first meeting, of April 1974, was devoted to 'Classification in ecology', and consisted of four papers and a workshop session led by Cormack and Williamson. The second meeting, of November 1974, dealt with 'Population dynamic models', and the third took the form of a session on 'sampling' during the winter meeting of that year.

Meanwhile, the annual 'York workshops' had continued to meet independently, attracting interest well beyond the two universities which acted as hosts. The number of people wishing to take part had increased substantially. A particular feature of the meetings was the relatively large proportion of papers given by graduate students. At the conclusion of the workshop held in July 1976, it was decided to ask the British Ecological Society and Biometrics Society to take over the two-day workshop as part of the functions of the Mathematical Ecology Group. This time, both societies agreed; the meeting of September 1977 was rated 'a complete success'[17].

The contribution made by each specialist group to its respective field of ecology depended very largely on the calibre and keenness of the group secretary and comparatively few other persons who arranged and co-ordinated activities. It was always envisaged that groups would come and go. In the words of the meetings secretary, the specialist groups, like all good ecological systems, 'should have a death-rate as well as a birth-rate'. In fact, none was disbanded. The emphasis was still on forming more specialist groups—the Mires Research

Group in 1975, the Freshwater Ecology Group in 1983, and the Education Group in 1985.

Concern as to 'the further splintering of the Society' gradually gave way to fears of another sort. Unless the Society accorded recognition to the spontaneous groups of ecologists seeking formal status, there was a risk of their looking elsewhere for that recognition. Matters had come full circle. At the Annual General Meeting of 1983, the president, Tony Bradshaw, spoke of how the Society found the activities of the specialist groups especially heartening. The groups offered the kind of meetings, workshops, and field events which *members* seemed to want. They provided 'a flexible system for grass roots involvement', and were something to be encouraged.

4.4

The Society and the Environmental Revolution

At long last, the dreams of the founder members of the British Ecological Society seemed to be coming true. There were signs of ecologists beginning to exert an influence in what was called 'the public policy decision-making process'. Ecologists were being invited to take part in appraisals of current land-use practices, and to help draw up proposals for the future. Not only did the increasing membership and prosperity give the Society the confidence to become a more outward-looking body, but the desire, and indeed the pressures, to do so became well-nigh irresistible.

The term 'environmental revolution' came to be used more and more to describe the scale on which traditional attitudes towards the use and management of the environment were being challenged. Mick Southern caught some of the exhilaration and frustration of the period in his presidential address of 1969, in which he depicted ecology as being at a 'crossroads'. As with any science, ecology had encountered many crossroads in its time, but this crossing was especially important. Globally, there was increasing awareness of the crisis facing civilization—too many people, and too little food and space. There was public debate as to whether the right response was to seek some radical expansion of the exploitation of the world's renewable resources, or to advocate a rapid deceleration in the increase of human population so as to bring man and resources back into balance at the limits of exploitation. As Southern (1970) remarked, the challenge for the ecologist was to put forward a third alternative, namely to create a different kind of balance where man and resources were held at a level *below* that of the maximum.

How far did ecologists have both the necessary motivation and equipment to press for this alternative? There were cynics who argued that ecology was too important a subject to be left to ecologists, but, as Southern argued, who else but ecologists had any real inkling of how an optimum balance between man and resources might be achieved? The aim must be to 'sell' ecology, not by any tricky salesmanship, but by intensified research, the teaching of basic principles, and by 'bruiting around the applied repercussions of these principles on management and conservation'. As the meetings of the Society made clear, there were ecologists keen to take on the challenge of applying ecology to the resolution of the world's problems.

The first steps towards consciously deciding how far the Society should go in becoming a more outward-looking body were taken during discussion of the annual accounts of January 1969. A subcommittee was set up to review the future role of the Society, with Palmer Newbould (one of the vice-presidents) as chairman, and the president, secretaries, treasurers, and Frederick Whitehead as members. Following a couple of meetings, Newbould drafted a paper drawing attention to 'the growing impact of ecological thinking in society and public affairs', particularly among planners, landscape architects, and journalists. If the Society were to exploit this new situation, there had to be not only a much greater awareness of these changes but a greater willingness on the part of the Society to develop an outward-looking and assertive approach to participating in public debate and controversy.

The importance of achieving these goals was set out in a leaflet sent to all members of the Society at the end of 1969. Newbould wrote a commentary in the first issue of *Bulletin* emphasizing how the purpose of the Society had always been 'to promote and foster the study of ecology in its widest sense'. As more and more references were made to ecology in public debate, it was particularly important that the Society should make itself heard. Unless the society was prepared to exert itself, ill-informed comment and imperfect advice would pass unchallenged, from whatever quarter they came, and the public reputation of ecology as a scientific discipline would be tarnished.

If the Society was to influence the public perception of environmental issues, a great deal of rethinking was required on the part of the Society. It was no longer enough to act as a forum for discussing topical issues. As Newbould wrote, 'now that ecology has broken into the mass media and the quality of the environment become a political issue', a much more positive stance was called for. Where the Society had something important to say, the message had to reach a much wider audience than hitherto. The Society's recent report on the role of the ecologist in rehabilitating industrial landscapes would, for example, have attracted a lot more publicity if it had been 'jazzed up as a glossy publication with a brightly coloured cover and photographs'. Whilst some members of the Society might be appalled by such an approach, Newbould contended that this was the only way of successfully competing for public attention.

The subcommittee believed the most effective way of breaking free from the mould of the past, and taking up the challenges of the present and future, was for the Society to establish a public affairs committee (or public relations committee) explicitly for that purpose. Its task would be to promote information about ecology, stressing its relevance to such issues as land-use planning and pollution control. The committee would speak on behalf of the Society on issues of public importance, secure better press coverage for the Society's activities, and generally advise Council on issues of public concern.

The germ of the idea for a public affairs committee can be traced back to a letter written by Palmer Newbould in July 1963, which recounted how the Ecological Society of America had formed a public affairs committee, designed to represent the Society

> on matters of public interest in which the knowledge and experience of ecologists are important for conservation and judicious use of our natural resources.

By having such a committee, the society would also be able to react promptly to requests from governmental and quasi-governmental agencies and elsewhere for the kinds of advice that ecologists were best qualified to give. Whilst there was no desire to 'crib everything the Americans do', Newbould believed there was a strong case for the British Ecological Society setting up something comparable[18].

There could be no disguising the dichotomy of viewpoints held within the Society. There were members who believed the activities of the new committee would give an added sense of purpose to the Society's pre-eminence in the field of British ecology. According to Newbould, many members had long wanted the Society to play a more active role in investigating ecological issues of public concern—it was only the way in which the Society was organized that had prevented their voices being heard earlier. There could, however, be no denying the strong feelings held by other members, who argued that the Society should only be concerned with strictly scientific issues and that it would be extremely difficult, in practice, to draw 'a clear line between political issues and scientific questions'. If the Society became involved in making value judgements, its scientific integrity would be undermined.

In the event, the public affairs committee emerged as a rather different animal from the one first perceived. At the Council meeting of April 1970, it was decided to change its name to the Ecological Affairs Committee, and to reduce its status to that of a subcommittee. It was emphasized that the subcommittee could only make its decisions known to the general public through Council. In the event of a statement having to be made on a very urgent matter, the president or other officers were empowered to act on behalf of Council. It did not become a 'full' committee until Council adopted the so-called 'federal' structure in 1974.

4.4.1 *The Cow Green reservoir and the Teesdale flora*

Council members believed they were more than justified in adopting a cautious stand over the setting up of an ecological affairs subcommittee. After all, they were hardly novices in the world of public affairs. During the late 1960s, Council had experienced at first hand the considerable difficulties which a scientific body was bound to face in fighting large-scale developments likely to inflict damage on the natural environment. Even if a way could be found of consulting the membership at every twist and turn in the debate over some 'burning' environmental issue, there was thought to be little chance of members reaching a consensus.

An example of the dangers of becoming involved in issues of public debate occurred in November 1967, when Peter Greig-Smith received a letter, in his capacity as editor of the *Journal of Ecology*, seeking the Society's views on the use of defoliants in the war in Vietnam. In acknowledging that the letter would have to be considered by Council, Eric Duffey wrote that it would hardly 'be proper for the Society to get mixed up in political issues of this magnitude', whatever the personal feelings of Council members. If any scientific body was to take up the question, it had to be the Royal Society[19].

Council found it much more difficult to decide how far the Society should become involved in opposing large-scale developments in the British landscape, particularly where they impinged directly on sites of obvious biological importance. By the late 1960s, there was an increasing number of these schemes—the most famous being the proposal to construct a reservoir in Upper Teesdale National Nature Reserve. The issues raised by the proposal, and the prominent part played by ecologists in the public debates, made so great an impact on the Society and British ecologists at large that it is worth recounting in some detail the circumstances that led up to the reservoir scheme.

The story begins in 1956, when the Tees Valley and Cleveland Water Board announced its intention to construct a reservoir at either Cow Green or Dine Holm in Upper Teesdale. As the Conservancy's regional officer confessed to Pearsall and other botanists at the time, there was no alternative to flooding part of the Upper Teesdale SSSI if the Water Board was 'to conform to the tenets of good land use and not flood land of high productive capacity'. The economic return from sheep farming on the two proposed reservoir sites was negligible[20].

Following visits to the affected sites, Pearsall, Nicholson, Clapham, and Pigott agreed that the construction of a reservoir at Dine Holm would inflict considerable damage on the flora. Pearsall and Nicholson publicly challenged the wisdom of establishing industries making heavy demands for water in what was one of the drier parts of the country. If employment considerations dictated such a location, ways had to be found of using seawater. If there were no alternatives to an impounding scheme, at least in the short term, Pearsall believed that development at the Cow Green site would cause the least damage to the SSSI.

Following the Board's decision to develop the Dine Holm site, a letter signed by 14 eminent botanists was published in *The Times* during February 1957, deploring the threatened incursion into an internationally acclaimed area of scientific interest. According to David H. Valentine, the professor of botany at Durham, the area below Cauldron Snout was probably the worst possible site for a reservoir in the Tees valley from the scientific point of view. In due course, the Conservancy, as well as the county council, came out in opposition to the scheme. The Water Board decided to proceed instead with a Private Bill to construct a reservoir on a lower tributary of the Tees, at Balderhead.

During the summer of 1964, the Water Board was suddenly faced with the need to double the output of the Teesdale catchment, following the announcement of plans by Imperial Chemical Industries (ICI) to construct two of the largest ammonia plants in the world over a period of two years. The only feasible course of action was to construct another river-regulating reservoir on the upper reaches or a tributary of the River Tees. Following exploratory discussions with officers of the Conservancy, the Water Board believed the construction of a reservoir in Cow Green would arouse the least opposition. By May 1965, the consulting engineers had satisfied themselves that a reservoir of only 770 acres was required, of which only 20 acres were likely to cover part of the SSSI. The Water Resources Board confirmed that it would be the cheapest and quickest site to develop. In the meantime, however, one of the Conservancy's staff, Derek Ratcliffe, had also surveyed the area, and reported that, because of the highly diversified complex of plant species present, even the loss of 20 acres would represent a serious loss of scientific interest. The report, together with the strong representations made by an increasing body of botanists, left the Conservancy with no choice but to announce in July 1965 that it would oppose the development of the Cow Green site.

In February 1965, a further letter was published in the *The Times*, signed by 14 eminent botanists, who drew attention to the extraordinary assemblages of rare species. A Teesdale Defence Committee was formed by the Botanical Society of the British Isles (BSBI), which included representatives of the Durham and Northumberland Naturalists' Trust. A leading role was played by Margaret Bradshaw of the extramural department at Durham. Harry Godwin circularized botanists, inviting them to support the Defence Committee. It was at first intended that a copy of the Committee's appeal brochure should be sent to members of the British Ecological Society with their copies of the *Journal of Ecology*, but when it missed the mailing date Council agreed to the Defence Committee using the Society's current membership list for that purpose.

Soon after the Private Bill had been introduced to Parliament, seeking powers to construct the reservoir and ancillary works, the BSBI wrote to the British Ecological Society, asking if it would sign the formal petition opposing the Bill. By doing so, it would add considerable weight to the opposition. At a Council meeting of January 1966, a few members expressed doubts as to whether 'the scientific case for opposing the reservoir was conclusive, and also whether the site had the international status claimed for it'. In the words of the minutes taken at the meeting, 'other members argued strongly that the scientific case for conserving the Teesdale flora was overwhelming and that if the Society did not give its support to this, it would have difficulty in finding reasons to do so for any other area'. The Council agreed that the president should sign the parliamentary petition, opposing the Bill.

In the event, a joint parliamentary petition was presented by the BSBI, British Ecological Society, Council for Nature, Linnean Society, SPNR, and three naturalists' trusts. Almost every amenity and naturalists' society of any standing supported the move. Whilst the petitioners recognized that adequate supplies of water were of vital importance to industry, the destruction of 'so splendid an heritage' could not be sanctioned under any circumstances. Upper Teesdale was 'an irreplaceable open air laboratory, containing unparalleled scientific riches which include a remarkable complex of plant communities'.

The Bill was first introduced to the House of Commons, where it was given the customary unopposed second reading and referred to a Select Committee for detailed consideration. In evidence to the Committee, opponents of the Bill contended that, instead of resorting to the short-term device of building a dam at Cow Green, the water authorities should make greater use of underground aquifers and 'the scope for using sea-water and supplies from other catchment areas'. If it was ultimately necessary to construct a reservoir, an alternative site should be found. Legal counsel for the promoters argued that Cow Green was the best site for the reservoir. Its quantifiable value for water supply purposes far outweighed the unquantifiable scientific and aesthetic value of anything that would be destroyed. The scientific importance of the site had, in any case, been exaggerated[21].

Among the expert witnesses called by the legal counsel for the opponents of the Bill, Margaret Bradshaw emphasized how the plant assemblages were unique. Harry Godwin responded to the many misconceptions of ecological research revealed in the submissions made by the promoters of the Bill and their witnesses. The scientific potential of the area did not arise from the study of individual rare plants, but rather from the grouping of the species in relation to the different types of gravelly flushes, peaty areas, sugar limestone, and eroded sugar limestone, each with its different drainage, slope, and exposure. In Godwin's words, 'you have a naturally made experiment in which there is a continuous variation of conditions' from the top of the slopes to the bottom. The effect of the reservoir would be to remove the bottom slice of the sequence. The offer of £100 000 on the part of ICI for a 'crash' programme of research prior to the flooding of the area overlooked the long-term nature of experiments and the continued improvement of research techniques[22].

Another of the expert witnesses was Donald Pigott, who had recently been appointed professor of biological sciences in the new University of Lancaster. Although he had initially distanced himself from the controversy, he was persuaded to become involved by the argument that, as the author of the only detailed scientific description of the flora, any standing back might be interpreted as opposition to the case for conservation. Some 10 years earlier, he had published in the *Journal of Ecology* an account of the different plant assemblages, together with a reappraisal of the various hypotheses that had been put forward to explain the richness of the Upper Teesdale flora. In every case, the species could be traced back to several limited types of habitat which, although now probably much less extensive, continued to provide congenial conditions for the distinctive flora (Pigott 1956). In his oral evidence to the Select Committee, Pigott explained how it was impossible to study 'a population without studying it in the system which controls that population'. It was only by having recourse to 'natural systems' that the scientist could discover whether his models of the different phenomena and processes were complete and reliable.

It was not enough to emphasize the purely scientific aspects. Witnesses were confronted with what seemed to be an obsessive questioning on the *direct* material utility of studying the Teesdale flora. Both Godwin and Pigott responded by emphasizing how a study, which began as pure science, might come to assume economic significance. Population genetics practised on such communities as those at Teesdale were not fundamentally different from those used in the selection of agricultural races of grass. In Pigott's words, nobody regarded guinea-pigs as very important, and yet, together with white rats, they had proved to be the most useful animals in medical research. One had to choose whatever plants were the most convenient scientifically, working on the premise that, without extrapolating beyond one's safe limits, there was something of general application to be learned about plants as a whole.

For the ecologist, there was a demoralizing feeling that, even if the arguments could have been better marshalled, they would have made little difference. The problems of cross-examination reflected the lack of any real understanding of ecological concepts among laymen, or indeed lawyers. Insofar as they gave the fate of plants any real consideration, they were concerned almost exclusively with uniqueness and rarity. The Bill was approved by the Select Committee and then by the House of Commons after a debate on a blocking motion at the Report Stage. As Gregory (1971) remarked, 'it was emotion and sentiment, rather than a cool and open-minded appraisal of the issues involved, that dictated the attitude of most MPs'.

As part of its preparations for the passage of the Bill through the House of Lords, the Teesdale Defence Committee urged the British Ecological Society, together with other leading organizations and distinguished biologists, to write to the Press, voicing their opposition. Advantage was taken of the Society's field meeting at Beinn Eighe in July 1966 to bring together five members of Council to discuss the request and, following a telephone call to the president, the text of a letter was approved and sent to *The Times*, expressing the Society's concern at the great loss to science if 'this remarkable series of plant communities is not saved from destructive development'. The Bill was the 'most important test case' to confront the country since the National Parks and Access to the Countryside Act of 1949, which had established the international reputation of the United Kingdom 'for progressive thinking on wildlife conservation in relation to modern land use'[23].

The first most members of Council knew of the letter was its appearance in *The Times*. At least one member of Council raised the question as to how far the initiative taken by the president and only five members of Council complied with the rules of the Society and created 'a dangerous precedent'. In his response, Eric Duffey argued that no set of rules could cover every eventuality, and that situations were bound to arise where decisions had to be taken quickly. This was one such case. The president's signing of the letter amounted to no more than an extension of Council's earlier agreement to support the Petition.

After a debate lasting nearly five hours, the Bill was given its second reading in the House of Lords and referred to a Select Committee, which heard much the same evidence as that given to the earlier Committee in the House of Commons. In his evidence, Pigott emphasized the scientific/cultural value of the vegetation by comparing it with a work of art. No one would contemplate knocking down the Chapel of Henry VIII in Westminster Abbey in order to widen the road, simply because a wider road would be more useful than the Chapel. For the ecologist, a particularly important event was the visit made by the Lords' Select Committee to Cow Green. As two of the members commented to Godwin and Pigott, it was only then that they clearly understood what was at stake, and how extremely difficult it had been for witnesses to describe the complexity of the vegetation.

The Lords' Select Committee concluded that, in view of the obvious need for a scheme, and the fact that only 20 acres of sugar limestone would be inundated, it was unreasonable to prevent the reservoir at Cow Green being built. Following a debate on the third reading, the Bill was approved without division, and received the Royal Assent (Gregory 1971).

For the British Ecological Society and ecologists in general, it was a poor consolation to know that things could never be quite the same again—that promoters of civil engineering schemes would have to take far greater account of ecological considerations if they were to avoid prolonged argument and adverse publicity. The immediate feeling was one of acute disappointment. There was even less enthusiasm among Council members and officers for another confrontation of that kind.

In the certain knowledge that there would be further reservoirs and other proposals for large-scale development, the primary consideration for the Society was how to respond next time. Was it better to fight each scheme as it came along, or to give greater weight to trying to change the whole climate of public thinking? Was it possible—or indeed appropriate—for a Society that prided itself on its scientific integrity to become so involved in the 'politics of decision-making' and, if so, how could those representing the Society be certain that they were executing the wishes of the membership?

4.4.2 The 'Split'

There was a curious paradox about the early 1970s. European Conservation Year had been a great success. In a speech to the Australian Conservation Foundation in April 1970, the Duke of Edinburgh spoke of a sudden and explosive concern for the environment. There was suddenly 'a massive and passionate concern for everything in and to do with nature and the pollution of the environment'. The change was so great as to warrant the term 'environmental revolution' (Prince Phillip 1978).

As so often happens with revolutions, all was not what it seemed. Beneath the surface, there was a deep sense of unease and frustration. The Society and its members had played only a very small part in generating the revolution. The words 'ecology' and 'environment' had been hijacked by those whose principal expertise was jumping on bandwagons. Far from bringing out the strengths of ecology, in its scientific sense, the shifts in public perception seemed to presage a further period of upheaval and strain, distracting many first-class minds from research and teaching.

No matter how exciting the challenges of the environmental revolution might be, there could be no forgetting the constraints within which the Society operated. As Eric Duffey wrote, in the *Bulletin* for December 1971, 'our past, to a considerable extent, still proclaims our future'. There was a risk of the Society overextending itself, and appearing to be 'a splendid head, pronouncing good intentions', without the body capable of fulfilling these intentions. The Society's membership was in any case overwhelmingly professional and most unlikely to support any decisive leap into the public arena on conservation issues alone. The expertise of the membership embraced a whole range of applications, including medicine, public health, agriculture, horticulture, fisheries, water supply biology, sewage disposal, and animal husbandry. There were not only members who used their expertise to safeguard rare and threatened organisms, but others who sought to reduce the numbers of overpopulous pest species. Whilst the latter activity was no longer such a fashionable topic for press releases, the support given to the Society's technical symposia and their published proceedings was a tangible indication of the importance which many members attached to these other forms of applied ecology.

It might have been easier for Council and its officers to develop a more coherent and positive role in nature conservation if the future organization of that movement had been clearer. At a time when the words 'environmental revolution' seemed to be on everyone's lips, there was every prospect that the Society's greatest post-war achievement, the Nature Conservancy, would be abolished. The circumstances in which this came about, and the implications for both the Society and the careers of many of its members, highlighted the complex and unexpected repercussions of the increasing interest being taken by Government in conservation and environmental issues. Not a few ecologists looked back with longing to the halcyon days of the 1950s and '60s, before 'conservation' and 'ecology' became so fashionable among politicians and the Press.

An important turning point for the Nature Conservancy had been in the early 1960s when those in influential positions in government began to perceive the role of the Conservancy in much wider terms than simply the designation of National Nature Reserves and scheduling of Sites of Special Scientific Interest. Landowners, who had resented the intrusion of the Conservancy into the countryside, and who expressed their hostility over such controversies as the future use of the island of Rhum, now began to realize the merits of having a body which could act as a kind of rural policeman, investigating, for example, the side-effects of pesticides, as they affected the game-covert, fox-earth and countryside generally. During its first 10 years, the Nature Conservancy had never been free from threats of abolition. Now it was confronted by misgivings and misunderstandings of another sort.

Reflecting the rise in public interest in environmental issues, questions were increasingly asked as to the adequacy of the powers given to such bodies as the

Nature Conservancy. How far would it be sensible to subsume the Conservancy into some larger body? In 1959, the Minister for Science had invited his Advisory Council on Scientific Policy to review the balance of scientific effort. The Council concluded that a new research council was required, with both the central responsibility and resources to cover all aspects of resource conservation (Lord President of the Council 1960, 1963).

In 1961, the Prime Minister had appointed a separate Committee of Enquiry into the arrangements for government-sponsored civil-science (under the chairmanship of Sir Burke Trend). The Committee endorsed the concept of a new natural resources research council in its report of October 1963. (Prime Minister 1963). There followed, in July 1964, the announcement by the new Secretary of State for Education and Science that a natural resources research council would be established, but that the word 'environment' would be substituted for that of 'resources' in order to emphasize the wide-ranging terms of reference of the new council. Under the Science and Technology Act of 1965, the Natural Environment Research Council (NERC) was set up, with the responsibility of encouraging and supporting research in the earth sciences and ecology, disseminating knowledge and advice on matters related to those fields, and establishing, maintaining, and managing nature reserves. The Act transferred the statutory responsibilities of the Nature Conservancy to NERC. From thenceforth, the Conservancy had the status of 'a charter committee' of the new Council.

Despite the marked advantages of being part of the Natural Environment Research Council (NERC), it became increasingly difficult for the Nature Conservancy to maintain its position as *the* official spokesman on nature conservation. The slowing down in the growth rate of the science budget provided further strain, and matters came to a head when structural changes were made in the way NERC was organized. The benefits of restoring the Conservancy's former independence were keenly debated and, in some quarters, advocated.

Meanwhile, in the spring of 1971, the Government had commissioned the Central Policy Review Staff (popularly known as the Think Tank) to review the whole range of Government research and development, and particularly the way in which it was financed and organized. The head of the Think Tank, Lord Rothschild, recommended that the funding of applied research and development should be placed on a customer–contractor basis. The appropriate Government departments should themselves decide what type and level of research and development they required. In order to provide the departments with sufficient funds to act as customers in contracting the work, part of the research councils' budget would be transferred (Lord Privy Seal 1971).

At the Council meeting of the British Ecological Society in January 1972, Eric Duffey was asked to draw up a comprehensive response to the Rothschild recommendations, for submission to the Government's Chief Scientific Adviser. It proved a difficult task. As Richard S. Clymo pointed out, there was a danger of adding to the kind of 'woolly generalized statements' that were getting ecology a poor reputation. In drafting the submission, it was important to ask how much notice was likely to be taken of the Society, bearing in mind it had little responsibility as a body for actually promoting or carrying out research in its field of interest, which, in any case, probably amounted to no more than 3% of the research council's budget. The overriding need was to put across points of significance which would otherwise not be adequately represented. In Clymo's view, these related to the fact that ecological problems were often 'not specifiable by the customer', or indeed by the contractor at first. The development and solution of such problems occurred over very long time-scales[24].

The eventual submission was an inevitable compromise, which would have been even longer if all the suggestions of Council members had been incorporated. It argued that not only was there no need for the major changes contemplated by the Rothschild Report, but many of the proposals were ill-conceived in respect of the conduct of ecological research. The Society was particularly critical of the way in which the Report denied the existence of strategic research, classifying all research as either basic science or applied research and development. As the president, Jack Harley, commented, strategic research was essentially 'orientated basic research', something which 'most of us are engaged upon'. It characterized the bulk of the work of such research stations as those of Rothamsted, East Malling, and Merlewood[25].

Experience indicated that the research worker was the best-qualified person to identify the more pressing requirements in ecological research and development. There would be an inevitable tendency for customer bodies to commission research of a short-term, narrowly based nature. In its submission, the Society emphasized how the research councils had a unique understanding of the complexity of many of the issues involved, and the time and resources needed to carry out the relevant research. It was inherent in the training and outlook of the ecologist engaged in research that he accepted 'the responsibility for considering all the consequences in space and time of activities, both his own and those of others, most likely to have important effects on the environment'.

The concept of the customer–contractor relationship, as set out in the Rothschild Report, was largely endorsed by the Government in a White Paper of July 1972 (Lord Privy Seal 1972). This came as no surprise—the real bombshell for those concerned with ecology and nature conservation was a further section of the White Paper, dealing *inter alia* with the Nature Conservancy. This described how the duality of functions in the Conservancy had led to 'stresses difficult to resolve within the present framework'. In the words of a Government Minister, the marriage of the conservancy girl into the research family, solemnized by the Science and Technology Act of 1965, had not been a happy one. If the original purposes of setting up the Conservancy in 1949, and NERC in 1965, were to be fulfilled, a fresh start had to be made. The Government had decided that the Conservancy's reserves, and the staff needed to run them, should become the responsibility of a new Nature Conservancy Council (NCC), which would be an independent statutory body, appointed by the Secretary of State for the Environment, and financed by a grant-in-aid. The funds for managing the reserves and for commissioning applied research would be transferred from NERC to the Department of the Environment. The balance of the Conservancy's budget, together with the stations and research staff, would remain in NERC[26].

In defending the decision to separate the conservation and research functions of the Conservancy, Government Ministers stressed that the Conservancy was only one of many agencies that had to be fitted into the infrastructure of Government. As part of the much wider changes in the organization of Government in 1970, the Department of the Environment had been created with the explicit purpose 'of placing in one Government Department the total approach to the environment'. It was only logical that the new department should include the official body responsible for nature conservation. By the same token, it was important to retain the research side of the Nature Conservancy in NERC, which had a complementary role to the Department of the Environment in the field of environmental research[27].

The 'Split' was announced in a White Paper, as opposed to a Green Paper, and accordingly no formal submissions were invited from the British Ecological Society, and none was made. Indeed the minutes of Council make no reference to the considerable outcry that greeted the White Paper. Whilst everyone agreed that changes were needed, scarcely any ecologist believed the solution was to tear asunder the research and executive functions of the Conservancy. There was a period of frantic lobbying while the 'bill of divorce' was being prepared. It was largely to no avail. The 'Split' and the abolition of the Nature Conservancy were enacted in a Nature Conservancy Council Bill, which was supported in principle by the Opposition and received the Royal Assent in July 1973. The stations and 150 staff remaining in the NERC became part of a new Institute of Terrestrial Ecology.

4.4.3 *Putting across the ecologists' message*

The overriding priority for most members of the Society had always been to win and retain the respect of their peers in the scientific community. Although never entirely overlooked, it was not until the 1970s that ecologists began to take conscious steps towards improving their public, as opposed to their scientific, image. The most striking point to come out of the debate on *A Blueprint for Survival* was the need for the Society to put across its own perception of ecology much more vigorously. There had to be a much more positive use made of every medium of communication available.

This more assertive stance arose out of two kinds of fear. The first centred on the growing misuse of the word 'ecology' which, by the 1980s, had even taken on political overtones, being linked with 'green' politics. Unless urgent steps were taken, there seemed a real possibility of the public coming to associate every ecologist with those who prophesied a future of unrelieved gloom and doom, or who advocated in the name of ecology a whole gamut of panaceas for restoring the 'health of the planet'. Such a public image of ecology would do nothing to quell the second, and much more immediate, fear, namely the consequences of the drastic reductions which were being made by Government in research and teaching, arising from cut-backs in public expenditure.

At the same time, the Society itself came under attack. Although trivial in itself, the incident provided further evidence, if such were needed, of the penalties for the Society of being branded as an academic body, out of touch with the preoccupations of the modern world. A leader in the popular magazine, *Animals*, criticized the Society for being so reluctant to give ecological advice. At a time when 'we look to ecologists for help in trying to avoid disasters resulting from our actions', it was, in the words of the leader column, dispiriting to discover that members of the Society were too busy, frightened of committing themselves, or simply reluctant to step down from the ivory tower of learning and dirty their hands with 'the sort of practical day-to-day involvement with actual problems that should be the fundamental concern of any ecologist worth the name'. Although Palmer Newbould, as chairman of the Society's Ecological Affairs Subcommittee, had no difficulty in writing a letter to the magazine, refuting the ill-informed criticisms, it was a warning of much worse to come if the Society was seen to be faltering in its development of a 'social conscience' (Anon. 1970)[28].

It was no longer enough to point out the errors in the various arguments being put forward, or to warn of the dangers of becoming mixed up with 'the politics and sociology of the ecology of the human environment'. As the great publicity accorded to the publication of *A Blueprint for Survival* made clear, there had to be a much more positive response. The *Blueprint* first appeared in the magazine *The Ecologist* in January 1972. It was compiled by 'a small team of people', alarmed by 'the extreme gravity of the global situation', and failure of Governments to take the necessary corrective measures. The aim of *A Blueprint for Survival* was to 'herald the formation of the Movement for Survival, a coalition of organizations seeking to establish 'a new philosophy of life'. A more stable society was envisaged, in which there would be the minimum of disruption to ecological processes, the maximum conservation of materials and energy, and a human population where recruitment equalled loss (Anon. 1972).

In one of its regular reports to Council, the Ecological Affairs Subcommittee warned of how it might be unwise for 'professional ecologists' to ignore *A Blueprint for Survival*, in spite of its limitations. The subcommittee organized two evening sessions at the Society's winter meeting of January 1973, to discuss ways of making 'more accurate ecological information' available. One session took the form of a debate on the motion 'that this Society should now abandon the blackboard in favour of the placard'. The motion was proposed by Palmer Newbould who argued that the word 'ecology' was now so misunderstood and misrepresented among the general public that the Society had a duty to put the record straight by taking an active part in public discussion, and seeking to influence children and students through the educational system. The motion was opposed by Jack Harley, who agreed that the Society should be concerned with environmental affairs, but only in respect of the facts. The Society should 'not take sides in controversial matters'.

The most substantial point to emerge from the debate was the need to make much greater use of the Press, radio and television. The Press had always been regarded with some suspicion. At its meeting of January 1969, Council sought to retain the privacy of the Society's meetings by reaffirming that reporters should only be admitted if contributors had been given prior notice and any report was approved by the Society before publication. At the same time, Council recognized the need for a more positive approach. Whilst it was far too premature to contemplate employing a paid press officer, it was agreed that the appointment of an honorary public relations officer was long overdue. The writer and ornithologist, Bruce Campbell, was appointed to the position. For the first time, press releases were sent to a large number of scientific journalists, drawing attention to the Society's winter meeting of 1972. They seemed to have little effect.

Something more was needed. Approaches were made to '40 prominent members of the Society', inviting them to become the Society's press officer. They all declined. At its meeting of April 1973, the Ecological Affairs Subcommittee concluded that there was no alternative to appointing a press officer for, say, £500 per annum. From discussions with a leading environmental journalist, it became clear that such a press officer would have to spend at least 200 hours a year dealing with the Society's affairs, including attendance at relevant meetings, examining the contents of journals, and consulting members. For this, payment of at least £1500 would be required. After considerable discussion as to whether the expenditure of 'this large sum of money' was warranted, a motion was put to Council by the president, contemplating the employment of such an officer for two years. There were nine votes in favour and six against, with three abstentions.

In announcing his own appointment in the *Bulletin* of December 1974, the journalist and environmental correspondent, Jon Tinker, outlined his three lines of attack, namely to issue brief press releases on selected papers published in the Society's journals, putting journalists in contact with those members who had agreed to act as spokesmen on various topics, and to publicize 'the more outward-looking meetings of the Society'. The overriding need was to convince journalists that the Society was a source of news and information. As Tinker reminded members, he could not create the news himself nor comment on the Society's behalf—he was an intermediary, dependent on the co-operation of members.

Reference to the Society, whether in the Press or on radio and television, was something of a lottery. On the basis of the preconference abstracts, Tinker asked for advance texts or summaries of 14 papers which were to be given at the winter meeting of 1975. Only eight authors responded. Press releases were prepared on four of the papers. Two made a considerable impact. After giving his paper, John Farrar (who left a few days later for a post in Dar-es-Salaam) spent the afternoon walking up and down under arc lamps beside a small stand of *Pinus sylvestris* in the grounds of the Royal Holloway College, talking about SO_2 damage to Scots pine. He appeared in a five-minute interview on BBC television late that night.

In a report written at the end of his term of employment, Jon Tinker criticized the Society for not taking the value of communication between the professional ecologist and the public seriously enough. Much greater use should be made of *Bulletin* for developing public relations; a conscious effort should be made to plan the Society's meetings with the needs of the press consultant in mind. For their part, members of the Ecological Affairs Subcommittee agreed on the need to provide the general public with reliable information on the Society's activities, and 'to demonstrate that ecologists were serious people', but it would be wrong to place so much emphasis on Press/public liaison as to 'let the tail wag the dog'. Publicity was not the main purpose of the Society's meetings.

After some delay, a new press consultant was appointed for a one-year period—a freelance journalist. Instead of producing stories for direct consumption, a greater priority was given to promoting a more general understanding of what was meant by 'ecology', and what ecologists did. Rather than sending out press handouts, longer articles would be written, based on interviews with researchers. The approach met with little greater success. There continued to be a lack of sufficiently controversial or newsworthy material to attract the interest of press correspondents and their editors.

Rather than seeking to influence the media in every possible way, the Ecological Affairs Committee concluded that it would be much more realistic to concentrate on one limited objective, namely to inject more ecological items into the scientific columns of the better papers. To achieve this, an ecologist with press contacts might prove to be better than a journalist with ecological leanings. In 1980, the publicity officer of the Nature Conservancy Council, Jean Ross, agreed to take on the additional responsibility of acting as press consultant to the Society, abstracting suitable material from its journals and symposia volumes, the *Bulletin*, and relevant minutes and papers of the various committees. It met with far greater success than previous efforts.

As other learned bodies and disciplines were to find, synthesis and reason have little place in selling newspapers or boosting television ratings—they are much more the attributes of the backroom approach, and it was this approach which the Ecological Affairs Committee came to adopt more and more during the 1980s. There was plenty of headway to make up. Members of the Committee were shocked to discover the extent of ignorance and prejudice even within Government departments and statutory bodies. They came across more than one senior Civil Servant, who had dismissed the Society out of hand as 'a political party of greenish hue' or an advocate of some alternative lifestyle to do with organic farming. The commonest misconception of all was the assumption that the Society was concerned only with nature conservation.

There was scope for improving liaison even with those bodies with whom the society had had long-standing links. As chairman of the Ecological Affairs Committee, Michael Usher inititiated a series of discussions with key personnel in the Nature Conservancy Council, Countryside Commission, and Wildlife Link. The director of the Forestry Commission was invited to attend part of the meeting of the Ecological Affairs Committee in February 1985. Among the topics discussed was the scope for making representation through the Commission's regional advisory committees, and the development of closer contacts with research staff in the Commission. A meeting with representatives of the Ministry of Agriculture, later that year, focused on future changes in farming practice, the ecological content of training courses for officers of the Agricultural Development Advisory Service (ADAS), and the problems of environmental research against a background of changing methods.

Such meetings not only provided an opportunity for the Society to make representations to some of the most powerful land-owning and -managing agencies in the countryside, but they often enabled the Society to learn at first hand what was happening in respect of research funding. At a meeting with the secretary of the Natural Environment Research Council, Society representatives discussed the Council's recently published corporate plan, in which ecology had fared so poorly in relation to some other disciplines within the Council's purview. The secretary of the Council stressed the need for ecologists to co-operate within strongly integrated subject areas, so that they could compete more effectively for funds against other sciences, where such integrated groups already existed.

4.5

A More Outward-Looking Society

It has never been easy to distinguish those activities which developed as a direct consequence of there being a British Ecological Society and those which merely surfaced for whatever reason under the Society's aegis. Although its journals and meetings bore eloquent witness to the labours of many members who, at various times, played a prominent role in the Society, the Society could not necessarily claim any direct credit for their works. In this respect the Society was no different from other natural history societies and field clubs (Doogue 1985), or indeed any voluntary body established to *serve* the needs of its membership.

The Society's precise role in promoting ecology became even more blurred during the Society's third quarter-century, as ecologists became more closely concerned with matters of public concern, and sought further sources of funding. Ecologists traditionally operated through their respective university departments, research institutes, and the various academic funding bodies. It was in this way that resources were obtained for research, and responsibilities for teaching and supervision were fulfilled. To this extent, ecologists shared with most of their academic colleagues an 'ivory-towered existence'. It was an existence that came under increasing scrutiny during the Society's third quarter-century. Not only were there external pressures encouraging, and indeed forcing, ecologists to look for additional sources of funding, but ecologists themselves were becoming much more active in finding new outlets for their expertise.

4.5.1 *The consultant ecologist*

There was nothing new about members of the Society assisting in the protection of sites, or giving their opinion as to the use and management of land and natural resources generally. Oliver had called for just that kind of consultative work in the first years of the Society's existence. Neither was there anything new in the fact that so few members accepted the challenge. The main difference between the past and present situation was the much greater scope in the 1970s for taking up a consultative role. It was an indication of the impact of the environmental revolution, and the repercussions of such disasters as the wreck of the oil tanker, the *Torrey Canyon*, on the rocks of the Scilly Isles in 1967, that Governments, landowners and land-users alike were now consciously seeking ecological advice.

240

The Society's first response to the overflowing postbag of enquiries was to compile lists of members' expertise. The first was drawn up in direct response to pleas from the honorary officers for help in answering enquiries from the general public. A notice printed in the *Bulletin* for June 1974 elicited only 14 offers of help. A specific request was made in 1976 for assistance in answering enquiries on pollution and conservation, mainly from schoolchildren engaged on project work. Sixteen members responded to a further appeal for help in 1980.

Difficulties were soon encountered in handling the requests for more detailed information. The National Trust asked for ecological guidance on how far it was necessary to cull the seal population of the Farne Islands[29]. The most obvious way of securing specialist advice at short notice was to compile a register of members' expertise, and to use this as the basis for convening small working parties. As a first step, the September issue of the *Bulletin* in 1974 included an edge-punched card, devised by John H. Lawton of the biology department at York, on which members were invited to record their interests under a series of general headings, covering habitat types, organisms, techniques, processes, and other interests. Some 85 specific interests were identified. By the following spring, over half the United Kingdom membership, and a quarter of overseas members, had returned a card[30]. In the event, the index was never used. The punch card system proved cumbersome to operate, and the details recorded were soon out of date.

Among the many recommendations made by Newbould's subcommittee on the future role of the Society had been a proposal to publish a list of members willing to serve in a private capacity as part-time ecological consultants. A questionnaire was included in each copy of the *Bulletin* for December 1970. A sufficiently large response was received for the names and addresses of 79 persons to be published in June 1972, grouped under 16 topics, which included conservation and management, air pollution, pest control, and quantitative methods. Special attention was drawn to those persons with experience in tropical ecology. A disclaimer made it clear that the Society could take no responsibility for the manner in which any consultancy work was carried out, including the scale of fees charged.

Few ecologists were likely to find consultancy work by simply having their names included in the Society's list. They had to find more practical ways of demonstrating the relevance of their expertise and experience. David Streeter of the School of Biology at Sussex wrote of how the Society itself could play an important part by encouraging members to tackle practical problems in the environment. Philip Grime of the botany department at Sheffield suggested that the *Bulletin* could be used to notify fellow-ecologists of sites where scientific work was needed, either because they were about to be destroyed, or because a strong scientific case was needed to preserve them. In cases of extreme urgency, the Society might provide funds for the necessary work. An important by-product of the Society's active involvement might be to establish such survey

work as a legitimate research activity in the eyes of some university departments.

In direct response to these suggestions, Council announced in the *Bulletin* of
March 1973 that a sum of up to £300 per annum would be made available for
financing 'ecological surveys of areas of scientific and natural history interest'
threatened or potentially threatened by development. Among the first grants to
be given was one of £100 for the employment of a field assistant in establishing a
series of permanent monitoring sites around Loch Lomond, where the littoral
fauna was threatened by the construction of a barrage, SO_2 pollution, and public pressures. During 1975, a grant of £350 was given for a survey of the site of
the proposed Empingham Reservoir in Rutland.

The success of such projects depended on more than money. Barry Gold-
smith of the botany department at University College, London, commented, in
the *Bulletin* for March 1973, that such site assessments were fraught with difficulty. How was it possible to weigh up different characteristics of ecosystems?
What was the relative importance of species richness, habitat diversity, extent,
stability, and naturalness? What was the purpose of—who would benefit
from?—such assessments? Was the priority to improve amenity, provide gene
banks, or preserve rare species? An opportunity to compare different
approaches occurred in 1976–7 when surveys were made of properties belonging to the National Trust in Kent, Norfolk, and Yorkshire. The Trust provided
a grant of £2000 to cover the expenses of the three surveyors nominated by the
Society, and their five assistants. The results were sufficiently encouraging for
the Trust to set up a conservation branch, with a view to making similar surveys
of other properties.

Very often, one of two situations arose. Either the Society was ignored altogether, or it was expected to give an instant response to requests for advice on
highly complex issues. The only way to combat these embarrassing situations
was to make a positive effort to 'spot issues coming up'. In an appeal for help,
printed in the *Bulletin* for December 1971, the Ecological Affairs Subcommittee
invited suggestions from members as to the 'subject areas where the Society
might formulate an opinion, and official bodies, working parties and co-
ordinating committees to whom such opinions might be expressed'. As the
notice stressed, the Society had to be selective. It could not become involved in
purely local issues, except where they raised 'general principles of national import'.

In response to these various appeals for information, one correspondent
drew attention to 'the very considerable threats to the Scottish coasts posed by
North Sea Oil development'. Sites of Special Scientific Interest had already been
damaged at Nigg and Dunnet Bays. As well as identifying those parts of high
biological interest, it was perhaps even more important 'to find sites of un-
importance', where development might be encouraged, 'leaving the more important sites to be inspected at greater leisure'. Arising from discussions on
Council, in which George M. Dunnet, the professor of zoology at Aberdeen,

took a prominent part, it was suggested that, as a first step, 'an objective statement of the type of problems which were occurring' should be printed in the
Bulletin.

In the event, the statement took the form of a paper prepared by the North
Tayside branch of the Conservation Society, to which comments were
appended by George Dunnet and J. B. Kenworthy, of the botany department at
Aberdeen. A major problem was to define and disentangle the economic, social,
political, and ecological aspects. Three types of representation could be made.
Individual ecologists could become closely involved with local development by
discussing new proposals with the local planning authorities. Secondly, the
Society might convene a symposium, which would clarify terminology, assess
the importance of scientifically established information in the planning process,
and bring together information from other parts of the world. Thirdly, the
Society might press for an extension of legislative control over the activities of
the oil industry.

The pre-eminence of the Nature Conservancy as an employer of ecologists
had given rise in some quarters to a false impression that ecologists were unable
to turn their hand to anything else. Not only was this untrue, but it had the
effect of precluding them from some of the most exciting challenges in environmental planning. In the *Bulletin* for June 1976, Rawden Goodier of the Nature
Conservancy Council drew attention to the implications of discussions taking
place in the European Economic Community about whether member states
should require an environmental impact statement to be made prior to consent
being given for developments affecting the quality of the human environment.
Such a requirement was imposed on federal agencies under the Environmental
Policy Act of 1969 in the United States.

The great increase in environmental awareness had opened up unprecedented opportunities. The local authority ecologists were the first practitioners
of a new kind of amenity–land planning, having to find instant answers to problems which other ecologists had hardly discovered. There was more than an element of irritation in another contribution to the *Bulletin*, written by John G.
Kelcey, an ecologist with the development corporation responsible for the
building of the new city of Milton Keynes. Although local authorities and industry were beginning to employ ecologists, the ecological community showed
every sign of being caught unprepared. As one moved 'from the general principles towards implementation, so the ecological problems became more and
more acute and the information more scarce'. Drawing on his experience at
Milton Keynes, Kelcey believed the only way ecologists could win the respect of
their peers in local government and industry was for there to be 'a radical
change in the entrenched attitudes of traditional ecologists, a re-ordering of
priorities and a change in the speed, methods and type of ecological research
currently being undertaken'.

One effect of the widening of employment opportunities was to call into
question the use of the term 'professional' among ecologists. Writing in the

Bulletin, J. F. Benson urged ecologists to follow the lead of engineers, town planners and architects, who used the term to describe only the *practice* of the subject, usually as a consultant. University teachers and officers in research institutes might be professional teachers and researchers, but they were not necessarily professional ecologists. It was more than a question of semantics: professional qualifications were at the heart of the whole career structure in local government. Ecologists were placed at a distinct disadvantage through their inability to put forward any kind of qualification denoting a minimum academic standard and practical experience gained. The British Ecological Society was the obvious body to help make good the deficiency.

It did not take long for the Ecological Affairs Committee to conclude that the Society itself could not act as the validatory body. It was considered inappropriate for 'a learned society' to become so closely involved in public affairs. The obvious alternative was for the Society to play a key role in establishing a professional institute of ecology. In discussing the feasibility of such a move, Council recognized, at its meeting of January 1979, that close account would have to be taken of recent initiatives taken by the Institute of Biology, which had set up an environmental division. At a meeting with representatives of the division, the president, together with Michael Usher and Roger Smith, learned of how it was intended to set up a register of environmental biologists. A vigorous vetting procedure would place great emphasis on practical experience. As the representatives of the Institute of Biology pointed out, the participation of the British Ecological Society in the assessment of applicants could play an important part in achieving greater recognition for the register. For the Society, the publication of the register would remove any need to revise and update its own list of consultants.

Throughout all these discussions, Council and the Ecological Affairs Committee were very conscious of the fact that the question had never been satisfactorily resolved as to how far the Society should express opinions on matters of public concern. It quickly came to the fore again in the mid-1980s, following the publication of a paper in the journal *Ambio* setting out the environmental consequences of nuclear war. For many ecologists, it was their first detailed insight into the wide-ranging atmospheric effects of a nuclear fireball, and the way in which a 'nuclear winter' might disrupt ecosystems on a global scale, bringing about irreversible, catastrophic change in the biosphere.

As a measure of its concern, the Society organized a one-day meeting in March 1985. Papers described the size of the world's nuclear arsenal and speculated on the effects of nuclear war on both terrestrial and marine ecosystems, and on human food production. A report on the meeting, printed in the *Bulletin* concluded by stressing how a preliminary assessment of the effects of a nuclear war was 'a legitimate scientific undertaking as people's natural concern about uncertain events lead them to address such questions'. People had a right to know the full implications of policies undertaken by Governments. The obvious way of securing the relevant evidence was through co-ordinated and disciplined modelling programmes, bearing in mind the parametric and structural uncertainties that were bound to arise.

In the same issue of the *Bulletin*, five members of the Society gave notice of their intention to table a resolution at the next Annual General Meeting, calling upon the Society to issue a statement drawing attention to the ecological consequences of nuclear war. The immediate pretext was the imminent publication of an international scientific study (under the aegis of the Scientific Committee on Problems of the Environment, SCOPE), dealing *inter alia* with this aspect. The New Zealand Ecological Society had already published a book setting out the ecological consequences of such a holocaust for that part of the globe.

A large part of the Annual General Meeting of December 1985 was taken up with a discussion of the motion, put forward by Tony Bradshaw (who had been chairman of the one-day conference), Michael Chadwick, Richard Law, Jonathan Silvertown and Jeff K. Waage. The motion had already been rejected unanimously by Council. As the president, Roy Taylor, told the meeting, even if passed, the motion could hardly be represented as the united view of the Society. The bulk of the membership had not been consulted. The most likely outcome for the honorary officers would be a deluge of correspondence from puzzled members. More seriously, there was danger that such a resolution, with its obvious political implications, would discourage Governments from seeking the advice of ecologists on those aspects where they did have some claim to special expertise.

On behalf of the proposers of the motion, Michael Chadwick insisted that, far from creating a precedent, the motion did no more than follow the example of the Society's founder, A. G. Tansley, who, with other members of the Society, had pressed 'the political case for conservation in Britain', and secured the establishment of the Nature Conservancy.

For another speaker, the principal concern was how far the motion would 'open the floodgate for other comparable motions to be brought annually on topics such as animal rights and "green politics"'. If passed, what would be the position of members who disagreed with such political judgements. Were they to resign and so lose their scientific contacts which were central to the purpose of the Society? On the motion being amended so as to make it explicitly clear that it reflected the view of 'the Annual General Meeting of the British Ecological Society', rather than the Society itself, the motion was carried by a show of hands.

4.5.2 *The sponsorship of education and research*

In the words of the revised constitution, the aim of the Society was

> to advance education and research in the subject of ecology as a branch
> of natural science and the dissemination of the results of such research.

Although there was nothing particularly novel about these sentiments, the Society became much more assertive in developing them during the 1970s.

Whilst the meetings and journals continued to play a central part in the life of the Society, increasing significance was attached to the sponsoring of education and research. Nine formal schemes for financial assistance had been introduced by the mid-1980s.

Almost every participant in the Society's debate on *A Blueprint for Survival* had stressed the need for greater understanding and appreciation of ecological subjects among children, young people, and teachers. Conscious of how little could be achieved, even if all the Society's resources were devoted to this cause, Council invited Charles Sinker, the director of the Field Studies Council, to form a small educational working group 'to explore the possibilities with the object of finding some limited sphere where they could expect a clear result in relation to input of resources'.

It was far from clear what kind of distinctive and practical contribution could be made by the Society. Summer schools, special lectures, and discussion programmes were 'a valuable and efficient means of promoting ecological education', but the Field Studies Council and naturalists' trusts were better placed to provide them, and had already acquired a considerable reputation for doing so. The use of reduced subscriptions to encourage sixth-formers and undergraduates to join the Society might lead only to disillusionment and a high wastage rate if the Society's publications continued to be so specialized. A proposal that the *Bulletin* might be expanded in content and scope, with short papers on teaching methods and perhaps a question-and-answer service, was withdrawn following strong opposition from the editor.

A week-long course organized for teachers was held at the Preston Montford Field Centre of the Field Studies Council in August/September 1975 on 'field teaching and modern ecological theory'. It was attended by 25 people; the five tutors were drawn from the Society and Field Studies Council. Further courses were organized. In 1977, a weekend refresher course for sixth-form and other post O-level teachers was focused on plant-feeding insects. The aim was to bring participants up to date with current research in this field of population ecology, and to illustrate how concepts could be put across through appropriate field exercises. A course on the 'Biological aspects of water pollution' in 1978 was followed a year later by three on quantitative plant ecology, decomposition, and field techniques in animal community ecology.

Financial assistance was offered to those attending these and other courses organized by the Field Studies Council and the Scottish Field Studies Association where teachers were unable to recoup the costs from their respective education authorities. The scheme was soon afterwards extended to cover the costs incurred by schools of extra equipment and facilities needed for carrying out local projects involving 'some constructive work' on the part of pupils.

A workshop was organized in September 1978, with the aim of assessing the training needs of students about to embark on a career tackling practical ecological and environmental problems. As Edward Newman, the rapporteur for the meeting, remarked, there were so many calls for help in solving ecological problems of major practical importance that it was far from easy to know how students should be prepared. Should universitites and polytechnics be altering their teaching to provide different kinds of expertise? Were they necessarily the best places to provide such training? What training were potential employers of ecologists looking for? Did they know themselves what they wanted?

Discussions ranged over the utility of ecology-plus-environmental science degrees, the value of one-year MSc courses and three-year PhDs, and the potential of refresher courses. A distinction was drawn between problem-solving (the use of ecological knowledge in managing an area of vegetation or controlling a pest), and decision-making, which involved the political skills of balancing essentially irreconcilable social, economic, and ecological demands. Much more attention had to be given to the skills of problem-solving.

It did not take many setbacks to convince Council that the main thrust of the Society's educational effort should continue to be in the field of higher and adult education. Both the opportunities and need for such an investment were set out in the *Bulletin* of late 1970 by Ian N. Healy (*1941–72*), of the zoology department at King's College, London. As he pointed out, the Society had been founded at a time when scientific activity, and especially biology, was the preserve of the privileged few, and cost very little public money—now there were many more scientists, and their work cost a great deal. Baying at the heels of these scientists was 'an educated public increasingly anxious to see the relevance of the scientists' activities to their own needs and interests'. A recent university extramural course on the environment had attracted 170 applicants.

Ian Healey put forward two proposals intended to exploit this potential audience. Both were intended for the kind of person who joined the county naturalists' trusts—the Kent Trust already had 3300 members. So far, the scientific community had given these bodies very little direct encouragement. As well as organizing meetings explicitly for this kind of person, a small number of bursaries should be made available so as to enable non-members to attend the Society's meetings. A pilot scheme, comprising six bursaries, was introduced in 1971.

It was the Society's own members who benefited most from bursaries and other forms of grant aid. As a result of rising costs, it was becoming increasingly difficult to attract the younger members to meetings. After lengthy discussions, Council agreed to make some contribution to the costs of those with limited means. It was announced in the *Bulletin* for June 1975 that £500 had been set aside. The scheme was extended and, in 1979, was replaced by another, whereby £750 were set aside for the winter meeting, £750 for the annual symposium, £500 for the summer meeting, and £250 for occasional and group meetings. Intended to meet subsistence costs only, the grants were available on a first come, first served, basis.

The specialist groups stood to gain much from this kind of support. Only three postgraduates attended the summer workshop of the Mathematical Ecology Group in 1978. As Michael Usher pointed out, the main aim of the workshop was to talk about work in progress, and a large graduate student input was essential. The obvious solution was to subsidize the expenses of those students reading papers or leading workshop sessions. The response to the offer of grant aid was encouraging. Five of the 11 papers at the workshop of 1979 were presented by postgraduates supported by a grant from the Society. In the words of Robert H. Smith, who succeeded Michael Usher as the Society's secretary of the Group, 'the high standard of both content and presentation more than justified the support'.

Rising costs were also curbing the activities of student societies. In September 1978, Council accepted a proposal by Edward Newman that financial support should be given to biological societies in universities and other institutions of higher education to cover the expenses of invited speakers. After lengthy discussion, a sum of £400 was set aside for an initial one-year period. Thirteen applications were received, and £232 awarded. Six societies applied for aid in the second year of the experiment. In a move to encourage more applications, the upper limit of £40 per application was removed in 1980.

In collaboration with the Nordic Council for Ecology, a scheme was devised whereby postgraduate students from Scandinavia received financial assistance in attending meetings and courses of the British Ecological Society. British students were given similar opportunities to meet leading ecologists in their field of study, and to discuss research problems with fellow students, in Scandinavia. Where the courses included field investigations, there was obviously the chance to see a little of the country visited.

As a way of encouraging research students to present papers at the Society's winter meeting, John Lawton put forward the idea of making an annual award to the student who presented the best paper, as judged on its scientific content and presentation. Considered an excellent idea, Council, at its meeting of April 1978, resolved to award a prize of £50 and a certificate. As one of the judges remarked of the 1979 competition, the standards of the six entries were so high as to outclass many of the professionals. There were three times that number of entries in 1980. The value of the prize was later increased to £100, and provision made for awards to highly commended candidates. The competition soon became one of the most successful features of the winter meeting, attracting 27 candidates in 1985 and 42 in 1986.

It was out of consideration for the training and educational elements that Council reversed another policy decision in 1978—this time to award grants to student expeditions. On the recommendation of the Education and Careers Committee, it agreed to make available a sum of £500 (raised to £750 in 1979) for assisting individual expeditions, up to a limit of £250 each, subject to their being approved by a panel of three members of the Society. The main criterion would be the extent to which the expeditions widened the ecological experience of the participants, rather than produced publishable results. As Tim C. Whitmore wrote later in the *Bulletin*, it was the purpose of the grants 'to help light a spark of interest in ecology outside the recipients' own country'. Since most of those to benefit would live in the United Kingdom, and the tropics were the most likely destination, the support was seen as another important way of bringing new talent to bear on tropical ecology. Six grants were made in 1979, and nine in 1980, the largest grant being £200 for an expedition to Kenya.

It was often a moot point whether education or research stood to gain more from these various types of sponsorship: the two activities impinged so closely on one another. As its new chairman, Michael Usher quickly realized that an obvious task for the Ecological Affairs Committee was to develop the concept of the small ecological project grants. The earlier scheme for assisting research on threatened habitats had never been given much publicity and involved very small sums of money. In the course of drawing up its budget for 1979/80, the Ecological Affairs Committee sought an allocation of £2500, which would enable it to respond more quickly to calls for assistance from those members carrying out 'small projects'. Council agreed to make £1000 available. A notice in the *Bulletin* made it clear that preference would be given to surveys of threatened habitats, those of special ecological interest, and sites with a history of ecological work. No grant was likely to exceed £250.

The scheme was to prove the most popular ever devised. The first grants were made for a survey of Jurassic limestone grassland in the North York Moors (£270), an investigation of the effects of air pollution on insectivorous birds in central Scotland (£75), and the ecological aspects of a study of the behaviour and ecology of the Houbara bustard in the Canary Islands (£150). The annual sums allocated for the scheme were increased in 1980 to £2000 and in the three subsequent years to £4000, £8000 and £15 000. At first there were some doubts as to whether the scheme was a legitimate use of the Society's funds. The standards adopted for approving projects were criticized. The rejection rate was, however, high. Of the first 300 applications, only 171 were successful. There were no misgivings on the part of the Ecological Affairs Committee. The scheme not only helped young ecologists, but experience indicated that it had a pump-priming effect.

Council had always insisted that it would be wrong for the Society to assume general responsibility for financing research. It was not until the early 1980s that serious consideration was given to providing such assistance, following a recommendation that £30 000 per annum should be allocated equally between the funding of a research fellowship and the equivalent of three research studentships. A dichotomy of views soon emerged, with some members drawing attention to the considerable prestige which the Society would gain from funding fellowships, and other members emphasizing how few persons would actually benefit. Rather than entering 'the research funding field normally occupied by the Research Councils', a better course for the Society might be to institute travelling fellowships, intended to help ecologists in whatever country they lived to travel abroad for research purposes. Not only would they involve the Society in far less supervision, but they could be discontinued much more easily, should the need arise.

The outcome was a decision by Council to offer a number of 'British Ecological Society Travelling Fellowships' of normally no more than £1500 each, intended to enable 'people active in ecological research' to visit other countries for the purposes of research. A sum of £15 000 per annum was set aside for an experimental period of two years. Considered a success, the scheme was put on a permanent footing at the April meeting of Council in 1984.

Once the Society had taken on the role of sponsoring educational and research ventures, there was no scope for turning back—indeed the different forms of sponsorship assumed an even greater role as other sources of funding contracted. The overall level of grants rose by over 70% between 1983 and 1984, and by a further 35% to £54 231 in 1985. The expedition grants increased by 14%. The support given to the travelling fellowships scheme rose from £6955 in its first year to £17 133 in 1985.

4.6

The Presidential Viewpoint

The most important election held by the Society was that of the president. He was chosen not only for his wide interest in ecology and contribution to a particular field of research, but also for his obvious concern for the Society and its role in public affairs. In order to give the president sufficient time to gain a firm grasp of 'current business and society work before taking the chair', the constitution was amended to enable him to be elected one year before the date of the vacancy and serve as the president-elect on Council and committees[31].

While he was president, Mick Southern suggested that the customary alternation of animal and plant ecologists might be expanded so as to include applied ecologists. Not everyone liked the idea. Tom Huxley, the director of the Countryside Commission for Scotland, protested that it was time for all such distinctions to be abolished. If the recent trend of establishing biology faculties in the universities continued, there would soon be 'many aspiring Presidents, who will have majored in biology, as opposed to botany or zoology'. Council took the 'safe' course of deciding that no change was needed for as long as 'a fairly liberal interpretation of botany and zoology' was made. Both were essential to 'a variety of applied biological sciences'[32].

4.6.1 *The outlook for ecologists*

Beginning in 1976, it became the custom for presidents to contribute 'a presidential viewpoint' to the *Bulletin*. It was entirely appropriate that the first should focus on what its author, Richard Southwood, called, the Society's prime duty, 'to foster the subject so that the decisions of the practioners can be soundly based'. Over the previous two decades, great progress had been made in using modern quantitative methods. Ecologists had been able to move from single-factor explanations to a more synoptic multidimensional view. Plant and animal ecologists had found much of interest in 'the other kingdom'.

There was, however, little evidence of ecology shaking off its traditional constraints of being perceived as 'a soft science' of little direct or obvious relevance to the more pressing problems of Government and industry. As Southwood pointed out, it was the complexity of the systems being studied by the ecologist that prevented laws and generalizations having 'the crisp simplicity and universal applicability' of, say, molecular biology. Among the physical sciences, meteorology came closest to sharing some of the complexities of ecology, and meteorology had long ago been recognized as a relevant and expensive science, and had been funded accordingly. Ecology had never crossed that threshold in public perception. For members of the British Ecological Society, it was far from clear whether the opportunities for such a breakthrough in the so-called environmental revolution had been wasted, or whether they had never been there in the first place.

It was right that the Society should do everything possible to create a climate in which a 'new breed of ecological practitioners' might be employed, as 'the counterpart of the engineer who makes his decisions on a knowledge of the laws of physics and mechanics', but, in doing so, Southwood warned of how it was important not to exaggerate what could be accomplished, and therefore bring the subject and profession into disrepute. As the great engineers of the past had discovered, there were penalties in proceeding before all the relevant expertise and experience had been acquired. The only protection, and it was only a partial one, was to draw a clear distinction between fact and opinion.

Whilst ecologists might speak of their desire, and indeed duty, to make 'their scientific expertise and understanding' more widely available, it was easy for the laymen to interpret the way in which they went about that task as betraying a lack of conviction or commitment. In his presidential viewpoint, printed in the *Bulletin* for February 1978, Peter Greig-Smith wrote of how it was perfectly proper for individual ecologists to express opinions on environmental matters, and to join pressure groups, but they had to be clear in their own minds, and make it abundantly clear to others, when they were speaking as ecologists and when as private citizens. Likewise, it would be improper for the Society to give 'a corporate opinion on practical affairs'. Any move in that direction would damage straightaway the Society's credibility on scientific issues. The Society's role had to be strictly limited to providing a forum in which members could discuss 'the contribution of ecological expertise to decision making'.

No matter how pertinent this advice might be for preserving the integrity of ecology as a science, it did little, at least in the short term, to attract public attention and patronage. Fearful of incurring the criticism of scientific colleagues, ecologists tended to play safe in drawing conclusions or making recommendations. As George Dunnet remarked, in his presidential viewpoint of August 1980, 'advice that is scientifically respectable may not be strong enough for developers or resource managers working under the usual constraints of time and money'. Having tried and found such ecological advice wanting, they might conclude that it could be dispensed with. Faced with so dangerous a contingency, Dunnet wrote of how 'it is up to *us* to ensure that decisions taken are consistent with our understanding of the ecological issues involved'.

There was, according to Dunnet, 'quite a lot going for us'. There were success stories of pollution control and the utilization and management of resources (for example in grazing systems and fisheries), but the question remained as to how much of the credit could be apportioned to the kind of ecologists who joined the British Ecological Society? How much of the attention of the media was really focused on the activities of such ecologists, and how much was on those who came from other disciplines and backgrounds, but who chose to work under the name of ecology?

Ecological terms were used ever more loosely. Part of the problem was that they were words in common use, which had been given a more or less precise meaning by ecologists. Although this had avoided the adoption of artificial jargon, it had done nothing to project the role of 'the professional ecologist'. Because the words also remained in everyday use, it was easy for non-ecologists to gain the impression that they understood the very complex ideas, encouraging them to make free and uncritical use of the terms. It was largely (but not entirely) the responsibility of 'the professional ecologist' to make clear what should be set aside strictly for ecologists working from a scientific standpoint.

As an illustration, Dunnet (1982) devoted the greater part of his presidential address to one of the more 'important current interfaces between ecologists and the public', namely monitoring and the making of environmental impact assessments. As a form of surveillance, monitoring took two forms, namely *repeated* measurements to define the characteristics of plants and animals, and secondly to define the concentration of selected chemical or physical characteristics of defined components of the environment for comparison with legally permitted or otherwise agreed levels. Not only was the design and execution of monitoring programmes a professional business, but so too should be the interpretation of the data collected, bearing in mind the need to take account of such factors as the extent, duration, progressiveness and reversibility of change, and the difficulties of assigning causes to the changes identified.

For a long time to come, it would be difficult to make firm and confident predictions on the basis of monitoring programmes. Drawing on his own experience in respect of breeding colonies of seabirds, and the problems inherent in gathering statistical data, Dunnet emphasized the need to be positive rather than negative, always making it clear how much confidence should be placed on the statements made. Ultimately, the quality of the ecologist's forecasts depended on a basic understanding of the relevant species, community and processes, as 'derived from good scientific research in either pure or applied ecology'.

For his presidential viewpoint of August 1982, Tony Bradshaw took the theme of 'achieving useful ecology'. Because the environment was all around and everybody had some experience of it, there was no difficulty in becoming an expert in the ecology of something. Within reason, this broadening of the ecological constituency was to be welcomed, but it gave rise to obvious frustration when those trained in ecology found it increasingly difficult to find a job, and other people tackled the many ecological problems waiting to be solved. Ecology had slipped so far out of the hands of ecologists that an offensive was needed, which positively presented 'an ecological approach to all problems where ecological principles and expertise' were appropriate.

In the new, rather tougher, world of the 1980s, when the recurrent message for all scientists was one of relevance, this meant grabbing every opportunity to use ecology in practical situations; being prepared to start low, by taking simple jobs or giving advice on simple matters; being persistent (by being persistent, the other side had to pay attention eventually if the arguments were good); accepting difficult problems where the evidence was far from adequate, and, finally, communicating effectively. All this applied as much to established ecologists, as to recent graduates. The former already had experience and status— they were more likely to be listened to.

At a time when doom and gloom were fashionable, it had to be recognized that the destruction of ecosystems by mining, quarrying, and other processes was an inevitable part of civilization, and that even if ecologists took little part in restoring such ecosystems, others would. Not only would 'all sorts of opportunities to practise the science of ecology be lost', but the restorative work would probably not be done as well as it might be. With the proper use of imagination, there was great scope for 'creative ecology in degraded terrestrial environments'. In his presidential address on 'The reconstruction of ecosystems', Bradshaw set out the opportunities for constructing on derelict sites not only the communities representing the early stages of succession, with their specialist floras and faunas, but examples of more advanced plant communities (Bradshaw 1983).

For Roy Taylor, in his viewpoint of August 1984, one of the most serious defects in applied ecology had been the growth of 'no-go' areas between the several applications, and most obviously within 'the seemingly unbridgeable gap between conservation and agriculture'. These two most highly developed and highly regarded aspects of applied ecology appeared to be locked in an interminable argument as futile as it was acrimonious. For too long, ecologists had allowed the notion to develop that they were part of the nature conservation movement, and sought to preserve the *status quo*. For just as long, the all-out quest to maximize food production had been allowed to stifle the expertise undoubtedly available, within both the research councils and universities, for investigating the wider ramifications of the increasing rates of change in the use and management of British farmland. It was a hiatus that encouraged agronomists and those working on natural populations to overlook or underestimate the relevance of one another's findings and techniques. Ecologists paid a particularly heavy price for generally neglecting the extensive population studies being carried out by agronomists on crop and weed plants in the trial plots of the various agricultural research institutions.

In fact, this particular 'no-go' area provided an outstanding example of the extent to which the research environment was in a state of flux. Less than two years later, in May 1986, Charles Gimingham, in his viewpoint, was able to refer to research initiated with the explicit aim of reconciling the interests of conservation, agriculture, and forestry. The change had come about not through any positive moves on the part of ecologists, but through the impatience of Parliament that so little was being done by the research councils and their 'customer' departments to bring about this reconciliation. It remained to be seen whether ecologists would make good use of the unprecedented opportunities to explore the agriculture/environment interface.

In practice, ecologists were left with little choice but to carry out 'useful' research. At a time of diminishing resources in the research councils and universities, any new venture in the biological sciences was likely to be financed by savings elsewhere. As Gimingham noted, the 'new, strong winds of change' were blowing in the direction of molecular and cell biology, genetics and biotechnology. These were exciting developments, opening up vast fields of new research, much of which might have very direct practical applications. Ecologists had to respond in kind. Not only was commissioned research more likely to attract public notice, but its proceeds would help to minimize further job losses and reductions in departmental budgets.

The role of the British Ecological Society was not to decry new fields of biological research, or particular applications of that research, but to ensure that advances continued to be broadly based. As well as accommodating cell biology and the other new ventures, a place had to be retained for ecology in university departments and research councils, where there was ample scope for both applied and basic research. To this end, the Society had to set an example by maintaining in its own meetings and publications a careful balance between the practical applications of ecology and the more traditional kinds of contribution, reporting the results of academic and apparently 'useless' research. Although the Society was unable to offer much in the way of material support, the less tangible support that came from the choice of appropriate themes for meetings, and the editorial policy of the various journals, was likely to increase in importance.

4.6.2 *The challenge of ecology*

Although no guidelines were laid down as to what a presidential address might contain, there was a presumption that it should be presented in the form of a scholarly review of some major field of research endeavour, presented in such a way as to be of interest to anyone with an interest in ecology. By 1980, over 30 such addresses had been given. John Harper suggested that they might be brought together in a volume. Having learned that the sales of a 500-page facsimile were likely to be as low as 600 copies, the project was abandoned. There were in any case doubts as to the historical value of republishing the addresses without very considerable editing.

Roy Taylor's address of 1984 had all the attributes of its genre. In his own words, it was a 'swan-song' that sought not only to illuminate a field of ecology, but to draw attention to the merits of a range of methods and techniques. Taylor recalled how he had become convinced that behaviour, expressed in terms of migration, was a key to understanding population dynamics. C. B. Williams and Roy French had already demonstrated the potential of an amateur observers' network for migrant Lepidoptera. Under Taylor's direction, a Rothamsted Insect Survey was established. Between 1969 and 1984, 364 daily records were made each year for a thousand species from over 100 sites. As Taylor (1986) remarked, it was likely to remain a unique venture. By illustrating what had been achieved in setting up the agricultural-pest warning system, which had made possible the exploration of several fundamental concepts of population structure, the presidential address gave some indication of what would be sacrificed if ecologists chose, or were forced, to abandon long-term research altogether.

Presidents of the Society rarely allowed their audiences to forget the enormous debt of gratitude they owed to the pioneers of ecology. In his presidential address of 1986, Charles Gimingham illustrated, by reference to the course of research on heathland ecology, how

> each individual builds on foundations previously laid by others; the questions asked and methods adopted are influenced by prevailing attitudes and available techniques.

From time to time, however, there was an ecologist whose approach was so fresh and different that his work set in motion many new trains of thought and enquiry. He joined the 'great names' of the science—those whose influence was out of all proportion to their number (Gimingham 1987).

One of these 'great names' was A. G. Tansley, the first president of the Society and chairman of its predecessor, the British Vegetation Committee. In his own presidential address of 1975, Clifford Evans recalled how Tansley, from an early date, had realized that the best way of relating knowledge of plant ecology to the main corpus of natural sciences was through detailed studies of the structure and functioning of individual plants in their natural surroundings. Bridges could then be built between these studies and those conducted in the laboratory. It was a challenging assignment. Evans (1976) cited his own experiences of measuring such adaptive mechanisms as respiration, first in the forests of Nigeria and then in the woodlands around Cambridge, as examples of the inaccessibility of plants to research in their natural surroundings.

Tansley did more than present a challenge of a practical nature. In its comprehensiveness, his concept of the ecosystem presented an intellectual challenge of the highest order. It was relevant to ecological studies of any place and at any time. The evolution of species had always gone on within ecosystems. Unless species were to disappear, each had to maintain a continuous state of adaptation to each other's activities. As Evans remarked, there was the strongest possibility that the mechanisms by which adaptation took place on an evolutionary time-scale were to be found within the ecosystem as a whole.

As Tansley himself discovered, terminology could be abused so easily. In the course of commenting on Pearsall's address to the golden jubilee conference of 1963, James Cragg voiced his own misgivings, referring to the loose way in which the word 'ecosystem' was used. He recalled how, when he became director of the Conservancy's Merlewood Research Station, he had been told that its main purpose was to study 'the whole ecosystem of natural woodlands'. It was 'a useful piece of shorthand', but, Cragg contended,

> I think a statement of this kind tends to be too much akin to the search for the philosopher's stone to give a sharp form to a research programme[33].

In his own presidential address of 1961, Cragg (1961) focused on the structure, maintenance and fate of populations, as they were revealed by studies of the relationships established between the individuals that comprised those populations and their respective environments. Many of his data were derived from studies of moorland animals on the Moor House National Nature Reserve. The word 'ecosystems' was scarcely mentioned.

Ecosystems continued to feature large in most presidential addresses. For Jack Harley, the deficiences arose from the manner in which ecologists studied the physiology of ecosystems. Because of their impatience to get results, too little time was spent on methodology. The penalties were all too clearly seen in studies of the role of fungi in terrestrial ecosystems. As Harley indicated in his presidential address of 1971, there were real difficulties in determining not only what species occurred in any ecological substrate and their state of vegetative activity, but of predicting from a study of fungi in culture what they might be doing ecologically. These were deficiencies that could only be overcome by fundamental investigation and accurate ecological observations (Harley 1972).

Ecosystem patterns and the environment were so complex as to defy all but the most resolute attempts to formulate a general framework—there were so many exceptions to any rule. Richard Southwood likened the position of ecologists to that of the inorganic chemist before the development of the periodic table, or to astronomy prior to the adoption of the Hertzsprung–Russell diagram relating the evolution of stars to their properties. Advances were, however, taking place. Over the previous decade, fresh light had been cast on the concept of the niche and the assembly of niches into an ecosystem. As Southwood stressed, in his presidential address of 1977, the development of a classification of ecological strategies was not only an aid to learning, but made it possible to test predictions against field observations. As with all classifications of nature, exceptions were easy to find, but the real challenge, the constructive work, was to use the exceptions to improve, modify, and even change the general framework.

In choosing the title 'Habitat, the templet for ecological strategies' for his presidential address, Southwood (1977) sought to express in quantitative terms the multitude of strategies that had evolved as trade-offs between costs and benefits in the processes of adapting to habitats. There were at least eight quantitative characteristics of a natural habitat, each of which had to be assessed against the organism's own dimensions in space and time. Whilst it was important not to visualize the habitat as being a rigid causal templet (or template in an engineering sense), Southwood believed the habitat features could be condensed into two axes, namely durational stability, which assessed spatial heterogeneity against time, and resource level and constancy, which expressed the temporal heterogeneity of the same space. Whilst such a two-dimensional treatment could never encapsulate all the complex situations of nature, it was much more realistic than the many previous attempts to organize ecological strategies along a single dimension.

Although much of the original stimulus to plant ecology had come from a desire to explain vegetation patterns, Peter Greig-Smith contended, in his address of 1979, that insufficient account had been taken of the almost universal patchiness of vegetation in the various attempts to integrate the approaches of community ecology and population ecology. Without taking full cognisance of the extent to which the dynamics of both plant and terrestrial animal populations took place against a highly heterogeneous background, there was even less chance of understanding how ecosystems functioned and of ultimately predicting and controlling changes within them.

One effect of the increasing application of numerical methods to the study of plant distribution had been to emphasize that randomness seldom occurred, even over very small areas. Greig-Smith (1979) sought to assess how far habitat differences could account for those patterns which were less intense and small-scale, and how far other mechanisms might cause or reinforce spatial heterogeneity. He concluded that although patchiness in the physical and chemical environment might determine patterns of all intensities and scales, a wide variety of other causes might produce pattern. With the possible exception in some instances of historically and chance-determined patterns, all of them would be mediated ultimately by changes in the environment.

In recounting the contributions made by successive presidents in their biennial addresses, there is a risk of implying that knowledge is acquired in a logical, incremental fashion. The truth is very different. The fitful and erratic progress made in ecology was nowhere better demonstrated than in the field of plant population dynamics. During the First World War, Tansley had published the results of Marsh's trials with *Galium* spp. grown in pure and mixed stands, describing the struggle for existence between them on different soils (Tansley 1917). The implications of the paper, namely that the biology of a species seen in isolation might take no account of its ecology, were for many years largely ignored. Attention was concentrated instead on the description of vegetation, the evolution of ecotypes, the physiology of adaptation, and later on productivity (Harper 1964, 1977).

It would be wrong to suggest that such figures as Tansley and Watt had no interest in competition, predation, disturbance, and diversity as they affected distribution and abundance patterns. On the contrary, their studies in the interwar period demonstrated a considerable understanding of these factors. They were, however, primarily concerned with the immediate causes for what they observed in the field. As Jeremy B. C. Jackson wrote, in his review of research on interspecific competition and species distribution (with the subtitle, 'The ghosts of theories and data past'), ecologists in the period before 1950 were little interested in 'the origin of life history patterns' (Jackson 1981).

The exception was Salisbury who, in one of the most prescient, yet over-looked, presidential addresses given to the British Ecological Society, focused on some 11 life history topics (Salisbury 1929). It was, according to Jackson (1981), 'a truly extraordinary paper', and yet it attracted so little attention—there was no reference to the address, or indeed to any of Salisbury's work, in Tansley's own presidential address to the British Ecological Society in 1939, reviewing the previous 25 years of British ecology (Tansley 1939b). Jackson wrote of how 'poor Salisbury was so far ahead of other ecologists in his under-standing of natural selection that they simply were not interested'.

The key figure in focusing the plant ecologist's attention on natural selection in the late 1950s and 1960s was John Harper, whose own early research had been orientated towards agricultural problems. For his presidential address to the British Ecological Society in 1967, Harper chose the title, 'A Darwinian approach to plant ecology'. In it, he set out to demonstrate how, by addressing themselves to the prime questions posed by Chapters 3 and 4 of the *Origin of Species* (Darwin 1859), plant ecologists could usefully occupy themselves for the next hundred years. Through repeated observations of marked individuals, it was possible to introduce measurements of population turnover into ecolo-gical studies in terms meaningful to the selection geneticist and the evolutionist. Stable vegetation would come to be seen as a state of continuous flux, in which the rates of turnover were critical characteristics of stability. The stage would be set for understanding the significance of the strategy of reproduction and of the life cycle itself, both of which were ecologically fascinating but neglected sub-jects for study (Harper 1967).

Plant population biology could no longer be regarded as a laggard. An American commentator went so far as to describe its revolutionary progress in the 1970s as 'the most important event in the field of ecology' (Antonovics 1980). In 1977, John Harper published his volume *Population Biology of Plants* setting out a demographic theory supported by a growing base of empirical data. It was, in the words of the review published in the *Journal of Ecology*, a valuable and timely synthesis of the work over the previous 20 years carried out by 'the Harperian School of Plant Ecology', and the more recent American school of plant demographers. The book made abundantly clear 'why the demographic theory developed for animal populations cannot be applied to plants, why plants have to be studied differently, and where the unanswered questions lie' (Solbrig 1979).

The course of development in plant population ecology illustrated how any publication, if it is to make an impact on ecological thinking, must win the re-spect (if not the wholehearted endorsement) of its readers and, equally import-ant, has to be regarded as timely in its appearance. The author may help to prepare the ground—introducing ideas and bodies of data in earlier publications, which may then be brought together and synthesized in some 'grand design'. At the same time, readers may be seeking, consciously or otherwise, some kind of theoretical construct. For these and other reasons, a monograph, *The Theory of Island Biogeography*, took its place in the distinguished line of American works to stir the imagination of British ecologists, extending back to Clements's volumes at the turn of the century. The monograph, published in 1967, was by Robert H. MacArthur and Edward O. Wilson of the biology departments at Princeton and Harvard respectively.

The ground had been well prepared. The increasing number of publications on species diversity, species–area phenomena, and character displacement was an indication of how far the earlier papers of MacArthur and Wilson had stimulated 'a marked enthusiasm and response on the part of many contemporary biologists'. As well as bringing much of this material under one cover, the monograph demonstrated both the challenge and the rewards of collaboration between an ecologist (MacArthur) and a taxonomist and zoogeographer (Wilson)—there could be no finer test of the ultimate unity of population biology. In the course of explaining their own equilibrium theory (by which the number of species on an island depended primarily on the balance of immigration and extinction rates) and examining the varied strategies available for colonization, the authors were in effect making 'biogeography more conducive to experimentation and more amenable to theoretical analysis'. In the words of one reviewer, the monograph was 'staking out a new era in the biological sciences' (MacArthur & Wilson 1967, Hamilton 1968).

Any appraisal of the course of research, of its achievements and missed opportunities, comes to focus eventually on the behaviour of ecologists themselves. Amyan Macfadyen introduced his presidential address by outlining how scientists were supposed to work, according to Popper, Kuhn, and other philosophers of science. Contemporary advances in host–parasitoid theory provided an insight into how far ecology might become an exact science, dealing with the falsification of (the assumptions of) a hypothesis by the crucial experiment. Nicholson and Bailey had put forward their original equation in the mid-1930s. Its assumption that the area of discovery of a parasite could be regarded as constant was challenged in a paper published in *Nature*, in which Hassell and Varley (1969) set out a new equation. Whilst still adopting a random search model, it took account of parasitic interference. There followed, some five years later, a paper in the *Journal of Animal Ecology* emphasizing the need to take greater account of spatial heterogeneity in assessing the numerical response of parasitoids (Cheke 1974).

No matter how ignorant or uncritical individual workers might be of research in other fields, Macfadyen argued that it was surely a function of members of the British Ecological Society to consider how far ecology reflected 'a coherent system of ideas and approaches', particularly in view of the strong criticisms that had sometimes been voiced. No single ecologist could comprehend all that was happening—each was tempted to bury his head in his own chosen patch of sand, and to ignore the rest. Ecology was so vastly extended, and yet, far from being able to focus on a more restricted body of theory, the most exciting developments in ecology required an even greater breadth of interest, as for example in the linking of ecology with advances in genetics.

Over two years elapsed between Macfadyen giving his address to the Annual Meeting in January 1973 and its publication in a much revised form. As Macfadyen (1975) wrote in a postscript, there were so many 'new ideas and new references of great importance'—evidence of how 'we seem to be entering a period now of deeper understanding and less confusion'. He had become even more convinced of the value of studying complex systems by means of appropriate quantitative models and by testing these against the realities of nature. Whether studying individual populations or an ecosystem, the ecologist was concerned with similar properties, namely control, stability, and variation. Whatever his special interests, it was necessary to be familiar not only with the basic methods of analysing these properties, but, in order to have access to possible analogies, with their application through the full range of ecological study.

The explosive increase in research activity had been reflected in the number of contributions to the Society's meetings and publications. As many as 56 papers were read at the winter meeting of 1977. Many parallel sessions had to be organized. In his presidential viewpoint, Peter Greig-Smith appealed to members, especially the younger members with careers in front of them, to resist the temptation of always choosing a meeting or session which appeared to be closest to their own interests. Narrowness was regrettable in any field of science, but particularly so in ecology, where insight so often came from initially unconsidered information and ideas.

As a forum in which botanists and zoologists, theorists and practical men, professionals and amateurs, met, the Society's most memorable meetings tended to be those when some controversy occurred, in which no party was keen to retreat from its well-prepared position. In the minds of some onlookers, and indeed participants, there was a growing and disturbing tendency for ecologists to go well beyond the normal give and take of intellectual controversy. In the words of Mick Southern, an increasingly aristocratic outlook had emerged, which regarded those carrying out highly sophisticated and experimental techniques as scientists, and the rest, to use an old gibe, as stamp collectors. This was to forget that science meant, quite literally, knowledge, and that more, not fewer, data would be required if ecologists were to gain a fuller understanding of the dynamic aspects of communities, with or without the aid of the computer.

With a little more humility, all ecologists would come to recognize that everyone was fundamentally interested in the existence, evolution, distribution, behaviour, abundance and interrelationship of plants and animals. Whether through studies in the laboratory or in the homes of the plants and animals, everyone had a part to play in acquiring 'this truly ecological knowledge', and therefore making it possible to explain what had been discovered (Southern

1970). The descriptive, historical, analytical (or reductionist), holistic, and practical approaches all had a part to play in any complete ecological investigation.
Those who argued otherwise failed to appreciate 'the necessary breadth of ecological science' (Gimingham 1987).

There was a timelessness about such exhortations. They recalled the fears
that beset William Smith in the latter days of the British Vegetation Committee.
Unless the claims of ecology were pursued with vigour, they would be dissi-
pated by apathy and disenchantment. Without humility, ecology would soon
fall 'into the hands of dogmatic unteachable people', who would make heretics
of all who dared to disagree with them. The British Ecological Society had
played an important part in steering British ecology clear of both hazards.
Through its very existence, the Society had denied the high ground to any particular aspect or grouping of ecologists. No matter what prominence was given
to a topic or person at a meeting or in one of the Society's publications, the turn
of others would assuredly come.

By its very title, the Society conferred a sense of unity and continuity. For 75
years, the Society had acted as a meeting point and publications outlet. If
required, it could serve as the mouthpiece for the overwhelming majority of
ecologists in Britain. Most crucially of all, the Society gave tangible form to a
science of great potential and relevance, but which still had to make its full
impact in academia and beyond. As the end of the 20th century drew near, and
with it the celebration of 100 years of British ecology, the need for such a body
as the 'BES' was as pressing as ever.

Appendix

Honorary Members of the British Ecological Society
Dates of Election

1913	W. G. Smith	1982	W. D. Billings
1934	L. Cockayne		R. Margalef
	H. C. Cowles		T. A. Rabotnov
1946	A. G. Tansley		V. M. Conway*
1955	V. E. Shelford		W. M. Curtis*
1958	W. H. Pearsall		R. E. Holttum*
	E. J. Salisbury		P. Saugman*
1959	C. Diver	1984	J. L. Harley
	C. S. Elton		R. O. Slatyer
	A. S. Watt		E. O. Wilson
1961	C. B. Williams	1986	J. L. Harper
1963	J. Braun-Blanquet		T. Kira
	H. Godwin	1987	H. G. Baker
	A. Hardy		
	A. J. Nicholson		
1964	E. G. Du Rietz		
	G. E. Hutchinson		
1965	V. S. Summerhayes		
1967	O. W. Richards		
1969	A. R. Clapham		
	H. Osvald		
	R. Tüxen		
1975	L. C. Birch		
1977	J. B. Cragg		
	K. Petrusewicz		
	P. W. Richards		
	F. Schwerdtfeger		
	H. N. Southern		
	S. Utida		
	H. Walter		
	V. Westhoff		
	V. C. Wynne-Edwards		
1978	H. Ellenberg		
	E. P. Odum		
	R. H. Whittaker		
	C. T. de Wit		
1979	J. W. G. Lund		
1980	T. B. Reynoldson		
	G. C. Varley		

*Previously an honorary associate member

263

Presidents of the British Ecological Society with numbers of members during their term of office

1913	A. G. Tansley	112	1952	C. B. Williams	768
1915	F. W. Oliver	111	1954	A. R. Clapham	841
1917	W. G. Smith	114	1956	G. C. Varley	934
1919	R. H. Yapp	138	1958	N. A. Burges	1010
1921	R. L. Praeger	151	1960	J. B. Cragg	1130
1923	F. E. Weiss	156	1962	P. W. Richards	1231
1926	T. W. Woodhead	170	1964	D. Lack	1441
1928	E. J. Salisbury	213	1966	J. L. Harper	1709
1930	F. E. Fritsch	255	1968	H. N. Southern	2023
1932	A. E. Boycott	242	1970	J. L. Harley	2381
1934	J. R. Matthews	322	1972	A. Macfadyen	2754
1936	W. H. Pearsall	341	1974	G. C. Evans	2875
1938	A. G. Tansley	350	1976	T. R. E. Southwood	2960
1940	C. Diver	335	1978	P. Greig-Smith	3580
1942	H. Godwin	363	1980	G. M. Dunnet	3512
1944	O. W. Richards	398	1982	A. D. Bradshaw	3789
1946	A. S. Watt	502	1984	L. R. Taylor	4171
1948	C. S. Elton	595	1986	C. H. Gimingham	4373
1950	W. B. Turrill	727	1988	R. J. Berry	

Secretaries and Treasurers of the British Ecological Society

SECRETARIES		TREASURERS	
F. Cavers	1913–16	H. Boyd Watt	1913–37
E. J. Salisbury	1917–30	A. S. Watt	1938–49
H. Godwin	1931–47	V. S. Summerhayes	1939–57
G. C. Varley	1938–41	C. E. Hubbard	1950–7
L. A. Harvey	1942–9	F. H. Whitehead	1958–66
A. R. Clapham	1948–50	J. L. Harper	1958–65
G. C. Varley	1950–3	R. A. French	1966–77
R. E. Hughes	1951–5	T. R. E. Southwood	1967–8
E. D. Le Cren	1954–63	R. Snaydon	1969–74
C. H. Gimingham	1956–60	S. McNeill	1975–9
P. Greig-Smith	1961–3	M. P. Hassell	1978–80
G. T. Goodman	1964–8	B. D. Turner	1980–
E. A. G. Duffey	1964–74	R. P. Gemmell	1981–6
M. J. Chadwick	1969–73	R. A. Benton	1987–
J. A. Lee	1974–80		
R. R. Askew	1975–7		
J. M. Cherrett	1978–85		
A. J. Davy	1981–6		
A. J. C. Malloch	1986–		
P. J. Edwards	1987–		

Editors of Publications of
the British Ecological Society

JOURNAL OF ECOLOGY	
F. Cavers	1913–16
A. G. Tansley	1917–37
W. H. Pearsall	1938–47
H. Godwin	1948–57
D. E. Coombe	1954–63
P. W. Richards	1958–63
P. Greig-Smith	1964–8
G. F. Asprey	1969–71
A. J. Willis	1969–75
P. J. Grubb	1972–7
C. H. Gimingham	1975–8
J. H. Tallis	1976–80
R. S. Clymo	1978–82
B. Hopkins	1980–5
B. Moss	1981–
J. A. Lee	1983–
J. White	1986–

JOURNAL OF APPLIED ECOLOGY	
A. H. Bunting	1964–8
V. C. Wynne-Edwards	1964–8
J. P. Dempster	1969–72
J. Warren Wilson	1969
H. W. Woolhouse	1970–4
T. H. Coaker	1973–7
R. W. Snaydon	1975–9
M. J. Way	1978–9
G. R. Sagar	1980–1
J. A. Wallwork	1981–3
T. M. Roberts	1983–6
W. C. Block	1984–
J. Miles	1987–

JOURNAL OF ANIMAL ECOLOGY	
C. S. Elton	1933–51
A. D. Middleton	1933–9
D. Chitty	1940–51
H. C. Gilson	1952–62
O. W. Richards	1963–7
J. B. Cragg	1963–6
K. H. Mann	1967
H. N. Southern	1968–75
E. Broadhead	1969–72
L. R. Taylor	1973–
J. M. Elliott	1976–

BULLETIN	
M. J. Chadwick	1970–3
M. B. Usher	1970–1
J. H. Lawton	1972–8
A. H. Fitter	1979–84
M. R. Young	1985–

FUNCTIONAL ECOLOGY	
P. Callow	1987–
J. Grace	1987–

SYMPOSIA VOLUMES	
E. Broadhead	1983–

Notes

KEY

BES MSS denotes files preserved by the British Ecological Society

BSC MSS, Tansley MSS denotes the collection of personal papers of Professor A. G. Tansley, deposited in the Botany School, Cambridge.

CRO, Pearsall MSS denotes the collection of personal papers of Professor W. H. Pearsall, deposited in the Cumbria Record Office, Kendal.

PRO denotes material preserved in the Public Record Office, Kew.

RBG MSS denotes material preserved in the Royal Botanic Gardens, Kew.

TMM, Woodhead MSS denotes the collection of personal papers of Dr T. W. Woodhead, deposited in the Tolson Memorial Museum, Huddersfield.

PART 1

1 BES MSS, MB 1.
2 BSC MSS, Tansley MSS.
3 BSC MSS, Tansley MSS.
4 BSC MSS, Tansley MSS.
5 BES MSS, MB 2.
6 BES MSS, MB 2.
7 PRO, T9/38, 840.
8 TMM, Woodhead MSS; BES MSS, MB 2.
9 BES MSS, MB 2.
10 BES MSS, MB 2
11 Botany School Library, Cambridge, E 108.1.
12 BES MSS, MB 2.
13 BES MSS, MB 2.
14 CRO, Pearsall MSS.
15 BSC MSS, Tansley MSS; TMM, Woodhead MSS.
16 BSC MSS, Tansley MSS.
17 TMM, Woodhead MSS.
18 BES MSS, MB 2.
19 TMM, Woodhead MSS.
20 BSC MSS, Tansley MSS.

PART 2

1 RGB MSS, Misc. Corresp., Tansley.
2 BES MSS, MB 2.
3 BES MSS, MB 2.
4 RGB MSS, Misc. Corresp., Oliver.
5 BES MSS, MB 3.
6 British Geological Survey, Edinburgh, MSS, LSA 100, 1–9.
7 BSC MSS, Tansley MSS.
8 BSC MSS, Tansley MSS.
9 CRO, Pearsall MSS.
10 CRO, Pearsall MSS.
11 CRO, Pearsall MSS.
12 CRO, Pearsall MSS.
13 CRO, Pearsall MSS.
14 BES MSS, MB 2.
15 BES MSS, MB 3.
16 BES MSS, MB 3.
17 RBG MSS, Misc. Corresp., Erdtman.
18 BES MSS, MB 3.
19 BES MSS, MB 6.
20 Scott Polar Research Institute, Cambridge, MS 1459/1.
21 BES MSS, MB 3.
22 Elton Library, Oxford, Elton MSS.
23 BES MSS, MB 5.
24 PRO, CO 758, 66/6.
25 Elton Library, Oxford, Elton MSS.
26 Royal Society of London, Council and Zoology Committee minutes.
27 CRO, Pearsall MSS.
28 BES MSS, MB 3.
29 BES MSS, MB 4.
30 BES MSS, MB 3.
31 Botany Department, University College, London, *Blakeney Point in 1920–3*, being the Joint Management and Scientific Report.
32 BES MSS, MB 3.
33 BES MSS, MB 3.
34 BES MSS, MB 3.
35 C. L. Coles (1981) *The Game Conservancy. Some History*. The Game Conservancy, Fordingbridge (unpublished manuscript).

36 Destructive Imported Animals Act, 1932, 22 George V, ch. 12.
37 BSC MSS, Tansley MSS.

PART 3
1 CRO, Pearsall MSS.
2 PRO, MAF 114, 203.
3 PRO, MAF 44, 32.
4 PRO, MAF 41, 1484–5.
5 BES MSS, MB 5 and 6.
6 BES MSS, MB 5 and 6.
7 Conference on Nature Preservation in Post-War Reconstruction memorandum 1 (1941) and memorandum 2 (1942).
8 Royal Society for Nature Conservation archives, Box 34.
9 Conference on Nature Preservation in Post-War Reconstruction memorandum 3 (1943), memorandum 4 (1943), memorandum 5 (1945) and memorandum 6 (1945).
10 PRO, HLG 92, 49.
11 PRO, CAB 87, 10.
12 PRO, HLG 93, 49.
13 PRO, HLG 93, 48.
14 PRO, CAB 132, 9 and 10.
15 Scottish Record Office, DD 12, 875.
16 BES MSS, SF/1/28 and 35.
17 BES MSS, SF/1/19.
18 BES MSS, SF/1/30.
19 BES MSS, SF/1/21 and 24.
20 BES MSS, SF/1/20.
21 BES MSS, SF/1/5.
22 BES MSS, SF/1/24.
23 BES MSS, SF/1/24.
24 BES MSS, SF/1/3.
25 BES MSS, SF/1/8.
26 BES MSS, SF/1/4.
27 BES MSS, SF/1/15.
28 BES MSS, SF/1/18.
29 BES MSS, SF/1/5.
30 CRO, Pearsall MSS.
31 BES MSS, MB 6.
32 BES MSS, MB 5.
33 BES MSS, MB 3.

34 RBG, Salisbury MSS 34.
35 CRO, Pearsall MSS.
36 BES MSS, MB 5.
37 *The Times*, 19 October 1964.
38 CRO, Pearsall MSS.
39 CRO, Pearsall MSS.
40 CRO, Pearsall MSS.

PART 4
1 BES MSS, SF/1/19.
2 BES MSS, SF/1/32.
3 BES MSS, JAPE, JPD 1.
4 BES MSS, SF/1/32.
5 BES MSS, SF/1/39.
6 BES MSS, SF/1/41
7 BES MSS, SF/1/33.
8 BES MSS, SF/11/29.
9 BES MSS, SF/1/32.
10 BES MSS, SF/11/39.
11 BES MSS, SF/11/17.
12 BES MSS, SF/11/29.
13 BES MSS, SF/11/15.
14 BES MSS, SF/11/16.
15 BES MSS, SF/11/9.
16 BES MSS, MEG, SF/1.
17 BES MSS, MEG, SF/2 and 3.
18 BES MSS, SF/1/32.
19 BES MSS, SF/1/33.
20 CRO, Pearsall MSS.
21 BES MSS, SF/1/33.
22 House of Lords Record Office, Commons Select Committee, Private Bills, 1966, printed evidence, vol. III.
23 BES MSS, SF/1/30; *The Times*, 13 July 1966.
24 BES MSS, SF/11/7A.
25 BES MSS, SF/1/42.
26 Hansard, Commons, *859*, 457.
27 Hansard, Commons, *859*, 460–1.
28 BES MSS, SF/11/7A.
29 BES MSS, SF/11/7A.
30 BES MSS, SF/11/38.
31 BES MSS, SF/11/4.
32 BES MSS, SF/1/33.
33 CRO, Pearsall MSS.

References

Adams, C. C. (1913). *Guide to the Study of Animal Ecology.* Macmillan, New York.

Adamson, R. S. (1912). An ecological study of a Cambridgeshire woodland. *J. Linn. Soc. Bot.* **40**, 339–87.

Adamson, R. S. (1927). The plant communities of Table Mountain. *J. Ecol.* **15**, 278–309.

Allen, D. E. (1976). *The Naturalist in Britain: a Social History.* Allen Lane, London.

Allen, D. E. (1986). *The Botanists. A History of the Botanical Society of the British Isles.* St Pauls Bibliographies, Winchester.

Anderson, D. (1984). Depression, dust bowl, demography, and drought: the colonial state and soil conservation in East Africa during the 1930s. *Afr. Affairs* **83**, 321–43.

Andrewartha, H. G. (1961). *Introduction to the Study of Animal Populations.* Methuen, London.

Andrewartha, H. G. & Birch L. C. (1954). *The Distribution and Abundance of Animals.* University Press, Chicago.

Anonymous (1901). Robert Smith. *Ann. Scot. Nat. Hist.* **37**, 1–2.

Anonymous (1903). An experiment in ecological surveying. *New Phytol.* **2**, 167–8.

Anonymous (1904). A second experiment in ecological surveying. *New Phytol.* **3**, 200–4.

Anonymous (1913). Notes and comments. *Naturalist, Hull,* 277–8.

Anonymous (1925). *Spitsbergen Papers. Vol. I, Scientific Results of the First Oxford University Expedition to Spitsbergen.* University Press, Oxford.

Anonymous (1929). *Spitsbergen Papers. Vol. II, Scientific Results of the Second and Third Oxford University Expeditions to Spitsbergen.* University Press, Oxford.

Anonymous (1934). Obituary. Marion Isabel Newbigin. *Scott. Geogr. Mag.* **50**, 331–2.

Anonymous (1935). Discussion on the origin and relationship of the British flora. *Proc. Roy. Soc. Lond. B* **118**, 197–241.

Anonymous (1939). Obituary. Dr Sydney Herbert Long. *Trans. Norfolk Norwich Nat. Soc.* **14**, 484–6.

Anonymous (1941a). The function of applied biology in war time. *Ann. Appl. Biol.* **28**, 170–7.

Anonymous (1941b). Biologial Flora of the British Isles. *J. Ecol.* **29**, 356–61.

Anonymous (1942). Biologists in war-time. *Nature, Lond.* **149**, 227–9.

Anonymous (1943). Memorandum on nomenclature and taxonomy in the Biologial Flora. *J. Ecol.* **31**, 93–6.

Anonymous (1944a). Ecological principles involved in the practice of forestry. *J. Ecol.* **32**, 83–115.

Anonymous (1944b). Nature conservation and nature reserves. *J. Ecol.* **32**, 45–82.

Anonymous (1946). Check list of British vascular plants. *J. Ecol.* **33**, 308–47.

Anonymous (1950–1). William Munn Rankin. *Proc. Bournemouth Nat. Sci. Soc.* **41**, 61–2.

Anonymous (1965). Obituary. Dr Robert S. Adamson. *S. Afr. J. Sci.* **61**, 443.

Anonymous (1970). Comment. *Animals* **13**, 147.

Anonymous (1972). A blueprint for survival. *Ecologist* **2** (*1*), 44pp.

Antonovics, J. (1980). The study of plant populations. *Science* **208**, 587–9.

Ashby, E. (1935). The quantitative analysis of vegetation. *Ann. Bot.* **49**, 779–802.

Babington, C. C. (1860). *Flora of Cambridgeshire.* Van Voorst, London.

Bailey, E. (1952). *Geological Survey of Great Britain.* Murby, London.

Bainbridge, R., Evans, G. C. & Rackham, O. (1966). *Light as an Ecological Factor.* Blackwell Scientific Publications, Oxford.

Baker, J. G. (1863). *North Yorkshire*. Longman, Green, London.

Baker, J. G. (1903). Reviews and book notices. *Naturalist, Hull*, 221–3 and 377–9.

Baker, J. R. (1939). Counterblast to Bernalism. *New Statesman and Nation* 18, 174–5.

Baker, J. R. (1976). Julian Sorell Huxley. *Biographical Memoirs of Fellows of the Royal Society* 22, 207–38.

Balls, W. L. & Brimble, L. J. F. (1955). Professor F. J. Lewis. *Nature, Lond.* 176, 237–8.

Barkman, J. (1981). Reinhold Tüxen. *Vegetatio* 48, 87–91.

Bernal, J. D. (1939). *The Social Function of Science*. Routledge, London.

Blackman, F. F. & Tansley, A. G. (1905). Ecology in its physiological and phyto-topographical aspects. *New Phytol.* 4, 199–203 and 232–53.

Blackman, F. F., Blackman, V. H., Keeble, F., Oliver, F. W. & Tansley, A. G. (1917). The reconstruction of elementary botanical teaching. *New Phytol.* 16, 241–52.

Blackman, G. E. (1935). A study by statistical methods of the distribution of species in grassland associations. *Ann. Bot.* 49, 749–77.

Blackman, G. E. (1942). The Biology War Committee. *Nature, Lond.* 149, 234–5.

Blackman, G. E. & Rutter, A. J. (1946). Physiological and ecological studies in the analysis of plant environment. I. The light factor and the distribution of the bluebell in woodland communities. *Ann. Bot.* 10, 361–90.

Blackman, G. E. & Rutter, A. J. (1947). Physiological and ecological studies in the analysis of plant environment. II. The interaction between light intensity and mineral nutrient supply in the growth and development of the bluebell. *Ann. Bot.* 11, 125–58.

Boardman, P. (1978). *The Works of Patrick Geddes*. Routledge & Kegan Paul, London.

Bolòs, O. de (1982). Josias Braun-Blanquet. *Vegetatio* 48, 193–6.

Boulger, G. S. (1902). The preservation of our indigenous flora. *S. East Nat.* 28–35.

Bower, F. O. (1918). Botanical Bolshevism. *New Phytol.* 17, 105–7.

Bower, F. O. (1919). *Botany of the Living Plant*. Macmillan, London.

Boycott, A. E. (1934). The habitats of land mollusca in Britain. *J. Ecol.* 22, 1–38.

Boycott, A. E. (1936). The habitats of fresh-water mollusca in Britain. *J. Anim. Ecol.* 5, 116–86.

Bradshaw, A. D. (1983). The reconstruction of ecosystems. *J. Appl. Ecol.* 20, 1–17.

Braun-Blanquet, J. (1932). *Plant Sociology: the Study of Plant Communities*. McGraw-Hill, London.

Brenchley, W. E. (1920). *Weeds of Farm Land*. Longman, Green, London.

Brenchley, W. E. & Adam, H. (1915). Recolonisation of cultivated land allowed to revert to natural conditions. *J. Ecol.* 3, 193–210.

Brenan, J. P. M. (1975). Victor Summerhayes: obituary. *Bull. Brit. Ecol. Soc.* 6 (*1*), 5–6.

Broadhead, E. & Wapshere, A. J. (1966). *Mesopocus* populations on larch in England—the distribution and dynamics of two closely-related coexisting species of Psocoptera sharing the same food resource. *Ecol. Monogr.* 36, 327–88.

Brock, S. E. (1914). Ecological relations of bird-distribution. *Br. Birds* 8, 30–44.

Brock, S. E. (1921). Bird-associations in Scotland. *Scott. Nat.* 11–21 and 49–58.

Brooks, F. T. (1942). Arthur William Hill. *Obituary Notices of Fellows of the Royal Society* 4, 87–100.

Brooks, F. T. & Chipp, T. F. (1931). *Report and Proceedings of the Fifth International Botanical Congress*. University Press, Cambridge.

Bunting, A. H. & Wynne-Edwards, V. C. (1964). Editorial. *J. Appl. Ecol.* 1, 1–2.

Burdon-Sanderson, J. S. (1893). Inaugural address. *Nature, Lond.* 48, 464–72.

Burges, A. (1951). The ecology of the Cairngorms. III. The *Empetrum-Vaccinium* zone. *J. Ecol.* 39, 271–84.

Burges, A. (1960). Time and size as factors in ecology. *J. Ecol.* 48, 273–85.

Burges, A. & Drover, D. P. (1953). The rate of podzol development in the sands of the Woy Woy district, NSW. *Austr. J. Bot.* 1, 83–94.

Burkill, I. H. (1931–2). Thomas Ford Chipp. *Proc. Linn. Soc.* 144, 169–74.

Burnett, J. H. (ed.) (1964). *The Vegetation of Scotland*. Oliver & Boyd, London.

Burtt, B. D. (1942). Some East African vegetation communities. *J. Ecol.* **30**, 65–146.

Butcher, R. W. (1927). A preliminary account of the vegetation of the river Itchen. *J. Ecol.* **15**, 55–65.

Butcher, R. W. (1933). Studies on the ecology of rivers. I. On the distribution of macrophytic vegetation in the rivers of Britain. *J. Ecol.* **21**, 58–91.

Buxton, P. A. (1923). *Animal Life in Deserts. A Study of the Fauna in Relation to the Environment.* Arnold, London.

Buxton, P. A. (1924). The temperature of the surface of deserts. *J. Ecol.* **12**, 127–34.

Buxton, P. A. (1926). Applied entomology. *Nature, Lond.* **117**, 623–4.

Buxton, P. A. (1932). Climate in caves and similar places in Palestine. *J. Anim. Ecol.* **1**, 152–9.

Buxton, P. A. (1955). *The Natural History of Tsetse Flies.* Lewis, London.

Carey, A. E. & Oliver, F. W. (1918). *Tidal Lands: a Study of Shore Problems.* Blackie, London.

Carr-Saunders, A. M. (1922). *The Population Problem: a Study in Human Evolution.* Clarendon, Oxford.

Carson, R. (1963). *Silent Spring.* Hamish Hamilton, London.

Cavers, F. (1910). The inter-relationships of the Bryophyta. *New Phytol.* **9**, 81–112.

Champion, H. G. (1933). Regeneration and management of Sal. *Indian Forest Rec.* **19**, 159 pp.

Cheke, R. A. (1974). Experiments on the effect of host spatial distribution on the numerical response of parasitoids. *J. Anim. Ecol.* **43**, 107–13.

Chipp, T. F. (1927). The Gold Coast forest. A study in synecology. *Oxford Forest Memoir* 7.

Chitty, D. & Southern, H. N. (eds) (1954). *Control of Rats and Mice.* Clarendon Press, Oxford.

Clapham, A. R. (1936). Over-dispersion in grassland communities and the use of statistical methods in plant ecology. *J. Ecol.* **24**, 232–51.

Clapham, A. R. (1956). Autecological studies and the 'Biological flora of the British Isles'. *J. Ecol.* **44**, 1–11.

Clapham, A. R. (1971). William Harold Pearsall. *Biographical Memoirs of Fellows of the Royal Society* **17**, 511–40.

Clapham, A. R. (1980). Edward James Salisbury. *Biographical Memoirs of Fellows of the Royal Society* **26**, 503–41.

Clapham, A. R. & Godwin, H. (1948). Studies of the post glacial history of British vegetation. VIII. Swamping surfaces in peats of the Somerset Levels. IX. Prehistoric trackways in the Somerset Levels. *Phil. Trans. Roy. Soc. Lond. B* **233**, 233–73.

Clapham, A. R., Tutin, T. G. & Warburg, E. F. (1962). *Flora of the British Isles.* University Press, Cambridge.

Clements, F. E. (1905). *Research Methods in Ecology.* University of Nevada, Lincoln.

Clements, F. E. (1912). Phytogeographical excursion in the British Isles. VIII. Some impressions and reflections. *New Phytol.* **11**, 177–9.

Clements, F. E. (1916). *Plant Succession. An Analysis of the Development of Vegetation.* Carnegie Institution, Washington.

Cockayne, L. (1898). On the burning and reproduction of sub-alpine scrub and its associated plants. *Trans Proc. N.Z. Inst.* **31**, 398–419.

Cockayne, L. (1911a). Observations concerning evolution, derived from ecological studies in New Zealand. *Trans Proc. N.Z. Inst.* **44**, 1–50.

Cockayne, L. (1911b). *Report on the Dune-Areas of New Zealand, their Geology, Botany and Reclamation.* Department of Lands, Wellington.

Cockayne, L. (1921). *The Vegetation of New Zealand.* Engelmann, Leipzig.

Cockayne, L. & Allan, H. H. (1927). The bearing of ecological studies in New Zealand on botanical taxonomic conceptions and procedure. *J. Ecol.* **15**, 234–77.

Cockayne, L. & Calder, J. W. (1932). The present vegetation of Arthur's Pass as compared with that of thirty-four years ago. *J. Ecol.* **20**, 270–83.

Colgan, N. & Scully, R. W. (1898). *Contributions towards a Cybele Hibernica, being Outlines of the Geographical Distribution of Plants in Ireland. Second Edition, Founded on the Papers of the Late Alexander Goodman More.* Ponsonby, Dublin.

Colinvaux, P. A. (1973). *Introduction to Ecology*. Wiley, London.

Colinvaux, P. A. (1982). Towards a theory of history: fitness, niche and clutch of *Homo sapiens. J. Ecol.* **70**, 393–412.

Colinvaux, P. A. (1983). *The Fates of Nations. A Biological Theory of History*. Penguin, Harmondsworth.

Collett, R. (1911–12). *Norgen Pattedyr*. Forlagt af H. Ascheboug, Kristiania, 136–62.

Collins, T. (1985). *Floreat Hibernia. A Bio-Bibliography of Robert Lloyd Praeger*. Royal Dublin Society, Dublin. Historical Studies in Irish Science and Technology, 5.

Conway, V. M. (1933). Further observations on the saltmash at Holme-next-the-Sea, Norfolk. *J. Ecol.* **21**, 263–7.

Conway, V. M. (1947). Ringinglow Bog, near Sheffield. Part I. Historical. *J. Ecol.* **34**, 149–81,

Conway, V. M. (1953). Other sites of special scientific interest. *Proc. Linn. Soc.* **165**, 94.

Conwentz, H. (1909). *The Care of Natural Monuments*. University Press, Cambridge.

Conwentz, H. (1913). Furstlich Hohenzollernsches Naturschutzgebiet im Bohmerwald. *J. Ecol.* **1**, 161–2.

Conwentz, H. (1914). On national and international protection of nature. *J. Ecol.* **2**, 109–22,

Cook, R. E. (1977). Raymond Lindeman and the trophic–dynamic concept in ecology. *Science* **198**, 22–6.

Cooper, W. S. (1926). The fundamentals of vegetational change. *Ecology* **7**, 391–413.

Cooper, W. S. (1935). Henry Chandler Cowles. *Ecology* **16**, 281–3.

Cowles, H. C. (1899). The ecological relations of the vegetation on the sand dunes of Lake Michigan. *Bot. Gaz.* **27**, 95–117, 167–202, 281–308 and 361–91.

Cowles, H. C. (1901). The physiographic ecology of Chicago and vicinity; a study of the origin, development and classification of plant societies. *Bot. Gaz.* **31**, 73–108 and 145–82.

Cowles, H. C. (1911). The causes of vegetative cycles. *Bot. Gaz.* **51**, 161–83.

Cowles, H. C. (1912). The International Phytogeographical Excursion. IV. Impressions of the foreign members of the party. *New Phytol.* **11**, 25–6.

Cragg, J. B. (1961). Some aspects of the ecology of moorland animals. *J. Anim. Ecol.* **30**, 205–33.

Crampton, C. B. (1906). Fossils and conditions of deposit, a theory of coal formation. *Trans. Edinb. Geol. Soc.* **9**, 73–92.

Crampton, C. B. (1911). *The Vegetation of Caithness Considered in Relation to the Geology*. Committee for the Survey and Study of British Vegetation, London.

Crampton, C. B. (1912). The geological relations of stable and migratory plant formations. *Scott. Bot. Rev.* **1**, 1–61.

Crampton, C. B. (1913). Ecology the best method of studying the distribution of species in Great Britain. *Proc. Roy. Phys. Soc. Edinb.* **19**, 22–36.

Crampton, C. B. & Carruthers, R. G. (1914). *The Geology of Caithness*. HMSO, Edinburgh.

Crampton, C. B. & MacGregor, M. (1913). The plant ecology of Ben Armine. *Scott. Geogr. Mag.* **29**, 169–92 and 256–66.

Crisp, D. J. (1964). *Grazing in Terrestrial and Marine Environments*. Blackwell Scientific Publications, Oxford.

Crump, W. B. (1931). In memoriam. Charles Edward Moss. *Naturalist, Hull*, 55–9.

Crump, W. B. & Crossland, C. (1904). *The Flora of the Parish of Halifax*. Halifax Scientific Society, Halifax.

Cullingworth, J. B. (1975). *Environmental Planning. Vol. I. Reconstruction and Land Use Planning, 1939*–1947. HMSO, London.

Currie, R. I. (1984). The marine biological associations. *Biologist* **31**, 245–9.

Dahl, E. (1956). *Rondane. Mountain Vegetation in South Norway and its Relation to the Environment*. Aschehoug, Oslo.

Darling, F. F. (1937). *A Herd of Deer*. University Press, Oxford.

Darling, F. F. (1947). *Natural History in the Highlands and Islands*. Collins, London.

Darwin, C. (1859). *On the Origin of Species*. Watts, London (reprint 1950).

Davies, W. (1936). The grasslands of Wales—a survey. In *A Survey of the Agricultural and Waste Lands of Wales* (ed. Stapledon, R. G.), pp. 13–107. Faber, London.

Davis, T. A. W. & Richards, P. W. (1933). The vegetation of Moraballi Creek, British Guiana. *J. Ecol.* **21**, 350–84.

Davy, J. B. (1938). *The Classification of Tropical Woody Vegetation Types.* Imperial Forestry Institute, Oxford, 13.

Davy, J. B. (1939). Bernard Dearman Davy. *Proc. Linn. Soc.* **151**, 234–6.

Day, W. R. (1946). Ecology and the study of climate. *Nature, Lond.* **157**, 827–9.

Diver, C. (1936). The problem of closely related species and the distribution of their populations. *Proc. Roy. Soc. Lond. B* **121**, 62–5.

Diver, C. & Good, R.D'O. (1934). The South Haven peninsula survey: general scheme of the survey. *J. Anim. Ecol.* **3**, 129–32.

Donoughue, B. & Jones, G. W. (1973). *Herbert Morrison. Portrait of a Politician.* Weidenfeld & Nicholson, London.

Doogue, D. (1986). Getting started. In *Reflections and Recollections. 100 Years of the Dublin Naturalists Field Club* (ed. anonymous), pp. 54–76. Dublin Naturalists Field Club, Dublin.

Dower, J. (1945). *National Parks in England and Wales.* HMSO, London, Cmd 6628.

Drude, O. (1896). Deutschlands Pflanzengeographie: ein geographisches Charakterbild der Flora von Deutschland. Verlag von J. Engelhorn, Stuttgart.

Dublin, L. I. (1950). Alfred J. Lotka. *J. Am. Statist. Ass.* **45**, 138–9.

Duff, A. G. & Lowe, P. D. (1981). Great Britain. In *Handbook of Contemporary Developments in World Ecology* (eds Kormondy E. J. & McCormick J. F.), pp. 141–56. Greenwood Press, Westpoint.

Duffey, E. (1970). Captain Cyril Diver. *Bull. Brit. Ecol. Soc.* **1**, 2–3.

Duffey, E. & Watt, A. S. (eds) (1971). *The Scientific Management of Animal and Plant Communities for Conservation.* Blackwell Scientific Publications, Oxford.

Duggar, B. M. (1929). *Proceedings of the International Congress of Plant Sciences.* Geo Banta, Wisconsin, I, 629–41 and 643–6.

Dunnet, G. M. (1982). Ecology and everyman. *J. Anim. Ecol.* **51**, 1–14.

Edelsten, H. McD. (1950–1). John Claud Fortescue Fryer. *Obituary Notices of Fellows of the Royal Society* **7**, 95–106.

Elton, C. S. (1924a). Field zoology. *Sch. Sci. Rev.* **6**, 90–5.

Elton, C. S. (1924b). Periodic fluctuations in the number of animals: their causes and effects. *Brit. J. Exp. Biol.* **2**, 119–63.

Elton, C. S. (1927). *Animal Ecology.* Sidgwick & Jackson, London.

Elton, C. S. (1929a). The relation of animal numbers to climate. *Conference of Empire Meteorologists. Agricultural Section* (ed. anonymous), pp. 121–7. HMSO, London.

Elton, C. S. (1929b). Ecology, animal. *Encyclopaedia Britannica, 14th Edition* **7**, 915–24.

Elton, C. S. (1930). *Animal Ecology and Evolution.* Clarendon, Oxford.

Elton, C. S. (1931). The study of epidemic diseases among wild animals. *J. Hyg., Camb.*, **31**, 435–56.

Elton, C. S. (ed.) (1933a). *Matamek Conference on biological cycles.* Matamek Factory, Canadian Labrador.

Elton, C. S. (1933b). *The Ecology of Animals.* Methuen, London.

Elton, C. S. (1933c). *Exploring the Animal World.* Allen & Unwin, London.

Elton, C. S. (1942). *Voles, Mice and Lemmings: Problems in Population Dynamics.* Clarendon Press, Oxford.

Elton, C. S. (1943). The changing realms of animal life. *Pol. Sci. Learn.* **2**, 7–11.

Elton, C. S. (1946). Competition and the structure of ecological communities. *J. Anim. Ecol.* **15**, 54–68.

Elton, C. S. (1949). Population interspersion: an essay on animal community patterns. *J. Ecol.* **37**, 1–23.

Elton, C. S. (1966). *The Pattern of Animal Communities.* Methuen, London.

Elton, C. S. & Miller, R. S. (1954). The ecological survey of animal communities. *J. Ecol.* **42**, 460–96.

Elton, C. S., Ford, E. B., Baker, J. R. & Gardner, A. D. (1931). The health and parasites of a wild mouse population. *Proc. Zool. Soc. Lond.*, 657–721.

Elton, C. S., Davis, D. H. S. & Findlay, G. M. (1935). An epidemic among voles on the Scottish border in the spring of 1934. *J. Anim. Ecol.* **4**, 277–88.

Erdtman, G. (1924). Studies in the micropalaeontology of postglacial deposits in northern Scotland and the Scotch Isles. *J. Linn. Soc. Bot.* **46**, 449–504.

Erdtman, G. (1929). Some aspects of the post-glacial history of British forests. *J. Ecol.* **17**, 112–26.

Evans, G. C. (1939). Ecological studies on the rain forest of Southern Nigeria. II. The atmospheric environmental conditions. *J. Ecol.* **27**, 436–82.

Evans, G. C. (1976). A sack of uncut diamonds: the study of ecosystems and the future resources of mankind. *J. Ecol.* **64**, 1–39.

Evans, W. (1919). The late Captain Sydney E. Brock. *Scott. Nat.* 27–8.

Faegri, K. (1981). Rolf Nordhagen. *Watsonia* **13**, 358.

Farrow, E. P. (1916). On the ecology of the vegetation of Breckland. II. Factors relating to the relative distributions of *Calluna*-heath and grass-heath in Breckland. *J. Ecol.* **4**, 57–64.

Farrow, E. P. (1917a). On the ecology of the vegetation of Breckland. III. General effects of rabbits on the vegetation. *J. Ecol.* **5**, 1–18.

Farrow, E. P. (1917b). On the ecology of the vegetation of Breckland. V. Observations relating to competition between plants. *J. Ecol.* **5**, 155–72.

Farrow, E. P. (1925). *Plant life on East Anglian Heaths, being Observational and Experimental Studies of the Vegetation of Breckland.* University Press, Cambridge.

Farrow, E. P. (1942). *A Practical Method of Self-Analysis.* Allen & Unwin, London.

Fenton, E. W. (1937). The influence of sheep on the vegetation of hill grazings in Scotland. *J. Ecol.* **25**, 424–30.

Fisher, R. A., Corbet, A. S. & Williams, C. B. (1943). The relation between the number of species and the number of individuals in a random sample of an animal population. *J. Anim. Ecol.* **12**, 42–58.

Flahault, C. (1897). Essai, d'une carte botanique et forestière de la France. *Annls Géogr.* **6**, 289–312.

Fream, W. (1888). The herbage of old grass land. *J. Roy. Agric. Soc.* **24**, 415–47.

Fritsch, F. E. (1906). Problems in aquatic biology, with special reference to the study of algal periodicity. *New Phytol.* **5**, 149–69.

Fritsch, F. E. (1907a). The role of algal growth in the colonization of new ground and in the determination of scenery. *Geogrl J.* **30**, 531–48.

Fritsch, F. E. (1907b). A general consideration of the subaerial and fresh-water algal flora of Ceylon. *Proc. Roy. Soc. Lond. B.* **79**, 197–254.

Fritsch, F. E. (1907c). The subaerial and freshwater algal flora of the tropics: a phytogeographical and ecological study. *Ann. Bot.* **21**, 235–75.

Fritsch, F. E. (1927). Some aspects of the present-day investigation of protophyta. *Report of the 95th Meeting of the British Association for the Advancement of Science*, 176–90.

Fritsch, F. E. (1937). The early history of the Association. *Fifth Report of Freshwater Biological Association*, 33–41.

Fritsch, F. E. & Rich, F. (1907). Studies on the occurrence and reproduction of British freshwater algae in nature. I. Preliminary observations on Spirogyra. *Ann. Bot.* **21**, 423–36.

Fritsch, F. E. & Rich, F. (1909). Studies on the occurrence and reproduction of British freshwater algae in nature. II. Five years' observations. *Proc. Bristol Nat. Soc.* **2**, 27–54.

Fritsch, F. E. & Salisbury, E. J. (1914). *An Introduction to the Study of Plants.* Bell, London.

G., F. (1907). Sir John Scott Burdon-Sanderson. *Proc. Roy. Soc. Lond. B.* **79**, iii–xviii.

Gause, G. F. (1934). *The Struggle for Existence.* Williams & Wilkins, Baltimore.

Geddes, P. (1900). Robert Smith. *Scott. Geogr. Mag.* **16**, 597–9.

Gimingham, C. H. (1979). James Robert Matthews. *Watsonia* **12**, 274–5.

Gimingham, C. H. (1986). Dr Alexander Stuart Watt. *J. Ecol.* **74**, 297–300.

Gimingham, C. H. (1987). Harnessing the winds of change: heathland ecology in retrospect and prospect. *J. Ecol.* (in press).

Gimingham, C. H., Spence, D. H. N. & Watson, A. (1983). Ecology. *Proc. Roy. Soc. Edinb. B.* **84**, 85–118.

Gleason, H. A. (1922). The vegetational history of the Middle West. *Ann. Ass. Am. Geogr.* **12**, 39–85.

Gleason, H. A. (1926). The individualistic concept of the plant association. *Bull. Torrey Bot. Club* **53**, 7–26.

Gleason, H. A. (1953). Autobiographical letter. *Bull. Ecol. Soc. Am.* **34**, 40–2.

Godwin, H. (1923). Dispersal of pond floras. *J. Ecol.* **11**, 160–3.

Godwin, H. (1929a). The sub-climax and deflected succession. *J. Ecol.* **17**, 144–7.

Godwin, H. (1929b). The 'sedge' and 'litter' of Wicken Fen. *J. Ecol.* **17**, 148–60.

Godwin, H. (1930). Review. *J. Ecol.* **18**, 181–3.

Godwin, H. (1934). Pollen analysis. An outline of the problems and potentialities of the method. *New Phytol.* **33**, 278–305 and 325–58.

Godwin, H. (1940a). Studies of the post-glacial history of British vegetation. III. Fenland pollen diagrams. IV. Post-glacial changes of relative land- and sea-level in the English fenland. *Phil. Trans. Roy. Soc. Lond. B.* **230**, 239–303.

Godwin, H. (1940b). Pollen analysis and forest history of England and Wales. *New Phytol.* **39**, 370–400.

Godwin, H. (1941). Studies of the post-glacial history of British vegetation. VI. Correlations in the Somerset Levels. *New Phytol.* **40**, 108–32.

Godwin, H. (1943). Coastal peat beds of the British Isles and North Sea. *J. Ecol.* **31**, 199–247.

Godwin, H. (1946). The relationship of bog stratigraphy to climatic change and archaeology. *Proc. Prehist. Soc.* **12**, 1–11.

Godwin, H. (1951). Comments on radiocarbon dating for samples from the British Isles. *Am. J. Sci.* **249**, 301–7.

Godwin, H. (1952). Review of 'Flora of the British Isles'. *J. Ecol.* **40**, 407–9.

Godwin, H. (1956). *The History of the British Flora. A Factual Basis for Phytogeography.* University Press, Cambridge.

Godwin, H. (1957). Arthur George Tansley. *Biographical Memoirs of Fellows of the Royal Society* **3**, 227–46.

Godwin, H. (1958). Sir Arthur George Tansley. *J. Ecol.* **46**, 1–8.

Godwin, H. (1960). Radiocarbon dating and Quaternary history in Britain. *Proc. Roy. Soc. Lond. B.* **153**, 287–320.

Godwin, H. (1973). Obituary: tribute to four botanists. *New Phytol.* **72**, 1245–50.

Godwin, H. (1977). Sir Arthur Tansley: the man and the subject. *J. Ecol.* **65**, 1–26.

Godwin, H. (1978). *Fenland: its Ancient Past and Uncertain Future.* University Press, Cambridge.

Godwin, H. (1981). *The Archives of the Peat Bogs.* University Press, Cambridge.

Godwin, H. (1985a). Early development of The *New Phytologist.* *New Phytol.* **100**, 1–4.

Godwin, H. (1985b). *Cambridge and Clare.* University Press, Cambridge.

Godwin, H. & Clifford, M. H. (1938). Studies of the post-glacial history of British vegetation. I. Origin and stratigraphy of fenland deposits near Woodwalton, Hunts. II. Origin and stratigraphy of deposits in southern fenland. *Phil. Trans. Roy. Soc. Lond. B.* **229**, 323–406.

Godwin, H. & Conway, V. M. (1939). The ecology of a raised bog near Tregaron, Cardiganshire. *J. Ecol.* **27**, 313–63.

Godwin, H. & Godwin, M. E. (1933). Pollen analyses of fenland peats at St Germans, near King's Lynn. *Geol. Mag.* **70**, 168–80.

Godwin, H. & Mitchell, G. F. (1938). Stratigraphy and development of two raised bogs near Tregaron, Cardiganshire. *New Phytol.* **37**, 425–54.

Godwin, H. & Tansley, A. G. (1929). The vegetation of Wicken Fen. In *The Natural History of Wicken Fen* (ed. Gardiner J. S.), Part 5, pp. 385–446.

Godwin, H. & Turner, J. S. (1933). Soil acidity in relation to vegetational succession at Calthorpe Broad, Norfolk. *J. Ecol.* **21**, 235–62.

Godwin, H., Godwin, M. E. & Clifford, M. H. (1935). Controlling factors in the formation of fen deposits, as shown by peat investigations at Wood Fen, near Ely. *J. Ecol.* **23**, 509–35.

Godwin, H., Walker, D. & Willis, E. H. (1957). Radiocarbon dating and post-glacial vegetational history: Scaleby Moss. *Proc. Roy. Soc. Lond. B.* **147**, 352–66.

Goodall, D. W. (1952). Quantitative aspects of plant distribution. *Biol. Rev.* **27**, 194–245.

Goodland, R. J. (1975), The tropical origin of ecology: Eugene Warming's jubilee. *Oikos* **26**, 240–5.

Goodman, G. T. (ed.) (1967). Survey of the nature of the technical advice required when treating land affected by industry. *J. Ecol.* **55**, 27P–34P.

Goodman, G. T., Edwards, R. W. & Lambert, J. M. (eds) (1965). *Ecology and the Industrial Society*. Blackwell Scientific Publications, Oxford.

Gregor, J. W. (1964). Obituaries. Edward Wyllie Fenton. *Proc. Linn. Soc.* **175**, 89–90.

Gregory, R. (1971). *The Price of Amenity*. Macmillan, London, 132–202.

Greig-Smith, P. (1957). *Quantitative Plant Ecology*. Butterworths, London.

Greig-Smith, P. (1979). Pattern in vegetation. *J. Ecol.* **67**, 755–79.

Greig-Smith, P. (1982). A. S. Watt, FRS: a biographical note. In *The Plant Community as a Working Mechanism* (ed. Newman E. I.), pp. 9–10. Blackwell Scientific Publications for British Ecological Society, Special Publication 1.

Grinnell, H. W. (1940). Joseph Grinnell. *Condor* **42**, 3–34.

Grinnell, J. (1904). The origin and distribution of the Chestnut-backed chickadee. *Auk* **21**, 364–82.

Guillebaud, W. H. (1949). The late Dr M. C. Rayner. *Forestry* **22**, 241–4.

Gunn, M. & Codd, L. E. (1981). *Botanical Exploration of Southern Africa*. Balkema, Cape Town.

Haeckel, E. (1866). *Generelle Morphologie der Organismen*. Reimer, Berlin.

Hairston, N. G. (1964). Studies on the organization of animal communities. *J. Ecol.* **52** suppl., 227–39.

Hall, A. D. (1903). *The soil; an Introduction to the Scientific Study of the Growth of Crops*. Murray, London.

Hall, A. D. & Russell, E. J. (1911). *A Report on the Agriculture and Soils of Kent, Surrey and Sussex*. HMSO for the Board of Agriculture, London.

Hamilton, T. H. (1968). Biogeography and ecology in a new setting. *Science* **159**, 71–2.

Hardy, A. C. (1924). The herring in relation to its animate environment. Part I. The food and feeding habits. *Ministry of Agriculture and Fisheries, Fishery Investigations series*, **7** (3), HMSO, London.

Hardy, A. C. (1968). Charles Elton's influence in ecology. *J. Anim. Ecol.* **37**, 3–8.

Hardy, M. (1902). Botanical geography and the biological utilisation of the soil. *Scott. Geogr. Mag.* **18**, 225–36.

Hardy, M. (1905). *Equisse de la géographie et de la végétation des Highlands d'Ecosse*. Imprimerie Générale Lahure, Paris.

Hardy, M. (1906). Botanical survey of Scotland. *Scott. Geogr. Mag.* **22**, 229–41.

Hardy, M. (1913). *An Introduction to Plant Geography*. Clarendon Press, Oxford.

Hardy, M. (1920). *The Geography of Plants*. Clarendon Press, Oxford.

Harley, J. L. (1972). Fungi in ecosystems. *J. Anim. Ecol.* **41**, 1–16.

Harley, J. L. (1981). Geoffrey Emett Blackman. *Biographical Memoirs of Fellows of the Royal Society* **27**, 45–82.

Harley, J. L. (1984). Introduction to symposium. *New Phytol.* **98**, 1.

Harper, J. L. (ed.) (1960). *The Biology of Weeds*. Blackwell Scientific Publications, Oxford.

Harper, J. L. (1964). The individual in the population. *J. Ecol.* **52** *suppl.*, 149–58.

Harper, J. L. (1967). A Darwinian approach to plant ecology. *J. Ecol.* **55**, 247–70.

Harper, J. L. (1977). *Population Biology of Plants.* Academic Press, London.

Harper, J. L. Clatworthy, J. N., McNaughton, I. H. & Sagar, G. R. (1961). The evolution and ecology of closely related species living in the same area. *Evolution* **15**, 209–27

Harris, T. M. (1963). Hugh Hamshaw Thomas. *Biographical Memoirs of Fellows of the Royal Society* **9**, 287–99.

Harrisson, T. H. (1933). The Oxford University Expedition to Sarawak, 1932. *Geogrl J.* **82**, 385–410.

Harvey, L. A. (1945). Symposium on 'the ecology of closely allied species'. *J. Ecol.* **33**, 115–16.

Harvie-Brown, J. A. & Bartholomew, J. G. (1893). *Naturalist's Map of Scotland.* Geographical Institute, Edinburgh.

Harvie-Brown, J. A. & Buckley, T. E. (1895). *A Vertebrate Fauna of the Moray Basin.* Douglas, Edinburgh.

Hassell, M. P. (1983). George C. Varley. *Antenna* **7**, 121–2.

Hassell, M. P. & Varley, G. C. (1969). New inductive population model for insect parasites and its bearing on biological control. *Nature, Lond.* **233**, 1133–7.

Hedberg, O. (1973). Gunnar Erdtman in memoriam. *Svensk Bot. Tidskr.* **67**, 311.

Herbertson, A. J. (1897). The mapping of plant associations. *Scott. Geogr. Mag.* **13**, 537–41.

Hewitt, C. G. (1921). *The conservation of the wild life of Canada.* Charles Scribner's Sons, New York.

Hill, A. W. (1935). Leonard Cockayne. *Obituary Notices of Fellows of the Royal Society* **1**, 443–57.

Hill, T. G. (1909). The Bouche d'Erquy. *New Phytol.* **8**, 97–103.

Hingston, R. W. G. (1930). The Oxford University Expedition to British Guiana. *Geogrl J.* **76**, 1–24.

Hobhouse, A. (1947). *National Parks Committee.* HMSO, London, Cmd 7121.

Hope-Simpson, J. F. (1940a). On the errors in the ordinary use of subjective frequency estimations in grassland. *J. Ecol.* **28**, 193–209.

Hope-Simpson, J. F. (1940b). The utilization and improvement of chalk down pastures. *J. Roy. Agric. Soc.* **100**, 44–9.

Hope-Simpson, J. F. (1941). Studies of the vegetation of the English chalk. VIII. A second survey of the chalk grasslands of the South Downs. *J. Ecol.* **29**, 217–67.

Hopkins, B. (1955). The species–area relations of plant communities. *J. Ecol.* **43**, 409–26.

Horwood, A. R. (1910). The extinction of cryptogamic plants. *S. East Nat.*, 56–86.

Horwood, A. R. (1913). Vestigial floras. *J. Ecol.* **1**, 100–2.

Hubbard, C. E. (1971). William Bertram Turrill. *Biographical Memoirs of Fellows of the Royal Society* **17**, 689–712.

Hume, C. W. (1939). Instructions for dealing with rabbits. *UFAWS Monograph and Report* **4E**, 20 pp.

Hutchinson, G. E. (1978). *An Introduction to Population Ecology.* Yale University Press, New Haven.

Huxley, J. S. (1914). The courtship-habits of the great crested grebe. *Proc. Zool. Soc. Lond.* 491–562.

Huxley, J. S. (1942). *Evolution. The Modern Synthesis.* Allen & Unwin, London.

Huxley, J. S. (1947). *Conservation of Nature in England and Wales.* HMSO, London, Cmd 7122.

Huxley, J. S. (1970). *Memories.* Allen & Unwin, London.

Ivimey-Cook, R. B. & Proctor, M. C. F. (1966). The plant communities of the Burren, Co. Clare. *Proc. Roy. Ir. Acad. B* **64**, 211–301.

Jaccard, P. (1912). The distribution of the flora in the alpine zone. *New Phytol.* **11**, 37–50.

Jacks, G. V. & Whyte, R. O. (1938). *Erosion and Soil Conservation.* Imperial Bureau of Soil Science, Technical Communication, 36.

Jacks, G. V. & Whyte, R. O. (1939). *The Rape of the Earth. A World Survey of Soil Erosion.* Faber, London.

Jackson, J. B. C. (1981). Interspecific competition and species' distributions: the ghosts of theories and data past. *Am. Zool.* **21**, 889–901.

Jefferies, R. L. & Davy, A. J. (eds) (1979). *Ecological Processes in Coastal Environments.* Blackwell Scientific Publications, Oxford.

Jefferies, T. A. (1915). Ecology of the purple heath grass. *J. Ecol.* **3**, 93–109.

Jeffers, J. N. R. (ed.) (1972). *Mathematical Models in Ecology.* Blackwell Scientific Publications, Oxford.

Jenkin, T. J. (1919). *Pasture Studies: Some Results.* Jarvis & Foster, Bangor.

Jessen, K. (1949). Studies in late Quaternary deposits and flora-history of Ireland. *Proc. Roy Ir. Acad. B* **52**, 85–290.

Jessen, K. & Milthers, V. (1928). Stratigraphical and palaeontological studies of interglacial freshwater deposits in Jutland and north-west Germany. *Danmarks Geologiske Undersogelse, Copenhagen, 48.*

Jones, W. N. & Rayner, M. C. (1920). *A Text-book of Plant Biology.* Macmillan, London.

Kendall, C. E. Y. (1921–2). The mollusca of Oundle. *J. Conch., Lond.,* **16**, 240–4 and 248–51.

Kendall, C. E. Y. (1929). Ecology of the British land molluscs. *Naturalist, Hull,* 247–50 and 273–6.

Kendall, C. E. Y., Dean, J. D. & Rankin, W. M. (1909). On the geographical distribution of mollusca in south Lonsdale. *Naturalist, Hull,* 314–19, 354–9, 378–81 and 435–7.

Kendall, M. G. (1963). Ronald Aylmer Fisher. Biometrika **50**, 1–15.

Lack, D. (1933). Habitat selection in birds with special reference to the effects of afforestation on the Breckland avifauna. *J. Anim. Ecol.* **2**, 239–62.

Lack, D. (1944). Ecological aspects of species-formation in passerine birds. *Ibis* **86**, 260–86.

Lack, D. (1945a). The Galapagos finches. A study in variation. *California Academy of Science, Occasional Paper, 21.*

Lack, D. (1945b). The ecology of closely related species with special reference to cormorant and shag. *J. Anim. Ecol.* **14**, 12–16.

Lack, D. (1947). *Darwin's Finches.* University Press, Cambridge.

Lack, D. (1965). Evolutionary ecology. *J. Anim. Ecol.* **34**, 223–31.

Lack, D. (1971). *Ecological Isolation in Birds.* Blackwell Scientific Publications, Oxford.

Lack, D. (1973). Obituary. *Ibis* **115**, 421–31.

Lack, D. & Varley, G. C. (1945). Detection of birds by radar. *Nature, Lond.* **156**, 446.

Laing, R. M. (1936). Obituary. Leonard Cockayne. *Trans. Proc. Roy. Soc. N.Z.* **65**, 457–67.

Lambert, J. M. (ed.) (1967). *The Teaching of Ecology.* Blackwell Scientific Publications, Oxford.

Le Cren, E. (1979). The first fifty years of the Freshwater Biological Association. *Forty-Seventh Annual Report of the Freshwater Biological Association,* 27–42.

Le Cren, E. & Holdgate, M. W. (eds) (1962). *The Exploitation of Natural Animal Populations.* Blackwell Scientific Publications, Oxford.

Lees, F. A. (1888). *The Flora of West Yorkshire.* Lovell Reeve, London.

Levy, H. (1932). *The Universe of Science.* Watts, London.

Lewis, F. J. (1904a). Geographical distribution of vegetation of the basins of the rivers Eden, Tees, Wear and Tyne. *Geogrl J.* **23**, 313–31.

Lewis, F. J. (1904b). Geographical distribution of vegetation of the basins of the rivers Eden, Tees, Wear and Tyne. *Geogrl J.* **24**, 267–85.

Lewis, F. J. (1906a). The plant remains in the Scottish peat mosses. II. The Scottish Highlands. *Trans Roy. Soc. Edinb.* **45**, 335–60.

Lewis, F. J. (1906b). The history of the Scottish peat mosses and their relation to the glacial period. *Scott. Geogr. Mag.* **22**, 341–52.

Lewis, F. J. (1907a). The plant remains in the Scottish peat mosses. III. The Scottish Highlands and the Shetland Islands. *Trans. Roy. Soc. Edinb.* **46**, 33–70.

Lewis, F. J. (1907b). The sequence of plant remains in the British peat mosses. *Sci. Prog.* **2**, 307–25.

Lewis, F. J. (1908). The British Vegetation Committee's excursion to the west of Ireland. *New Phytol.* **7**, 253–60.

Lewis, F. J. (1910). The plant remains in the Scottish peat mosses. IV. The Scottish Highlands and Shetland. *Trans. Roy. Soc. Edinb.* **47**, 793–833.

Lewis, F. J. & Dowding, E. S. (1926). The vegetation and retrogressive changes of peat areas in central Alberta. *J. Ecol.* **14**, 317–41.

Libby, W. F. (1952). *Radiocarbon Dating.* University Committee on Publications in the Physical Sciences, Chicago.

Lindeman, R. (1942). The trophic–dynamic aspect of ecology. *Ecology* **23**, 399–418.

Lofthouse, R. (1887). The river Tees: its marshes and their fauna. *Naturalist, Hull*, 1–16.

Lord President of the Council (1960). *Annual Report of the Advisory Council on Scientific Policy, 1959–60.* HMSO, London, Cmnd 1167.

Lord President of the Council (1963). *Annual Report of the Advisory Council on Scientific Policy, 1962–63.* HMSO, London, Cmnd 2163.

Lord Privy Seal (1971). *A Framework for Government Research and Development.* HMSO, London, Cmnd 4814.

Lord Privy Seal (1972). *A Framework for Government Research and Development.* HMSO, London, Cmnd 5046.

Lotka, A. J. (1925). *Elements of Physical Biology.* Williams & Wilkins, Baltimore.

Lowe, P. D. (1976). Amateurs and professionals: the institutional emergence of British plant ecology. *J. Soc. Biblphy Nat. Hist.* **7**, 517–35.

Ludi, W. (ed.) (1952). *Die Pflanzenwelt Irlands.* Veroff. geobot. Inst., Zurich.

M., A. E. (1950). Obituary notice: Dr George Herbert Pethybridge. *Trans. Brit. Mycol. Soc.* **33**, 161–4.

M., J. E. & N., E. T. (1919). Clement Reid. *Proc. Roy. Soc. Lond. B* **90**, viii–x.

Macan, T. T. (1970). *Biological Studies of the English Lakes.* Longman, London.

MacArthur, R. H. & Wilson, E. O. (1967). *The Theory of Island Biogeography.* University Press, Princeton.

Macfadyen, A. (1975). Some thoughts on the behaviour of ecologists. *J. Anim. Ecol.* **44**, 351–63.

MacGillivray, M. (1832a). Remarks on the phenogamic vegetation of the river Dee, in Aberdeenshire. *Mem. Wernian Nat. Hist. Soc.* **6**, 539–56.

MacGillivray, W. (1832b). *The Travels and Researches of Alexander von Humboldt.* Oliver & Boyd, Edinburgh.

MacGillivray, W. (1855). *The Natural History of Dee side and Braemar.* Her Majesty's Command, London.

McIntosh, R. P. (1975). H. A. Gleason, 'individualistic ecologist'. *Bull. Torrey Bot. Club* **102**, 253–73.

McIntosh, R. P. (1985). *The Background of Ecology: Concept and Theory.* University Press, Cambridge.

Mackerras, I. M. (1969). Alexander John Nicholson. *J. Ent. Soc. Aust.* (*N.S.W.*) **6**, 57–60.

McVean, D. N. & Ratcliffe, D. A. (1962). *Plant Communities of the Scottish Highlands.* HMSO, London.

Malloch, A. J. C. (1976). An annotated bibliography of the Burren. *J. Ecol.* **64**, 1093–105.

Marsden-Jones, E. M. & Turrill, W. B. (1930). Report on the transplant experiments of the British Ecological Society at Potterne, Wilts. *J. Ecol.* **18**, 352–78.

Marsden-Jones, E. M. & Turrill, W. B. (1933). Second report on the transplant experiments of the British Ecological Society at Potterne, Wilts. *J. Ecol.* **21**, 268–93.

Marsden-Jones, E. M. & Turrill, W. B. (1945). Sixth report of the transplant experiments of the British Ecological Society at Potterne, Wiltshire. *J. Ecol.* **33**, 57–81.

Marsh, A. S. (1915). The maritime ecology of Holme next the Sea, Norfolk. *J. Ecol.* **3**, 65–73.

Martin, C. J. (1939). Arthur Edwin Boycott. *Obituary Notices of Fellows of the Royal Society* **2**, 561–71.

Matthews, J. R. (1914). The White Moss Loch: a study in biotic succession. *New Phytol.* **13**, 134–48.

Matthews, J. R. (1922). The distribution of plants in Perthshire in relation to 'age and area'. *Ann. Bot.* **36**, 321–7.

Matthews, J. R. (1929). Obituary notices. William Gardner Smith. *Trans. Proc. Bot. Soc. Edinb.* **30**, 175–8.

Matthews, J. R. (1937). Geographical relationships of the British flora. *J. Ecol.* **25**, 1–90.

Mayr, E. (1942). *Systematics and the Origin of Species from the Viewpoint of a Zoologist.* Columbia University Press, New York.

Merrett, P. (ed.) (1971). *Captain Cyril Diver. A Memoir.* Nature Conservancy, Wareham.

Metcalfe, G. (1950). The ecology of the Cairngorms. Part II. The mountain Callunetum. *J. Ecol.* **38**, 46–74.

Miall, L. C. (1897). *Thirty Years of Teaching.* Macmillan, London.

Michael, E. L. (1920). Marine ecology and the coefficient of association: a plea in behalf of quantitative biology. *J. Ecol.* **8**, 54–9.

Middleton, A. D. (1934). Periodic fluctuations in British game populations. *J. Anim. Ecol.* **3**, 231–49.

Middleton, A. D. (1942). *The Control and Extermination of Wild Rabbits.* Bureau of Animal Population, Oxford.

Mitchell, F. (1976). *The Irish Landscape.* Collins, London.

Moore, J. J. (1960). A re-survey of the vegetation of the district lying south of Dublin (1905–1956). *Proc. Roy. Ir. Acad. B* **61**, 1–36.

Moore, J. J. (1962). The Braun-Blanquet system: a reassessment. *J. Ecol.* **50**, 761–69.

Moore, J. J. (1981). Founders of phytosociology. *Bull. Brit. Ecol. Soc.* **12**, (*1*), 2–4.

Moore, N. W. (1962). The heaths of Dorset and their conservation. *J. Ecol.* **50**, 369–91.

Moore, N. W. (ed.) (1966). Pesticides in the environment and their effects on wildlife. *J. Appl. Ecol.* **3** *suppl.*

Moss, C. E. (1896). Why do flowers bloom in spring and in the autumn? *Halifax Nat.* **1**, 61–3.

Moss, C. E. (1898). Green scums. *Halifax Nat.* **3**, 79–81.

Moss, C. E. (1900a). Changes in the Halifax flora during the last century and a quarter. *Naturalist, Hull,* 165–72.

Moss, C. E. (1900b). Norland Clough. *Halifax Nat.* **5**, 40–5.

Moss, C. E. (1902–3). Moors of south-west Yorkshire. *Halifax Nat.* **7**, 88–94.

Moss, C. E. (1907). *Geographical Distribution of Vegetation in Somerset: Bath and Bridgwater District.* Royal Geographical Society, London.

Moss, C. E. (1910). The fundamental units of vegetation. *New Phytol* **9**, 18–53.

Moss, C. E. (1913). *Vegetation of the Peak District.* University Press, Cambridge.

Moss, C. E., Rankin, W. M. & Tansley, A. G. (1910). The woodlands of England. *New Phytol.* **9**, 113–49.

Munro, T. (1935). Note on musk-rats and other animals killed since the inception of the campaign against musk-rats. *Scott. Nat.* 11–16.

Murray, J. & Pullar, F. P. (1900). A bathymetrical survey of the fresh-water lochs of Scotland. *Scot. Geogr. Mag.* **16**, 193–234.

Murray, J. & Pullar, L. (eds) (1910). *Bathymetrical Survey of the Scottish Fresh-water Lochs During the Years 1897 and 1909.* Challenger Office, Edinburgh.

Newbigin, M. (1901). Sir John Murray's scheme for the investigation of the natural history of the Forth Valley. *Scott. Geogr. Mag.* **17**, 644–51.

Newbigin, M. (1913). Geography in Scotland since 1889. *Scott. Geogr. Mag.* **29**, 471–9.

Nicholson, A. J. (1933). The balance of animal populations. *J. Anim. Ecol.* **2**, 132–8.

Nicholson, A. J. & Bailey, V. A. (1935). The balance of animal populations. Part I. *Proc. Zool. Soc. Lond.*, 551–98.

Nicholson, E. M. (1926). *Birds in England. An Account of the State of our Bird-Life and a Criticism of Bird Protection.* Chapman & Hall, London.

Nicholson, E. M. (1977). Sir Landsborough Thomson. *Br. Birds* **70**, 384–7.

Nordhagen, R. (1940). *Norsk Flora.* Aschehoug, Oslo.

Nordhagen, R. (1964). Review. *New Phytol.* **63**, 119–22.

Oliver, F. W. (1906). The Bouche d'Erquy 1906. *New Phytol.* **5**, 189–95.

Oliver, F. W. (1907). The Bouche d'Erquy 1907. *New Phytol.* **6**, 244–52.

Oliver, F. W. (1912). The shingle beach as a plant habitat. *New Phytol.* **11**, 73–99.

Oliver, F. W. (1914). Nature reserves. *J. Ecol.* **2**, 55–6.

Oliver, F. W. (1917). President's address. *J. Ecol.* **5**, 56–60.

Oliver, F. W. (1919). 'No department the door of which should not be opened'. *New Phytol.* **18**, 56–8.

Oliver, F. W. (1920). *Spartina* problems. *Ann. Appl. Biol.* **7**, 25–39.

Oliver, F. W. (1925). *Spartina townsendii*; its mode of establishment, economic uses and taxonomic status. *J. Ecol.* **13**, 74–91.

Oliver, F. W. (1927). *An Outline of the History of the Botanical Department of University College, London.* University College, London.

Oliver, F. W. (1928). Nature reserves. *Trans. Norfolk Norwich Nat. Soc.* **12**, 317–22.

Oliver, F. W. & Salisbury, E. J. (1913). Vegetation and mobile ground as illustrated by *Suaeda fruticosa* on shingle. *J. Ecol.* **1**, 249–72.

Oliver, F. W. & Tansley, A. G. (1904). Methods of surveying vegetation on a large scale. *New Phytol.* **3**, 228–37.

Ovington, J. D. (1950). The afforestation of the Cublin Sands. *J. Ecol.* **38**, 303–19.

Ovington, J. D. (1951). The afforestation of Tentsmuir Sands. *J. Ecol.* **39**, 363–75.

Ovington, J. D. (1953). Studies of the development of woodland conditions under different trees. I. Soils pH. *J. Ecol.* **41**, 13–34.

Ovington, J. D. (1962). Quantitative ecology and the woodland ecosystem. *Adv. Ecol. Res.* **1**, 103–92.

Ovington, J. D. (1964). The ecological basis of the management of woodland nature reserves in Great Britain. *J. Ecol.* **52**, *suppl.*, 29–37.

Ovington, J. D. & Heitkamp, D. (1960). The accumulation of energy in forest plantations in Britain. *J. Ecol.* **48**, 639–46.

Pawson, H. C. (1960). *Cockle Park Farm.* Clarendon, Oxford.

Pearsall, W. H. (1917 and 1918). The aquatic and marsh vegetation of Esthwaite Water. *J. Ecol.* **5**, 180–202 and **6**, 53–74.

Pearsall, W. H. (1918). On the classification of aquatic plant communities. *J. Ecol.* **6**, 75–84.

Pearsall, W. H. (1920). The aquatic vegetation of the English Lakes. *J. Ecol.* **8**, 163–201.

Pearsall, W. H. (1921). The development of vegetation in the English Lakes, considered in relation to the general evolution of glacial lakes and rock basins. *Proc. Roy. Soc. Lond. B.* **92**, 259–84.

Pearsall, W. H. (1924). The statistical analysis of vegetation. *J. Ecol.* **12**, 135–9.

Pearsall, W. H. (1938). Soil types and plant ecology in Yorkshire. *Naturalist, Hull*, 57–64.

Pearsall, W. H. (1940a). Dr T. W. Woodhead. An appreciation. *Naturalist, Hull*, 157–8.

Pearsall, W. H. (1940b). A milestone in plant ecology. *J. Ecol.* **28**, 241–4.

Pearsall, W. H. (1959). The Freshwater Biological Association and its laboratory. *Adv. Sci.* **15**, 521–3.

Pearsall, W. H. (1964). The development of ecology in Britain. *J. Ecol.* **52**, *suppl.*, 1–12.

Percival, E. & Whitehead, H. (1929). A quantitative study of the fauna of some types of stream-bed. *J. Ecol.* **17**, 282–314.

Pethybridge, G. H. & Praeger, R. L. (1905). The vegetation of the district lying south of Dublin. *Proc. Roy. Ir. Acad. B* **25**, 124–80.

Phelps-Brown, H. (1967). Sir Alexander Morris Carr-Saunders. *Proc. Br. Acad.* **53**, 379–89.

Phillips, J. (1934–5). Succession, development, the climax and the complex organism: an analysis of concepts. *J. Ecol.* **22**, 554–71, **23**, 210–46 and 488–508.

Phillips, J. (1954). A tribute to Frederic E. Clements and his concepts in ecology. *Ecology* **35**, 114–15.

Pigott, C. D. (1956). The vegetation of Upper Teesdale in the north Pennines. *J. Ecol.* **44**, 545–86.

Pigott, C. D. & Walters, S. M. (1954). On the interpretation of the discontinuous distributions shown by certain British species of open habitats. *J. Ecol.* **42**, 95–116.

Pinder, J. (ed.) (1981). *Fifty Years of Political and Economic Planning*. Heinemann, London.

Poore, M. E. D. (1954). The principles of vegetation classification and the ecology of Woodwalton Fen. PhD (unpublished) thesis, University of Cambridge.

Poore, M. E. D. (1955a). The use of phytosociological methods in ecological investigations. I. The Braun-Blanquet system. *J. Ecol.* **43**, 226–44.

Poore, M. E. D. (1955b). The use of phytosociological methods in ecological investigations. II. Practical issues involved in an attempt to apply the Braun-Blanquet system. *J. Ecol.* **43**, 245–69.

Poore, M. E. D. (1955c). The use of phytosociological methods in ecological investigations. III. Practical applications. *J. Ecol.* **43**, 606–51.

Poore, M. E. D. (1956). The ecology of Woodwalton Fen. *J. Ecol.* **44**, 455–92.

Poore, M. E. D. & McVean, D. N. (1957). A new approach to Scottish mountain vegetation. *J. Ecol.* **45**, 401–39.

Pound, R. (1896). The plant-geography of Germany. *Am. Nat.* **30**, 465–8.

Pound, R. (1954). Frederic E. Clements as I knew him. *Ecology* **35**, 112–13.

Pound, R. & Clements, F. E. (1898a). The vegetation regions of the prairie province. *Bot. Gaz.* **25**, 381–94.

Pound, R. & Clements, F. E. (1898b). A method of determining the abundance of secondary species. *Minnesota Bot. Stud.* **1**, second series, 19–24.

Pound, R. & Clements, F. E. (1900). *The Phytogeography of Nebraska*. University of Nebraska, Lincoln.

Praeger, R. L. (1897). *Open-Air Studies in Botany: Sketches of British Wild-Flowers in their homes*. Griffin, London.

Praeger, R. L. (1901a). Irish topographical botany. *Proc. Roy. Ir. Acad.* **7**, third series.

Praeger, R. L. (1901b). Flowering plants as illustrated by British wild flowers. V. Dispersal and distribution. *Knowledge* **24**, 217–20.

Praeger, R. L. (1903). The flora of Clare Island. *Ir. Nat.* **12**, 277–94.

Praeger, R. L. (1904). The flora of Achill Island. *Ir. Nat.* **13**, 265–79.

Praeger, R. L. (1906). A conference on vegetation study. *Ir. Nat.* **15**, 1–5.

Praeger, R. L. (1908). The British Vegetation Committee in the west of Ireland. *Naturalist, Hull*, 412–16.

Praeger, R. L. (1923). Disperal and distribution. *J. Ecol.* **11**, 114–23.

Praeger, R. L. (1937). *The Way that I Went. An Irishman in Ireland*. Hodges, Figgis, Dublin.

Price, S. R. (1916). Captain A. S. Marsh. *J. Ecol.* **4**, 119–20.

Prime Minister (1963). *Committee of Enquiry into the Organisation of the Civil Service*. HMSO, London, Cmnd 2171.

Prince Philip (1978). *The Environmental Revolution: Speeches on Conservation, 1962–77*. Deutsch, London.

Ramsay, J. D. (1947). *National Parks*. HMSO, London, Cmd. 7235.

Ramsbottom J. (1931). Charles Edward Moss. *J. Bot., Lond.*, **69**, 20–3.

Rankin, W. M. (1910). The peat moors of Lonsdale. *Naturalist, Hull*, 119–22 and 153–61.

Ranwell, D. S. (ed.) (1967). Sub-committee report on landscape improvement advice and research. *J. Ecol.* **55**, 1P–8P.

Raunkiaer, C. (1934). *The Life Forms of Plants and Statistical Plant Geography*. Clarendon, Oxford.

Rayner, M. C. (1915). Obligate symbiosis in *Calluna vulgaris*. *Ann. Bot.* **29**, 97–132.

Rayner, M. C. (1918). Botany and the teaching of biology. *New Phytol.* **17**, 193–7.

Reid, C. (1899). *The Origin of the British Flora*. Dulau, London.

Research Study Group. (1961). *Toxic Chemicals in Agriculture and Food Storage*. HMSO, London.

Richards, O. W. (1926). Studies on the ecology of English heaths. III. Animal communities of the felling and burn successions at Oxshott Heath, Surrey. *J. Ecol.* **14**, 244–81.

Richards, O. W. (1930). The animal community inhabiting rotten posts at Bagley Wood, near Oxford. *J. Ecol.* **18**, 131–8.

Richards, O. W. (1940). The biology of the small white butterfly, with special reference to the factors controlling its abundance. *J. Anim. Ecol.* **9**, 243–88.

Richards, O. W. (1955). The ecology of populations. *J. Anim. Ecol.* **24**, 465–6.

Richards, P. W. (1936). Ecological observations on the rain forest of Mount Dulit, Sarawak. *J. Ecol.* **24**, 1–37 and 340–60.

Richards, P. W. (1939). Ecological studies on the rain forest of southern Nigeria. I. The structure and floristic composition of the primary forest. *J. Ecol.* **27**, 1–61.

Richards, P. W. (1952). *The Tropical Rain Forest. An Ecological Study*. University Press, Cambridge.

Richards, P. W. (1964). What the tropics can contribute to ecology. *J. Anim. Ecol.* **33**, 1–11.

Richards, P. W. (1979). Sir Edward Salisbury: an appreciation. *Bull. Brit. Ecol. Soc.* **10**, 30–1.

Richards, P. W., Tansley, A. G. & Watt, A. S. (1940). The recording of structure, life form and flora of tropical forest communities as a basis for their classification. *J. Ecol.* **28**, 224–39.

Riley, N. D. (1943). The interrelations of plants and insects; the place of both in the ecosystem. *Proc. Roy. Ent. Soc. Lond. C.* **8**, 39–55.

Ritchie, J. (1920). *The Influence of Man on the Animal Life in Scotland*. University Press, Cambridge.

Ritchie, J. (1949). *Nature Reserves in Scotland: Final Report*. HMSO, London, Cmd 7814.

Ross, R. (1954). Ecological studies on the rain forest of southern Nigeria. III Secondary succession in the Shasha Forest Reserve. *J. Ecol.* **42**, 259–82.

Rothschild, M. (1985). *Dear Lord Rothschild*. Hutchinson, London.

Royal Commission on Coastal Erosion. (1909). *Minutes of Evidence Accompanying the Second Report*. Cd 4461, vol. II, part II, QQ 15656–837.

Rübel, Von E. (1920). Über die Entwicklung der Gesellschafts-morphologie. *J. Ecol.* **8**, 18–40.

Russell, E. J. (1961). Reginald George Stapledon. *Biographical Memoirs of Fellows of the Royal Society* **7**, 249–70.

Rutter, A. J. & Whitehead, F. H. (eds) (1963). *The Water Relations of Plants*. Blackwell Scientific Publications, Oxford.

Rycroft, H. B. (1979). Professor R. H. Compton. *Veld and Flora* **65** (*3*), 74–5.

S., A. E. (1917). Sir John Murray. *Proc. Roy. Soc. Lond. B* **89**, vi–xv.

Sagar, G. R. (1985). Profile of John L. Harper. *Studies on Plant Demography. A Festschrift for John L. Harper* (ed. White J.), pp. xix–xxv. Academic Press, London.

Salisbury, E. J. (1912). The competition of furze and bracken, particularly on Harpenden Common. *Trans. Herts. Nat. Hist. Soc. Fld Club* **15**, 71–2.

Salisbury, E. J. (1916). The oak-hornbeam woods of Hertfordshire. Parts I and II. *J. Ecol.* **4**, 83–117.

Salisbury, E. J. (1917). The ecology of scrub in Hertfordshire: a study in colonization. *Trans. Herts. Nat. Hist. Soc. Fld Club* **17**, 53–64.

Salisbury, E. J. (1918). The oak-hornbeam woods of Hertfordshire. Parts III and IV. *J. Ecol.* **6**, 14–52.

Salisbury, E. J. (1920). The significance of the calcicolous habit. *J. Ecol.* **8**, 202–15.

Salisbury, E. J. (1921). Stratification and hydrogen-ion concentration of the soil in relation to leaching and plant succession with special reference to woodlands. *J. Ecol.* **9**, 220–40.

Salisbury, E. J. (1922a). Botany. *Sci. Prog.* **16**, 554.

Salisbury, E. J. (1922b). The soils of Blakeney Point: a study of soil reaction and succession in relation to the plant covering. *Ann. Bot.* **36**, 391–431.

Salisbury, E. J. (1923). The effects of coppicing as illustrated by the woods of Hertfordshire. *Trans. Herts. Nat. Hist. Soc. Fld Club* **18**, 1–21.

Salisbury, E. J. (1925). Note on the edaphic succession in some dune soils with special reference to the time factor. *J. Ecol.* **13**, 322–28.

Salisbury, E. J. (1928). A proposed biological flora of Britain. *J. Ecol.* **16**, 161–2.

Salisbury, E. J. (1929). The biological equipment of species in relation to competition. *J. Ecol.* **17**, 197–222.

Salisbury, E. J. (1933). Review. *New Phytol.* **32**, 75–6.

Salisbury, E. J. (1938). A. E. Boycott. An appreciation. *J. Anim. Ecol.* **7**, 395–6.

Salisbury, E. J. (1939). The study of British vegetation. *Nature, Lond.*, **144**, 305–7.

Salisbury, E. J. (1942). *The Reproductive Capacity of Plants. Studies in Quantitative Biology.* Bell, London.

Salisbury, E. J. (1952). Francis Wall Oliver. *Obituary Notices of Fellows of the Royal Society* **8**, 229–40.

Salisbury, E. J. (1954). Felix Eugene Fritsch. *Obituary Notices of Fellows of the Royal Society* **9**, 131–40.

Salisbury, E. J. (1963). William Bertram Turrill. *Proc. Bot. Soc. Br. Isl.* **5**, 194–6.

Salisbury, E. (1964). The origin and early years of the British Ecological Society. *J. Ecol.* **52** suppl., 13–18.

Salisbury, E. J. & Tansley, A. G. (1921). The Durmast oakwoods of the Silurian and Malvernian strata near Malvern. *J. Ecol.* **9**, 19–38.

Schimper, A. F. W. (1903). *Plant-Geography upon a Physiological Basis.* Clarendon Press, Oxford.

Schröter, C. (1904–8). *Das Pflanzenleben der Alpen.* Verlag A. Raustein, Zürich.

Scott, D. C. (1920). Charles Gordon Hewitt. *Proc. Trans. Roy. Soc. Can.* **14**, vi–viii.

Scudo, F. M. & Ziegler, J. R. (1978). *The Golden Age of Theoretical Ecology 1923–1940.* Springer-Verlag, Lecture Notes in Biomathematics, 22, Berlin.

Seward, A. C. (1917). Harold Henry Welch Pearson. *Proc. Roy. Soc. Lond. B* **89**, lx–lxvii.

Seward, A. C. (1931). *Plant Life Through the Ages.* University Press, Cambridge.

Shanks, A. (1943). Mr Hugh Boyd Watt. *Glasgow Nat.* **14**, 99–100.

Shaw, G. A. (1969). Francis Cavers. *Naturalist, Hull*, 59–62.

Sheail, J. (1975). The concept of national parks in Great Britain, 1900–1950. *Trans. Inst. Br. Geogr.* **66**, 41–56.

Sheail, J. (1976). *Nature in Trust: the History of Nature Conservation in Britain.* Blackie, Glasgow.

Sheail, J. (1981). *Rural Conservation in Inter-war Britain.* Clarendon Press, Oxford.

Sheail, J. (1982a). Underground water abstraction: indirect effects of urbanization on the countryside. *J. Hist. Geogr.* **8**, 395–408.

Sheail, J. (1982b). Wild plants and the perception of land-use change in Britain: an historical perspective. *Biol. Conserv.* **24**, 129–46.

Sheail, J. (1984a). The rabbit (exploited animals). *Biologist* **31**, 135–40.

Sheail, J. (1984b). Nature reserves, national parks, and post-war reconstruction in Britain. *Environ. Conserv.* **11**, 29–34.

Sheail, J. (1985). *Pesticides and Nature Conservation: the British Experience 1950–1975.* Clarendon Press, Oxford.

Sheail, J. (1986). Grassland management and the early development of British ecology. *Br. J. Hist. Sci.* **19**, 283–99.

Sheail, J. (1987). The extermination of the muskrat in inter-war Britain. *Arch. Nat. Hist.* (in press).

Sheail, J. & Adams, W. M. (eds) (1980). Worthy of preservation: a gazetteer of sites of high biological or geological value, identified since 1912. *University College, London, Discussion Paper in Conservation*, **28**.

Shelford, V. E. (1913). *Animal Communities in Temperate America as Illustrated in the Chicago Region. A Study in Animal Ecology*. Geographical Society of Chicago, 5, Chicago.

Shelford, V. E. (1915). Principles and problems of ecology as illustrated by animals. *J. Ecol.* **3**, 1–23.

Shelford, V. E. (1938). The organization of the Ecological Society of America 1914–19. *Ecology* **19**, 164–6.

Shimwell, D. W. (1971). *The Description and Classification of Vegetation*. Sidgwick & Jackson, London.

Sinnott, E. W. (1917). The 'age and area' hypothesis and the problem of endemism. *Ann. Bot.* **31**, 209–16.

Smith, H. S. (1935). The role of biotic factors in the determination of population densities. *J. Econ. Entomol.* **28**, 873–98.

Smith, J. M. (1974). *Models in Ecology*. University Press, Cambridge.

Smith, R. (1898). Plant associations of the Tay basin. *Trans. Proc. Perthsh. Soc. Nat. Sci.* **2**, 200–17.

Smith, R. (1899). On the study of plant associations. *Nat. Sci.* **14**, 109–20.

Smith, R. (1900a). Botanical survey of Scotland. I. Edinburgh district. *Scott. Geogr. Mag.* **16**, 385–416.

Smith, R. (1900b). Botanical survey of Scotland. II. North Perthshire district. *Scott. Geogr. Mag.* **16**, 441–67.

Smith, R. (1900c). On the seed dispersal of *Pinus sylvestris* and *Betula alba*. *Ann. Scot. Nat. Hist* 43–6.

Smith, W. G. (1902a). A botanical survey of Scotland. *Scott. Geogr. Mag.* **18**, 132–9.

Smith, W. G. (1902b). The origin and development of heather moorland. *Scott. Geogr. Mag.* **18**, 587–97.

Smith, W. G. (1903a). Botanical survey for local naturalists' societies. *Naturalist, Hull*, 5–13.

Smith, W. G. (1903b). Reviews and book notices. *Naturalist, Hull*, 57–60.

Smith, W. G. (1904). Botanical survey of Scotland. III and IV. Forfar and Fife. *Scott. Geogr. Mag.* **20**, 617–28.

Smith, W. G. (1905a). Botanical survey of Scotland. III and IV. Forfar and Fife. *Scott. Geogr. Mag.* **21**, 4–23, 57–83 and 117–26.

Smith, W. G. (1905b). Formation of a Committee for the Survey and Study of British Vegetation. *New Phytol.* **4**, 23–6.

Smith, W. G. (1905c). The Central Committee for the Survey and Study of British Vegetation. *New Phytol.* **4**, 254–7.

·Smith, W. G. (1906). Plant formations of the Dublin mountains. *Ir. Nat.* **15**, 126–30.

Smith, W. G. (1907). The Central Committee for the Survey and Study of British Vegetation. *New Phytol.* **6**, 103–7.

Smith, W. G. (1909a). The present position of botanical survey in Britain. *Trans. Proc. Bot. Soc. Edinb.* **24**, 53–9.

Smith, W. G. (1909b). The British Vegetation Committee. *New Phytol.* **8**, 203–6.

Smith, W. G. (1912a). The British Vegetation Committee. *New Phytol.* **11**, 99–103.

Smith, W. G. (1912b). *Anthelia*: an Arctic–Alpine plant association. *Trans. Proc. Bot. Soc. Edinb.* **26**, 36–44.

Smith, W. G. (1913). Raunkiaer's 'life-forms' and statistical methods. *J. Ecol.* **1**, 16–26.

Smith, W. G. (1918). The distribution of *Nardus stricta* in relation to peat. *J. Ecol.* **6**, 1–13.

Smith, W. G. (1919). President's address. *J. Ecol.* **7**, 110–16.

Smith, W. G. (1924). Obituary notices: Eugene Warming. *Trans. Proc. Bot. Soc. Edinb.* **29**, 113–16.

Smith, W. G. & Crampton, C. B. (1914). Grassland in Britain. *J. Agric. Sci. Camb.* **6**, 1–17.

Smith W. G. & Moss, C. E. (1903). Geographical distribution of vegetation in Yorkshire. *Geogrl J.* **21**, 375–401.

Smith, W. G. & Rankin, W. M. (1903). Geographical distribution of vegetation in Yorkshire. *Geogrl J.* **22**, 149–78.

Snodgrass, C. P. (1965). Editorial. A. J. Herbertson. *Scott. Geogr. Mag.* **81**, 143–4.

Solbrig, O. T. (1979). Review. *J. Ecol.* **67**, 386–8.

Southern, H. N. (1940). The ecology and population dynamics of the wild rabbit. *Ann. Appl. Biol.* **27**, 509–26.

Southern, H. N. (1959). Mortality and population control. *Ibis* **101**, 429–36.

Southern, H. N. (1970). Ecology at the cross-roads. *J. Anim. Ecol.* **39**, 1–11.

Southward, A. J. & Roberts, E. K. (1984). The Marine Biological Association 1884–1984. *Rep. Trans. Devon. Ass. Advmt Sci.* **116**, 155–99.

Southwood, T. R. E. (1977). Habitat, the templet for ecological strategies. *J. Anim. Ecol.* **46**, 337–65.

Southwood, T. R. E. (1987). Owain Westmacott Richards. *Biographical Memoirs of Fellows of the Royal Society* (in press).

Spencer, H. (1862) *First Principles.* Williams & Norgate, London.

Spencer, H. (1898–9). *The Principles of Biology*, pp. 3–410. Williams & Norgate, London.

Stamp, L. D. (1925). The aerial survey of the Irrawaddy delta forests (Burma). *J. Ecol.* **13**, 262–76.

Stamp, L. D. (1940). Review. *Geogrl J.* **95**, 57–8.

Stapledon, R. G. (1911). *The Effect of the Drought of 1911 on Cotswold Grass Land*, pp. 9–18. (Reprint from *Sci. Bull.* **3**.) Smith, Cirencester.

Stapledon, R. G. (1913). Pasture problems: drought resistance. *J. Agric. Sci. Camb.* **5**, 129–51.

Stapledon, R. G. (1914). Pasture problems: the responses of individual species under manures. *J. Agric. Sci. Camb.* **6**, 499–511.

Stapledon, R. G. (1928). Cocksfoot grass (*Dactylis glomerata*) ecotypes in relation to the biotic factor. *J. Ecol.* **16**, 71–104.

Stapledon, R. G. (1933a). Climate and the improvement of hill land. In *Four Addresses on the Improvement of Grassland* (ed. Stapledon R. G.), pp. 27–36. Welsh Plant Breeding Station, Aberystwyth.

Stapledon, R. G. (1933b). Land improvement. In *Four Addresses on the Improvement of Grassland* (ed. Stapledon R. G.), pp. 19–26. Welsh Plant Breeding Station, Aberystwyth.

Stapledon, R. G. (1939). *The Plough-up Policy and Ley Farming.* Faber, London.

Stapledon, R. G. & Hanley, J. A. (1927). *Grass Land. Its Management and Improvement.* Clarendon Press, Oxford.

Stapledon, R. G. & Jenkin, T. J. (1916). Pasture problems: indigenous plants in relation to habitat and sown species. *J. Agric. Sci. Camb.* **8**, 26–64.

Stearn, W. T. (1985). Marietta Pallis, ecologist and author. *Ann. Musei Goulandris* **7**, 157–73.

Steven, H. M. (1930). The bearing of some ecological conceptions on the formation of plantations. *Forestry* **4**, 26–33.

Stevenson, W. I. (1978). Patrick Geddes. *Geographers. Bibliographical Studies* (eds Freeman T. W. & Pinchemel P.), pp. 53–65. Mansell, London.

Stopes, M. C. (1918). An argument for morphology. *New Phytol.* **17**, 198–9.

Summerhayes, V. S. (1941). The effect of voles on vegetation. *J. Ecol.* **29**, 14–48.

Summerhayes, V. S. & Elton, C. S. (1923). Contributions to the ecology of Spitsbergen and Bear Island. *J. Ecol.* **11**, 214–86.

Summerhayes, V. S. & Elton, C. S. (1928). Further contributions to the ecology of Spitsbergen. *J. Ecol.* **16**, 193–268.

Summerhayes, V. S. & Williams, P. H. (1926). Studies on the ecology of English heaths. II. Early stages in the recolonization of felled pinewood at Oxshott Heath and Esher Common, Surrey. *J. Ecol.* **14**, 203–43.

Tansley, A. G. (1904). The problems of ecology. *New Phytol.* **3**, 191–200.

Tansley, A. G. (1905). Ecological expedition to the Bouche d'Erquy 1905. *New Phytol.* **4**, 192–4.

Tansley, A. G. (1906). Two recent ecological papers. *New Phytol.* **5**, 219–22.

Tansley, A. G. (1908). *Lectures on the Evolution of the Filicinean Vascular System.* Botany School, Cambridge.

Tansley, A. G. (1911a). The International Phytogeographical Excursion in the British Isles. *New Phytol.* **10**, 271–91.

Tansley, A. G. (ed.) (1911b). *Types of British Vegetation.* University Press, Cambridge.

Tansley, A. G. (1912a). The Scottish Botanical Review. *New Phytol.* **11**, 70–1.

Tansley, A. G. (1912b). The forest of Provence. *Gdnrs Chron.* **51**, 89–90, 112–13 and 131.

Tansley, A. G. (1913a). Primary survey of the Peak District of Derbyshire. *J. Ecol.* **1**, 275–85.

Tansley, A. G. (1913b). The aims of the new journal. *J. Ecol.* **1**, 1–4.

Tansley, A. G. (1913c). A universal classification of plant-communities. *J. Ecol.* **1**, 27–42.

Tansley, A. G. (1913–14). The International Phytogeogaphic Excursion in America in 1913. *New Phytol.* **12**, 322–36.

Tansley, A. G. (1914). Presidential address. *J. Ecol.* **2**, 194–203.

Tansley, A. G. (1916a). Albert Stanley Marsh. *New Phytol.* **15**, 81–5.

Tansley, A. G. (1916b). The development of vegetation. *J. Ecol.* **4**, 198–204.

Tansley, A. G. (1917). On competition between *Galium saxatile* and *Galium sylvestre* on different types of soil. *J. Ecol.* **5**, 173–9.

Tansley, A. G. (1919). Postscript. *New Phytol.* **18**, 108–10.

Tansley, A. G. (1920a). *The New Psychology and its Relation to Life.* Allen & Unwin, London.

Tansley, A. G. (1920b). The classification of vegetation and the concept of development. *J. Ecol.* **8**, 118–149.

Tansley, A. G. (1921). Recent text-books of botany. *New. Phytol.* **20**, 132–6.

Tansley, A. G. (1922a). Studies of the vegetation of the English chalk. II. Early stages of redevelopment of woody vegetation on chalk grassland. *J. Ecol.* **10**, 168–77.

Tansley, A. G. (1922b). The new Zurich–Montpellier school. *J. Ecol.* **10**, 241–8.

Tansley, A. G. (1923). *Practical Plant Ecology. A Guide for Beginners in Field Study of Plant Communities.* Allen & Unwin, London.

Tansley, A. G. (1924). Eug. Warming in memoriam. *Bot. Tidsskr.* **39**, 54–6.

Tansley, A. G. (1925). The vegetation of the southern English chalk. *Feschrift Carl Schröter* (ed. Brockmann-Jorosch H.), pp. 406–30. Veröffentlichungen des Geobotanischen Institutes Rübel, Zürich, 3.

Tansley, A. G. (1927). *The Future Development and Functions of the Oxford Department of Botany.* Clarendon Press, Oxford.

Tansley, A. G. (1929a). Obituary notice: William Gardner Smith. *J. Ecol.* **17**, 172–3.

Tansley, A. G. (1929b). Succession: the concept and its values. *Proceedings of the International Congress of Plant Science* (ed. Anonymous), pp. 677–86.

Tansley, A. G. (1931). Obituary notice: Charles Edward Moss. *J. Ecol.* **19**, 209–14.

Tansley, A. G. (1934). Review. *J. Ecol.* **22**, 318.

Tansley, A. G. (1935). The use and abuse of vegetational concepts and terms. *Ecology* **16**, 284–307.

Tansley, A. G. (1936). Prof. Frank Cavers. *Nature, Lond.* **137**, 1022.

Tansley, A. G. (1939a). Obituary. Carl Schröter. *J. Ecol.* **27**, 531–4.

Tansley, A. G. (1939b). British ecology during the past quarter-century: the plant community and the ecosystem. *J. Ecol.* **27**, 513–30.

Tansley, A. G. (1939c). *The British Islands and their Vegetation.* University Press, Cambridge.

Tansley, A. G. (1940a). Sigmund Freud. *Obituary Notices of Fellows of the Royal Society* **3**, 247–75.

Tansley, A. G. (1940b). Henry Chandler Cowles. *J. Ecol.* **28**, 450–2.

Tansley, A. G. (1942). *The Values of Science to Humanity.* Allen & Unwin, London.

Tansley, A. G. (1945). *Our Heritage of Wild Nature.* University Press, Cambridge.

Tansley, A. G. (1947a). The early history of modern plant ecology in Britain. *J. Ecol.* **35**, 130–7.

Tansley, A. G. (1947b). Frederic Edward Clements. *J. Ecol.* **34**, 194–6.

Tansley, A. G. (1948). Postscript. *J. Ecol.* **36**, 180.

Tansley, A. G. (1949). *Britain's Green Mantle.* Allen & Unwin, London.

Tansley, A. G. (1951). *What is Ecology?* Council for the Promotion of Field Studies, London.

Tansley, A. G. (1954). Some reminiscences. *Vegetatio* **5–6**, vii–viii.

Tansley, A. G. & Adamson, R. S. (1925). Studies of the vegetation of the English chalk. III. The chalk grasslands of the Hampshire-Sussex border. *J. Ecol.* **13**, 177–223.

Tansley, A. G. & Adamson, R. S. (1926). Studies of the vegetation of the English chalk. IV. A preliminary survey of the chalk grasslands of the Sussex Downs. *J. Ecol.* **14**, 1–32.

Tansley, A. G. & Chipp, T. F. (eds) (1926). *Aims and Methods in the Study of Vegetation.* British Empire Vegetation Committee and Crown Agents, London.

Tansley, A. G. & Fritsch, F. E. (1905). Sketches of vegetation at home and abroad. I. The flora of the Ceylon littoral. *New Phytol.* **4**, 1–17 and 27–55.

Taylor, E. G. R. (1939). A national atlas of Britain. *Nature, Lond.* **144**, 929–30.

Taylor, E. G. R. (1963). A welcome to the Atlas of Britain. *Geogrl Mag.* **36**, 447–8.

Taylor, L. R. (1986). Synoptic dynamics, migration and the Rothamsted Insect Survey. *J. Anim. Ecol.* **55**, 1–38.

Taylor, L. R. & Elliott, J. M. (1981). The first fifty years of the *Journal of Animal Ecology.* *J. Anim. Ecol.* **50**, 951–71.

Thomas, A. S. (1944). The vegetation of the Sese Islands, Uganda. *J. Ecol.* **29**, 330–53.

Thomas, A. S. (1941). The vegetation of the Karamoja district, Uganda. *J. Ecol.* **31**, 149–79.

Thomas, A. S. (1945). The vegetation of some hillsides in Uganda. *J. Ecol.* **33**, 10–43 and 153–72.

Thomas, A. S. (1960). Changes in vegetation since the advent of myxomatosis. *J. Ecol.* **48**, 287–306.

Thomas, A. S. (1963). Further changes in vegetation since the advent of myxomatosis. *J. Ecol.* **51**, 151–83.

Thomas, H. H. (1941). Albert Charles Seward. *Obituary Notices of Fellows of the Royal Society* **3**, 867–80.

Thomas, H. H. (1953). Federick Ernest Weiss. *Obituary Notices of Fellows of the Royal Society* **8**, 601–8.

Thomas, H. H. (1954–5). Francis John Lewis. *Proc. Linn. Soc.* **167**, 136–7.

Thomas, P. T. (1966). Prof. T. J. Jenkin. *Nature, Lond.* **210**, 12.

Thomson, A. L. (1926). *Problems of Bird-Migration.* Witherby, London.

Thomson J. A. (ed.) (1910). *Life of William MacGillivray*, pp. 114–58. Murray, London.

Thorpe, W. H. (1974). David Lambert Lack. *Biographical Memoirs of Fellows of the Royal Society* **20**, 271–93.

Thorpe, W. H. (1975). Sir Julian Sorell Huxley. *Ibis* **117**, 536–8.

Tobey, R. C. (1981). *Saving the Prairies: the Life of the Founding School of American Plant Ecology, 1895–1955.* University of California Press, Berkeley.

Trail, J. W. H. (1903). Suggestions towards the preparation of a record of the flora of Scotland. *Trans. Proc. Bot. Soc. Edinb.* **22**, 263–77.

Turrill, W. B. (1929). *The Plant-Life of the Balkan peninsula. A Phytogeographical Study.* Clarendon Press, Oxford.

Turrill, W. B. (1951). Some problems of plant range and distribution. *J. Ecol.* **39**, 205–27.

Turrill, W. B. (1958). John Christopher Willis. *Biographical Memoirs of Fellows of the Royal Society* **4**, 353–9.

Turrill, W. B. (1961). Eric Marsden-Jones. *Proc. Bot. Soc. Br. Isl.* **4**, 353–4.

Usher, M. B. & Williamson, M. H. (eds) (1974). *Ecological Stability.* Chapman & Hall, London.

Varley, G. C. (1947). The natural control of population balance in the knapweed gall-fly. *J. Anim. Ecol.* **16**, 139–87.

Varley, G. C. (1957). Ecology as an experimental science. *J. Anim. Ecol.* **26**, 251–61.

Vorontsov, N. N. & Gall, J. M. (1986). Georgyi Frantsevich Gause. *Nature, Lond.* **323**, 113.

Wadham, S. M. (1920). Changes in the salt marsh and sand dunes of Holme-next-the-Sea. *J. Ecol.* **8**, 232–8.

Wadsworth, R. M. (ed.) (1968). *The Measurement of Environmental Factors in Terrestrial Ecology.* Blackwell Scientific Publications, Oxford.

Waller, R. (1962). *Prophet of the New Age: the Life and Thought of Sir George Stapledon.* Faber, London.

Walton, J. (1922). A Spitsbergen salt marsh. *J. Ecol.* **10**, 109–21.

Warming, E. (1896). *Lehrbuch der Ökologischen: Pflanzengeographie: Ein Einfürung in die Kenntniss der Pflanzenvereine.* Borntraeger, Berlin.

Warming, E. (1909). *Oecology of plants: an Introduction to the Study of Plant Communities.* Clarendon Press, Oxford.

Warwick, T. (1934). The distribution of the muskrat in the British Isles. *J. Anim. Ecol.* **3**, 250–67.

Warwick, T. (1940). A contribution to the ecology of the musk-rat in the British Isles. *Proc. Zool. Soc. Lond. A* **110**, 165–201.

Watson, H. C. (1832). *Outlines of the Geographical Distribution of British plants; Belonging to the Division of Vasculares or Cotyledones.* Privately printed, Edinburgh.

Watt, A. S. (1919). On the causes of failure of natural regeneration in British oakwoods. *J. Ecol.* **7**, 173–203.

Watt, A. S. (1923). On the ecology of British beechwoods with special reference to their re-generation. *J. Ecol.* **11**, 1–48.

Watt, A. S. (1925). On the ecology of British beechwoods with special reference to their re-generation. II. The development and structure of beech communities on the Sussex Downs. *J. Ecol.* **13**, 27–73.

Watt, A. S. (1926). Yew communities of the South Downs. *J. Ecol.* **14**, 282–316.

Watt, A. S. (1931). Preliminary observations on Scottish beechwoods. *J. Ecol.* **19**, 137–57 and 321–59.

Watt, A. S. (1934). The vegetation of the Chiltern Hills, with special reference to the beech-woods and their seral relationships. *J. Ecol.* **22**, 230–70 and 445–507.

Watt, A. S. (1936). Studies in the ecology of Breckland. I. Climate, soil and vegetation. *J. Ecol.* **24**, 117–38.

Watt, A. S. (1937). Studies in the ecology of Breckland. II. On the origin and development of blow-outs. *J. Ecol.* **25**, 91–112.

Watt, A. S. (1938). Studies in the ecology of Breckland. III. The origin and development of the Festuco-agrostidetum on eroded sand. *J. Ecol.* **26**, 1–37.

Watt, A. S. (1940). Studies in the ecology of Breckland. IV. The grass heath. *J. Ecol.* **28**, 42–70.

Watt, A. S. (1947a). Pattern and process in the plant community. *J. Ecol.* **35**, 1–22.

Watt, A. S. (1947b). Contributions to the ecology of bracken. IV. The structure of the com-munity. *New Phytol.* **46**, 97–121.

Watt, A. S. (1955). Bracken versus heather, a study in plant sociology. *J. Ecol.* **43**, 490–506.

Watt, A. S. (1957). The effect of excluding rabbits from grassland B in Breckland. *J. Ecol.* **45**, 861–78.

Watt, A. S. (1961). Ecology. In *Contemporary Botanical Thought* (eds MacLeod A. M. & Cobley L. S.), pp. 115–31. Oliver & Boyd, Edinburgh.

Watt, A. S. (1964). The community and the individual. *J. Ecol.* **52** suppl., 203–11.

Watt, A. S. & Jones, E. W. (1948). The ecology of the Cairngorms. I. The environment and the altitudinal zonation of the vegetation. *J. Ecol.* **36**, 283–304.

Webb, D. A. (1950). The Ninth International Phytogeographical Excursion. Ireland. *Ir. Nat. J.* **10**, 1–8.

Webb, D. A. (1954). Is the classification of plant communities either possible or desirable? *Bot. Tidsskr.* **51**, 362–70.

Webb, S. (1903). The diminution and disappearance of the south-eastern fauna and flora. *S. East. Nat.* 48–60.

Weiss, F. E. (1908). The dispersal of fruits and seeds by ants. *New Phytol.* **7**, 23–8.

Weiss, F. E. (1909a). A preliminary account of the submerged vegetation of Lake Windermere as affecting the feeding ground of fish. *Mem. Proc. Manchr. Lit. Phil. Soc.* **53**, xi.

Weiss, F. E. (1909b). The dispersal of the seeds of the gorse and the broom by ants. *New Phytol.* **8**, 81–9.

Weiss, F. E. (1925). Plant structure and environment with special reference to fossil plants. *J. Ecol.* **13**, 301–13.

Weiss, F. E. (1929). Obituary notice: Richard Henry Yapp. *J. Ecol.* **17**, 405–8.

West, R. G. (1964). Inter-relations of ecology and Quaternary palaeobotany. *J. Ecol.* **52** *suppl.*, 47–57.

West, R. G. (1981). Palaeobotany and Pleistocene stratigraphy in Britain. *New Phytol.* **87**, 127–37.

West, R. G. (1986). Sir Harry Godwin. *New Phytol.* **103**, 1–3.

Wheeler, F. D. (1892). Presidential address. *Trans. Norfolk Norwich Nat. Soc.* **5**, 235–49.

White, F. B. (1890). The flora of river-shingles. *Scott. Nat.* **4**, 290–9.

White, F. B. (1895). The flowering plants. *Trans. Proc. Perthsh. Soc. Nat. Sci.* **2**, 50–9.

White, F. B. (1898). *The Flora of Perthshire.* Blackwood, Edinburgh.

White, J. (1982). A history of Irish vegetation studies. In *Studies on Irish Vegetation* (ed. White J.), pp. 15–42. Royal Dublin Society, Dublin.

White, J. (1985). The census of plants in vegetation. In *The Population Structure of Vegetation* (ed. White J.), pp. 33–88. Junk, Dordrecht.

Whittaker, E. T. (1941). Vito Volterra. *Obituary Notices of Fellows of the Royal Society* **3**, 691–729.

Whittaker, R. H. (1962). Classification of natural communities. *Bot. Rev.* **28** (*1*), reprint 239 pp.

Wigglesworth, V. B. (1956). Patrick Alfred Buxton. *Biographical Memoirs of Fellows of the Royal Society* **2**, 69–84.

Wigglesworth, V. B. (1982). Carrington Bonsor Williams. *Biographical Memoirs of Fellows of the Royal Society* **28**, 667–84.

Williams, C. B. (1926). Voluntary or involuntary migration of butterflies. *Entomologist* **59**, 281–8.

Williams, C. B. (1930). *The Migration of Butterflies.* Oliver & Boyd, Edinburgh.

Williams, C. B. (1943). Area and number of species. *Nature, Lond.* **152**, 264–7.

Williams, C. B. (1944). Some applications of the logarithmic series and the index of diversity to ecological problems. *J. Ecol.* **32**, 1–44.

Williams, C. B. (1947). The logarithmic series and its application to biological problems. *J. Ecol.* **34**, 253–72.

Williams, C. B. (1950). The application of the logarithmic series to the frequency of occurrence of plant species in quadrats. *J. Ecol.* **38**, 107–38.

Williams, C. B. (1951). Changes in insect populations in the field in relation to preceding weather conditions. *Proc. Roy. Soc., Lond. B* **138**, 130–56.

Williams, C. B. (1954). The statistical outlook in relation to ecology. *J. Ecol.* **42**, 1–13.

Williams, C. B. (1958a). *Insect Migrations.* Collins, London.

Williams, C. B. (1958b). Review. *J. Ecol.* **46**, 780–1.

Williams, C. B. (1964a). Some experiences of a biologist with R. A. Fisher and statistics. *Biometrics* **20**, 301–6.

Williams, C. B. (1964b). *Patterns in the Balance of Nature and Related Problems in Quantitative Ecology.* Academic Press, London.

Williams, W. T. (1965). Review. *J. Ecol.* **53**, 820.

Williams, W. T. & Lambert, J. M. (1959). Multivariate methods in plant ecology. I. Association-analysis in plant communities. *J. Ecol.* **47**, 83–101.

Williams, W. T. & Lambert, J. M. (1960). Multivariate methods in plant ecology. II. The use of an electronic digital computer for association-analysis. *J. Ecol.* **48**, 689–710.

Williamson, M. (1969). Review. *J. Anim. Ecol.* **38**, 464.

Willis, E. (1986). Professor Sir Harry Godwin. *J. Archaeol. Sci.* **13**, 303–4.

Willis, J. C. (1916). The evolution of species in Ceylon, with reference to the dying out of species. *Ann. Bot.* **30**, 1–23.

Willis, J. C. (1922). *Age and Area: a Study in Geographical Distibution and Origin of Species.* University Press, Cambridge.

Winch, N. J. (1825). *An Essay on the Geographical Distribution of Plants, through the Counties of Northumberland, Cumberland and Durham.* Charnley, Newcastle-upon-Tyne.

Woodhead, T. W. (1904). Notes on the bluebell. *Naturalist, Hull,* 41–8 and 81–8.

Woodhead, T. W. (1906). Ecology of the woodland plants in the neighbourhood of Huddersfield. *J. Linn. Soc. Bot.* **37**, 333–406.

Woodhead, T. W. (1908). Plant geography and ecology in Switzerland. *Naturalist, Hull,* 170–4.

Woodhead, T. W. (1915). *The Study of Plants. An Introduction to Botany and Plant Ecology.* Clarendon, Oxford.

Woodhead, T. W. (1919). Academic botany and the farm and garden. *New. Phytol.* **18**, 50.

Woodhead, T. W. (1923). Botanical survey and ecology in Yorkshire. *Naturalist, Hull,* 97–128.

Woodhead, T. W. (1924). The age and composition of the Pennine peat. *J. Bot. Lond.* **62**, 301–4.

Woodhead, T. W. (1929a). William Gardener Smith. *Naturalist, Hull,* 153–6.

Woodhead, T. W. (1929b). History of the vegetation of the southern Pennines. *J. Ecol.* **17**, 1–34.

Worster, D. (1979). *Dust Bowl. The Southern Plains in the 1930s.* Oxford University Press, New York.

Worster, D. (1985). *Nature's Economy: a History of Ecological Ideas.* University Press, Cambridge.

Worthington, E. B. (1983). *The Ecological Century: a Personal Appraisal.* Clarendon Press, Oxford.

Wynne-Edwards, V. C. (1959). The control of population-density through social behaviour: a hypothesis. *Ibis* **101**, 436–41.

Wynne-Edwards, V. C. (1962). *Animal Dispersion in Relation to Social Behaviour.* Oliver & Boyd, Edinburgh.

Wynne-Edwards, V. C. (1987). Sir Alister Hardy. *J. Anim. Ecol.* **56**, 365–7.

Yapp, R. H. (1908). Sketches of vegetation at home and abroad. IV. Wicken Fen. *New Phytol.* **7**, 61–81.

Yapp, R. H. (1909). On stratification in the vegetation of a marsh, and its relations to evaporation and temperature. *Ann. Bot.* **23**, 275–319.

Yapp, R. H. (1922). The concept of habitat. *J. Ecol.* **10**, 1–17.

Index